SCHOLAR Study Guide
Advanced Higher Phy

Authored by:
Julie Boyle (St Columba's School)
Chad Harrison (Tynecastle High School)

Reviewed by:
Grant McAllister (St Andrew's RC High School)

Previously authored by:
Andrew Tookey
Campbell White

Heriot-Watt University
Edinburgh EH14 4AS, United Kingdom.

First published 2019 by Heriot-Watt University.

This edition published in 2019 by Heriot-Watt University SCHOLAR.

Copyright © 2019 SCHOLAR Forum.

Members of the SCHOLAR Forum may reproduce this publication in whole or in part for educational purposes within their establishment providing that no profit accrues at any stage, Any other use of the materials is governed by the general copyright statement that follows.

All rights reserved. No part of this publication may be reproduced, stored in a retrieval system or transmitted in any form or by any means, without written permission from the publisher.

Heriot-Watt University accepts no responsibility or liability whatsoever with regard to the information contained in this study guide.

Distributed by the SCHOLAR Forum.

SCHOLAR Study Guide Advanced Higher Physics

Advanced Higher Physics Course Code: C857 77

ISBN 978-1-911057-78-9

Print Production and Fulfilment in UK by Print Trail www.printtrail.com

Acknowledgements

Thanks are due to the members of Heriot-Watt University's SCHOLAR team who planned and created these materials, and to the many colleagues who reviewed the content.

We would like to acknowledge the assistance of the education authorities, colleges, teachers and students who contributed to the SCHOLAR programme and who evaluated these materials.

Grateful acknowledgement is made for permission to use the following material in the SCHOLAR programme:

The Scottish Qualifications Authority for permission to use Past Papers assessments.

The Scottish Government for financial support.

The content of this Study Guide is aligned to the Scottish Qualifications Authority (SQA) curriculum.

All brand names, product names, logos and related devices are used for identification purposes only and are trademarks, registered trademarks or service marks of their respective holders.

Contents

1 Rotational Motion and Astrophysics — 1

1. Kinematic relationships 5
2. Angular motion 33
3. Rotational dynamics 73
4. Angular momentum 93
5. Gravitation 105
6. General relativity and spacetime 133
7. Stellar physics 161
8. End of section 1 test 191

2 Quanta and Waves — 195

1. Introduction to quantum theory 199
2. Wave particle duality 217
3. Magnetic fields and particles from space 235
4. Simple harmonic motion 255
5. Waves 277
6. Interference 305
7. Division of amplitude 315
8. Division of wavefront 331
9. Polarisation 341
10. End of section 2 test 355

3 Electromagnetism — 359

1. Electric force and field 361
2. Electric potential 379
3. Motion in an electric field 391
4. Magnetic fields 409
5. Capacitors 437
6. Inductors 461

7	Electromagnetic radiation	489
8	End of section 3 test	495

4 Investigating Physics — 503

1	Initial planning, using equipment and recording data	505
2	Measuring and presenting data	513
3	Evaluating findings	519
4	Scientific report	523

5 Units, prefixes, uncertainties and data analysis — 529

A	Units, prefixes and scientific notation	531
B	Uncertainties	545
C	Data analysis	555

Glossary — 568

Hints for activities — 578

Answers to questions and activities — 591

Rotational Motion and Astrophysics

1 Kinematic relationships	5
1.1 Introduction	7
1.2 Calculus methods	7
1.3 Deriving the equations for uniform acceleration	8
1.4 Uniform acceleration in a straight line	10
1.5 Graphical methods for uniform acceleration	15
1.6 Motion sensors	20
1.7 Graphical methods for non-uniform acceleration	23
1.8 Examples of varying acceleration	25
1.9 Extended information	28
1.10 Summary	29
1.11 End of topic 1 test	30
2 Angular motion	**33**
2.1 Introduction	35
2.2 Angular displacement and radians	35
2.3 Angular velocity and acceleration	38
2.4 Kinematic relationships for angular motion	40
2.5 Tangential speed and angular velocity	46
2.6 Centripetal acceleration	50
2.7 Centripetal force	55
2.8 Applications	60
2.9 Extended information	68
2.10 Summary	68
2.11 End of topic 2 test	69
3 Rotational dynamics	**73**
3.1 Introduction	75
3.2 Torque and moment	75
3.3 Newton's laws applied to rotational dynamics	78
3.4 Calculating the moment of inertia	83
3.5 Extended information	89
3.6 Summary	89
3.7 End of topic 3 test	90

4 Angular momentum — 93
- 4.1 Introduction — 94
- 4.2 Angular momentum — 94
- 4.3 Conservation of angular momentum — 94
- 4.4 Rotational kinetic energy — 97
- 4.5 Comparing linear and angular motion — 100
- 4.6 Extended information — 101
- 4.7 Summary — 102
- 4.8 End of topic 4 test — 102

5 Gravitation — 105
- 5.1 Introduction — 107
- 5.2 Newton's law of gravitation — 107
- 5.3 Weight — 111
- 5.4 Gravitational fields — 114
- 5.5 Gravitational potential and potential energy — 119
- 5.6 Escape velocity — 124
- 5.7 Extended information — 128
- 5.8 Summary — 128
- 5.9 End of topic 5 test — 129

6 General relativity and spacetime — 133
- 6.1 Introduction — 135
- 6.2 Comparison of special and general relativity — 135
- 6.3 The equivalence principle — 136
- 6.4 Understanding the consequences — 138
- 6.5 Show me the evidence — 140
- 6.6 Spacetime diagrams — 142
- 6.7 Worldlines — 144
- 6.8 The curvature of spacetime — 146
- 6.9 Black holes — 151
- 6.10 Extended information — 154
- 6.11 Summary — 155
- 6.12 End of topic 6 test — 156

7 Stellar physics — **161**

- 7.1 Introduction — 163
- 7.2 Properties of stars — 163
- 7.3 Star formation — 174
- 7.4 Stellar nucleosynthesis and fusion reactions — 175
- 7.5 Hertzsprung-Russell diagrams — 177
- 7.6 Stellar evolution and life cycles on H-R diagrams — 181
- 7.7 Extended information — 185
- 7.8 Summary — 186
- 7.9 End of topic 7 test — 187

8 End of section 1 test — **191**

Unit 1 Topic 1

Kinematic relationships

Contents

1.1	Introduction	7
1.2	Calculus methods	7
1.3	Deriving the equations for uniform acceleration	8
1.4	Uniform acceleration in a straight line	10
	1.4.1 Horizontal motion	10
	1.4.2 Vertical motion	12
1.5	Graphical methods for uniform acceleration	15
	1.5.1 Graphs for motion with constant velocity	15
	1.5.2 Graphs for motion with constant positive acceleration	16
	1.5.3 Graphs for motion with constant negative acceleration	17
	1.5.4 Example graphs	17
	1.5.5 Objects in freefall	19
1.6	Motion sensors	20
	1.6.1 Motion of bouncing ball	22
	1.6.2 Motion on a slope	23
1.7	Graphical methods for non-uniform acceleration	23
1.8	Examples of varying acceleration	25
1.9	Extended information	28
1.10	Summary	29
1.11	End of topic 1 test	30

Prerequisites

- Knowledge of the difference between vector and scalar quantities.
- Calculus - differentiation and integration.
- Some familiarity with the kinematic relationships would be useful.
- Understand the shape of displacement-time graphs, velocity-time graphs and acceleration-time graphs for constant acceleration.

Learning objective

By the end of this topic you should be able to:

- use calculus notation to represent velocity as the rate of change of displacement with respect to time;
- use calculus notation to represent acceleration as the rate of change of velocity with respect to time;
- use calculus notation to represent acceleration as the second differential of displacement with respect to time;
- derive the three kinematic relationships from the calculus definitions of acceleration and velocity:

$$v = u + at \qquad s = ut + \tfrac{1}{2}at^2 \qquad v^2 = u^2 + 2as;$$

- apply these equations to describe the motion of a particle with uniform acceleration moving in a straight line;
- state that the gradient of a graph can be found by differentiation;
- state that the gradient of a displacement-time graph at a given time is the instantaneous velocity;
- state that the gradient of a velocity-time graph at a given time is the instantaneous acceleration;
- state that the area under a graph can be found by integration;
- state that for a velocity-time graph the displacement can be found by integrating between the limits;
- state that for an acceleration-time graph the change in velocity can be found by integrating between the limits;
- use calculus methods to solve problems based on objects moving in a straight line with varying acceleration.

TOPIC 1. KINEMATIC RELATIONSHIPS

1.1 Introduction

This topic deals with linear motion. From the Higher course, you should already be familiar with the vector nature of the quantities displacement, velocity and acceleration. You may also recall the kinematic relationships describing the motion of objects with constant acceleration. We will now use calculus to derive these relationships, starting from the definitions of instantaneous acceleration and velocity.

We will then look at some examples of the use of motion sensors to study uniform acceleration in a straight line. We will also explore graphical methods for analysing motion, for both uniform and varying acceleration. Lastly, we will finish the topic by studying some specific examples of objects moving in a straight line with varying acceleration.

1.2 Calculus methods

The average velocity of an object can be found from the equation:

$$v_{\text{average}} = \frac{\Delta s}{\Delta t}$$

where Δs is the change in displacement and Δt is the change in time. However, if we want to find the instantaneous velocity, we must consider a very small time interval. In other words, we must find the derivative of displacement with respect to time. So we need to use calculus notation:

$$v = lim \frac{\Delta s}{\Delta t} \quad (\text{as } \Delta t \to 0)$$
$$v = \frac{ds}{dt}$$

Likewise, we could find the average acceleration at Higher by using the equation:

$$a_{\text{average}} = \frac{\Delta v}{\Delta t}$$

but to find instantaneous acceleration we will need to consider the rate of change of velocity with respect to time. In other words, we must differentiate velocity with respect to time. This gives:

$$a = lim \frac{\Delta v}{\Delta t} \quad (\text{as } \Delta t \to 0)$$
$$a = \frac{dv}{dt}$$

© HERIOT-WATT UNIVERSITY

UNIT 1. ROTATIONAL MOTION AND ASTROPHYSICS

Since $v = \frac{ds}{dt}$, the acceleration can also be written as:

$$a = \frac{d}{dt}\frac{ds}{dt} = \frac{d^2s}{dt^2}$$

This means that acceleration can either be expressed as the first derivative of velocity with respect to time or as the second derivative of displacement with respect to time:

$$a = \frac{dv}{dt} \quad \text{or} \quad a = \frac{d^2s}{dt^2}$$

It is worth noting that acceleration is a vector quantity and so an object will accelerate if either the magnitude of its velocity changes or the direction alters. In this topic we will solely explore objects moving in a straight line, but the next topic will consider the acceleration of an object moving in a circular path.

1.3 Deriving the equations for uniform acceleration

The equations describing motion with a constant acceleration can be derived from the definitions of acceleration and velocity.

We will start with the definition of instantaneous acceleration: Acceleration is the rate of change of velocity.

$$a = \frac{dv}{dt}$$

We are looking at motion where a is a constant. To find the velocity after time t, we will integrate this expression over the time interval from $t = 0$ to $t = t$.

$$\int_u^v dv = \int_{t=0}^t a\,dt = a \int_{t=0}^t dt$$

Carrying out this integration

$$[v]_u^v = a\,[t]_0^t$$
$$\therefore v - u = at$$
$$\therefore v = u + at$$

(1.1)

TOPIC 1. KINEMATIC RELATIONSHIPS

Equation 1.1 gives us the velocity v after time t, in terms of the acceleration a and the initial velocity u. Velocity is defined as the rate of change of displacement. We will now use this definition to derive the second of the kinematic relationships.

$$v = \frac{ds}{dt}$$

Since $v = \frac{ds}{dt}$ and $v = u + at$, we have:

$$\frac{ds}{dt} = u + at$$

Integrating over the time interval from 0 to t gives:

$$\int_{s=0}^{s} ds = \int_{t=0}^{t} (u + at)dt$$
$$\therefore [s]_0^s = \left[ut + \frac{1}{2}at^2\right]_0^t \qquad (1.2)$$
$$\therefore s = ut + \frac{1}{2}at^2$$

Equation 1.2 gives us the displacement s after time t, in terms of the acceleration and the initial velocity. To obtain the third kinematic relationship, we first rearrange Equation 1.1.

$$t = \frac{v - u}{a}$$

Substituting this expression for t into Equation 1.2 gives us

$$s = u\left(\frac{v-u}{a}\right) + \frac{1}{2}a\left(\frac{v-u}{a}\right)^2$$
$$\therefore 2as = 2vu - 2u^2 + v^2 - 2vu + u^2$$
$$\therefore 2as = -2u^2 + v^2 + u^2$$
$$\therefore v^2 = u^2 + 2as$$

(1.3)

We have obtained three equations relating displacement s, time elapsed t, acceleration a and the initial and final velocities u and v. Note that with the exception of t, these are all vector quantities.

© HERIOT-WATT UNIVERSITY

$$v = u + at \qquad s = ut + \frac{1}{2}at^2 \qquad v^2 = u^2 + 2as$$

Although we are dealing with vector quantities, we are only considering the special case of motion in a straight line. It is vital to ensure we assign the correct +ve or -ve sign to each of the quantities s, u, v and a.

It is also useful to remember that for an object moving in a straight line with uniform acceleration, the average velocity over a period of time is given by $\frac{(u+v)}{2}$.

The displacement can therefore be found from $s = \frac{(u+v)}{2} \times t$.

1.4 Uniform acceleration in a straight line

We are now going to use the kinematic relationships to describe the motion of a particle moving in one dimension.

1.4.1 Horizontal motion

The kinematic relationships can be used to solve problems of motion in one dimension with constant acceleration. In all cases we will be ignoring the effects of air resistance.

Example

A car accelerates from rest at a rate of 4.0 m s^{-2}.

1. What is its velocity after 10 s?
2. How long does it take to travel 72 m?
3. How far has it travelled after 8.0 s?

We can list the data given to us in the question

u = 0 m s^{-1} (the car starts from rest)
a = 4.0 m s^{-2}

1. We are told that the time elapsed t = 10 s and we wish to find v. So with u, a and t known and v unknown, we will use the equation $v = u + at$.

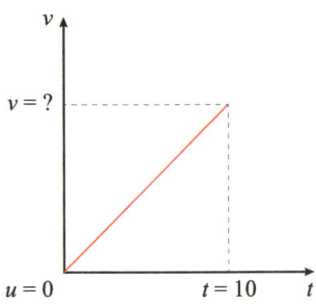

$$v = u + at$$
$$\therefore v = 0 + (4.0 \times 10)$$
$$\therefore v = 40 \text{ m s}^{-1}$$

2. In part 2 we know u, a and s, and t is the unknown, so we use $s = ut + \frac{1}{2}at^2$.

$u = 0$ m s^{-1}
$a = 4.0$ m s^{-2}
$s = 72$ m
$t = ?$

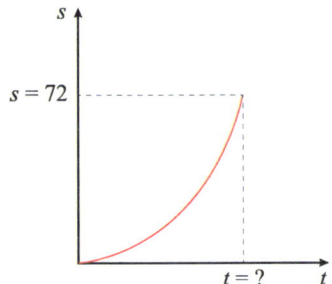

$$s = ut + \tfrac{1}{2}at^2$$
$$\therefore 72 = 0 + \left(\tfrac{1}{2} \times 4.0 \times t^2\right)$$
$$\therefore 72 = 2t^2$$
$$\therefore t^2 = 36$$
$$\therefore t = 6.0 \text{ s}$$

3. Finally in part 3 we are asked to find s, given u, a and t, so we will again use $s = ut + \frac{1}{2}at^2$, this time using different data.

$u = 0$ m s^{-1}
$a = 4.0$ m s^{-2}
$t = 8.0$ s
$s = ?$

$$s = ut + \tfrac{1}{2}at^2$$
$$\therefore s = 0 + \left(\tfrac{1}{2} \times 4 \times 8^2\right)$$
$$\therefore s = 128 \text{ m}$$

The same strategy should be used in solving all of the problems for constant acceleration. Firstly, list the data given to you in the question. This will ensure that you use the correct values when you perform any calculations, and should also make it clear to you which of the kinematic relationships to use. It is often useful to make a sketch diagram with arrows, to ensure that any vector quantities are being measured in the correct direction.

Horizontal motion   Go online

Suppose a car is being driven at a velocity of 12.0 ms^{-1} towards a set of traffic lights, which are changing to red. The car driver applies her brakes when the car is 30.0 m from the stop line. What is the minimum uniform deceleration needed to ensure the car stops at the line?

There is an online activity available which will provide further practice in this type of problem.

1.4.2 Vertical motion

When dealing with freely-falling bodies on Earth, the acceleration of the body is the acceleration due to gravity, $g = 9.8$ m s^{-2}. If any force other than gravity is acting in the vertical plane, the body is no longer in free-fall, and the acceleration will take a different value. Problems should be solved using exactly the same method we used to solve horizontal motion problems.

Example

A student drops a stone from a second floor window, 15 m above the ground.

1. How long does it take for the stone to reach the ground?
2. With what velocity does it hit the ground?

When dealing with motion under gravity, we must take care with the direction we choose as the positive direction. Here, if we take a as a positive acceleration, then v and s will also be positive in the downward direction.

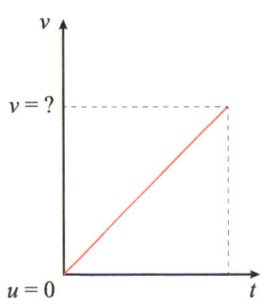 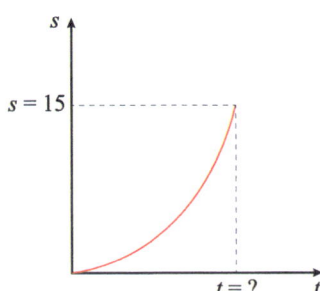

We are told that
$u = 0$ m s^{-1}
$a = g = 9.8$ m s^{-2}
$s = 15$ m

1. To find t, we use $s = ut + \frac{1}{2}at^2$.
   Putting in the appropriate values

$$s = ut + \tfrac{1}{2}at^2$$
$$\therefore 15 = 0 + \left(\tfrac{1}{2} \times 9.8 \times t^2\right)$$
$$\therefore 15 = 4.9 t^2$$
$$\therefore t^2 = \frac{15}{4.9}$$
$$\therefore t^2 = 3.06$$
$$\therefore t = 1.7 \text{ s}$$

2. To find v, we use $v^2 = u^2 + 2as$.
   Again, we put the appropriate values into this kinematic relationship

$$v^2 = u^2 + 2as$$
$$\therefore v^2 = 0 + (2 \times 9.8 \times 15)$$
$$\therefore v^2 = 294$$
$$\therefore v = 17 \text{ m s}^{-1}$$

Difficulties sometimes occur when the initial velocity is directed upwards, for example if an object is being thrown upwards from the ground. To solve such a problem, it is usual to take the vertical displacement as being positive in the upwards direction. The velocity vector is then positive when the object is travelling upwards, and negative when it is returning to the ground. In this situation the acceleration is always negative, as it is always directed towards the ground.

Quiz: Motion in one dimension

Go online

Useful data:

| Gravitational acceleration on Earth g | 9.8 m s^{-2} |

Q1: An object is moving with a uniform acceleration of 5 m s^{-2}. A displacement-time graph showing the motion of this object has a gradient which

a) increases with time.
b) decreases with time.
c) equals 5 m s^{-1}.
d) equals 5 m s^{-2}.
e) equals 0.

....

Q2: A car accelerates from rest with a uniform acceleration of 2.50 m s^{-2}. How far does it travel in 6.00 s?

a) 7.50 m
b) 15.0 m
c) 18.75 m
d) 45.0 m
e) 112.5 m

....

Q3: Neglecting air resistance, a stone dropped from the top of a building 125 m high hits the ground after

a) 1.25 s
b) 4.00 s
c) 5.05 s
d) 12.7 s
e) 25.5 s

....

Q4: A stone is dropped from a window 28.0 m above the ground. What is the velocity of the stone when it is 10.0 m above the ground?

a) 10.0 m s^{-1}
b) 14.0 m s^{-1}
c) 18.8 m s^{-1}
d) 23.4 m s^{-1}
e) 98.1 m s^{-1}

....

Q5: A diver jumps upwards with an initial vertical velocity of 3.00 m s⁻¹ from a diving board which is 8.00 m above the swimming pool. With what vertical velocity does he enter the pool?

a) 3.0 m s⁻¹
b) 3.6 m s⁻¹
c) 9.0 m s⁻¹
d) 13 m s⁻¹
e) 24 m s⁻¹

1.5 Graphical methods for uniform acceleration

We will look at how motion with constant acceleration can be represented in graphical form. We can use graphs to show how the acceleration, velocity and displacement of an object vary with time.

1.5.1 Graphs for motion with constant velocity

The graphs below all represent motion with a constant velocity.

Figure 1.1: Graphs for motion with constant velocity

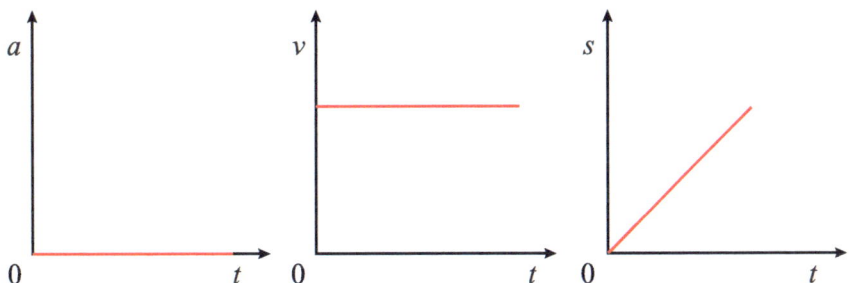

By inspecting the gradient of the displacement-time graph, we can see that the steepness of the graph is constant. This means that $\frac{ds}{dt}$, the instantaneous velocity, is constant.

The velocity-time graph is horizontal and so its gradient is zero. This tells us that $\frac{dv}{dt}$, the acceleration, is zero.

You will recall from Higher that acceleration can be found from a velocity-time graph by calculating the gradient of the line on the graph. The gradient of a straight line can be found from the equation:

$$\text{gradient} = \frac{y_2 - y_1}{x_2 - x_1}$$

For uniform acceleration this is equivalent to using the equation:

$$a = \frac{v - u}{t}$$

We will further explore the gradient method in section 1.7. There, we will look at varying acceleration, where the equation $a = \frac{v-u}{t}$ will no longer be valid.

You will also recall from Higher that the area under a velocity time graph is equal to the displacement. Similarly, it is worth noting that the area under an acceleration time graph is equal to the change in velocity.

Graphs for motion with constant velocity Go online

There is an online activity at this stage displaying more information on the graphs for motion with constant velocity.

1.5.2 Graphs for motion with constant positive acceleration

The graphs below all represent motion with a constant positive acceleration.

Figure 1.2: Graphs for motion with constant positive acceleration

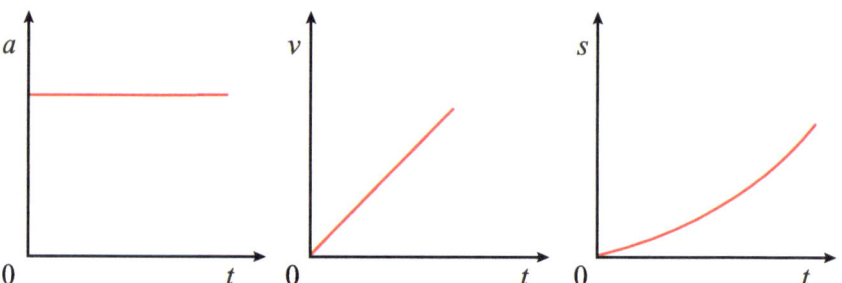

By inspecting the gradient of the displacement-time graph, we can see that the graph is getting steeper. This means that $\frac{ds}{dt}$, the instantaneous velocity, is increasing.

By inspecting the gradient of the velocity-time graph, we can see that the steepness of the graph is constant. This means that $\frac{dv}{dt}$, the acceleration, is constant.

Note that since acceleration and velocity are the same sign, the object is speeding up.

Graphs for motion with constant positive acceleration Go online

There is an online activity at this stage displaying more information on the graphs for motion with constant positive acceleration.

1.5.3 Graphs for motion with constant negative acceleration

The graphs below all represent motion with a constant negative acceleration.

Figure 1.3: Graphs for motion with constant negative acceleration

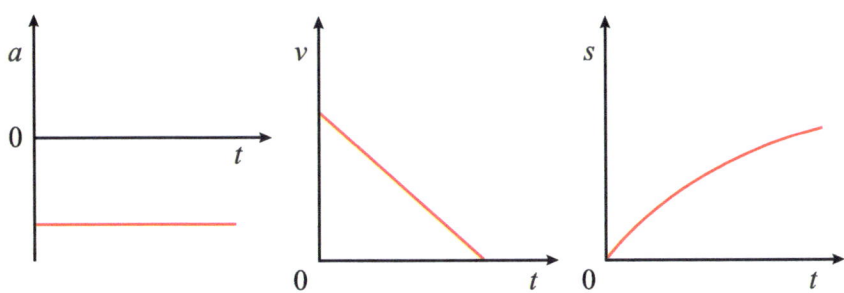

By inspecting the gradient of the displacement-time graph, we can see that the graph is getting less steep. This means that $\frac{ds}{dt}$, the instantaneous velocity, is decreasing.

By inspecting the gradient of the velocity-time graph, we can see that the steepness of the graph is constant and negative. This means that $\frac{dv}{dt}$, the acceleration, is constant and negative.

Note that since the acceleration and the velocity are the opposite sign, the object is slowing down.

Graphs for motion with constant negative acceleration Go online

There is an online activity at this stage displaying more information on the graphs for motion with constant negative acceleration.

1.5.4 Example graphs

Examples

1.

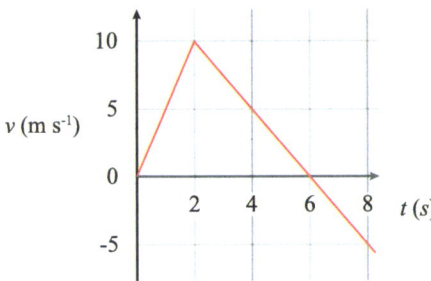

For the velocity - time graph shown, calculate:

1. The acceleration from 2.0 s to 8.0 s.
2. The displacement after 8.0 s.

1. acceleration = gradient of v-t graph = -15/6 = -2.5 m s⁻²
   or

$$a = \frac{v - u}{t}$$
$$a = \frac{-5 - 10}{6}$$
$$a = -2.5 \text{ m s}^{-2}$$

2. displacement = area under v-t graph = $1/2 \times 6 \times 10 - 1/2 \times 2 \times 5 = 25$ m

....

2.

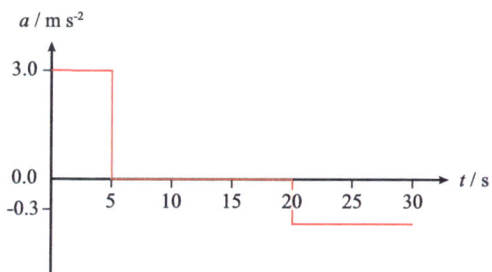

Find the change in velocity in:

1. The first 5 seconds.
2. Between 0 and 30 seconds.

1.
$$\text{change in velocity} = \text{area under } a - t \text{ graph}$$
$$= 3 \times 5$$
$$= 15 \text{ m s}^{-1}$$

2.
$$\text{change in velocity} = \text{area under } a - t \text{ graph}$$
$$= (3 \times 5) - (0.3 \times 10)$$
$$= 15 - 3$$
$$= 12 \text{ m s}^{-1}$$

TOPIC 1. KINEMATIC RELATIONSHIPS

| **Interactive example graphs** | Go online 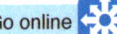 |

There is an online activity at this stage exploring these relationships.

1.5.5 Objects in freefall

The displacement of an object with constant acceleration can be found from

$$s = ut + \frac{1}{2}at^2$$

For a dropped object, the initial velocity is 0 m s^{-1} and the acceleration is the acceleration due to gravity, g. So the value of g can be evaluated from the gradient of the straight line graph of s against t^2.

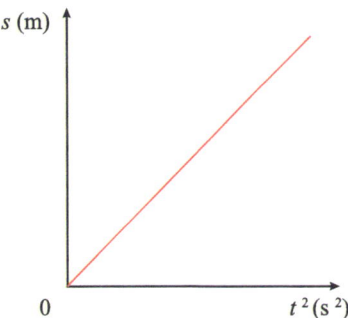

$$s = \frac{1}{2}gt^2$$

A graph of s against t^2 has gradient $\frac{1}{2}g$.

The acceleration due to gravity can also be found by considering the displacement-time graph. The gradient of the $s - t$ graph is increasing since the instantaneous velocity increases. The instantaneous velocity, v, at a given time, t, can be found by drawing a tangent to the graph and determining its gradient. To sketch a tangent to the graph, a straight line must be drawn such that it touches the curve only at one point.

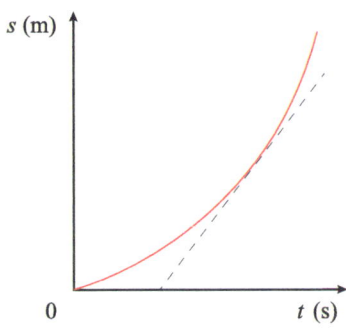

The acceleration due to gravity can then be found by substituting the values of v and t into the equation

$$a = \frac{v - u}{t}$$

1.6 Motion sensors

The displacement of an object can be measured using a motion sensor.

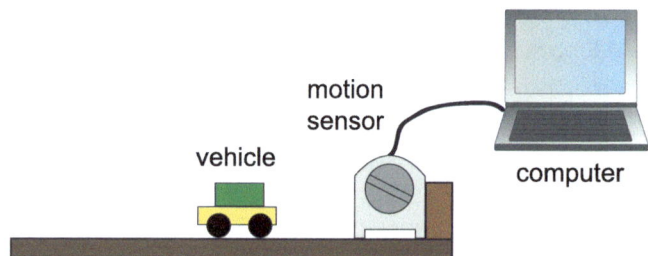

The motion sensor contains an ultrasound transmitter. This sends out pulses of high frequency sound that are reflected back to a special microphone on the sensor. The sensor measures the time for the sound pulses to return to the microphone and uses the speed of sound to calculate the distance of the object from the sensor. This is usually displayed on a computer screen as a displacement time graph.

Light gates are devices that can be used to measure how long it takes an object to pass a point.

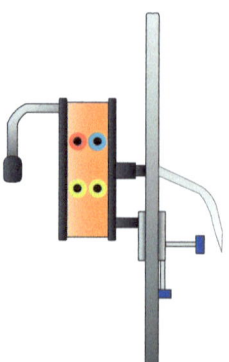

A beam of light passes from the box to a sensor on the arm of the light gate. The light gate is attached to a timer such as a datalogger. When an object breaks the beam the timer measures the amount of time the beam is blocked for. If the length of the object is known then the speed of the object can be calculated, this is often done by the timing device. The following set up can be used to measuring the acceleration of a vehicle on a slope.

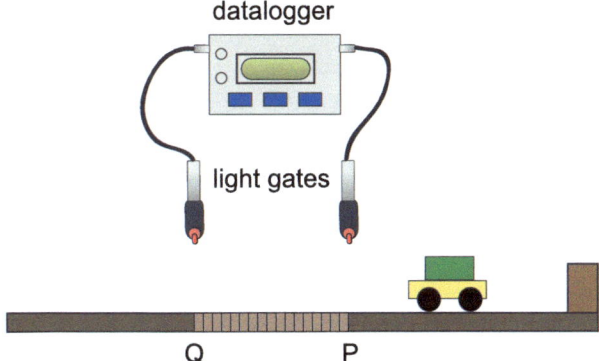

Two light gates are placed at P and Q and connected to the datalogger. The length of the car is measured an entered into the datalogger. When the car passes through the light gate at P the datalogger measures the time it takes to pass and then calculates its speed by dividing the length of the car by the time taken, this is the initial velocity, u. It then measures the time taken for the car to go from P to Q, t. Finally it measures the time taken to pass through the light gate at Q and then calculates its speed by dividing the length of the car by the time taken, this is the final velocity, v. The acceleration of the car can be calculated using equation $a = \frac{v-u}{t}$. The apparatus shown above can also be used to calculate the acceleration by another method.

If instead of measuring the time taken to go from light gate P to Q the distance between the two light gates, s, is measured then equation $v^2 = u^2 + 2as$ can be used to calculate the acceleration.

Acceleration can be measured with a single light gate as long as a 'double mask' is used.

The above mask is used as follows:
When the first section of the mask, A, passes through the light gate the datalogger measures the time to pass and then calculates the initial speed, u, by dividing the length of A by the time. The time for the gap to pass through the light gate, t, is recorded. When the final section, B, passes through the light gate the datalogger measures the time to pass and then calculates the final speed, v, by dividing the length of B by the time. The acceleration of the light gate can be calculated using equation $a = \frac{v-u}{t}$.

The motion of an object can also be analysed using video techniques. This has become increasingly important in the film and computer areas of the entertainment industry. The illustration below shows how motion capture is used to analyse the movement of a person.

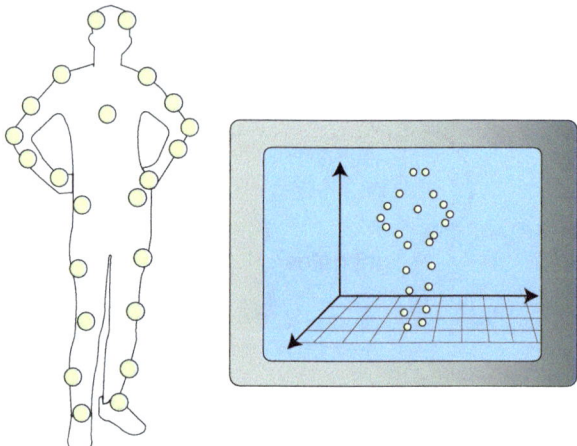

The person has a number of lights placed on their clothes and then stand against a dark background containing a grid. A video is taken of them as they move across the grid and then analysed frame by frame so that their motion can be reproduced in computer graphics. Similar techniques are used when analysing the motion of vehicles in crash test laboratories.

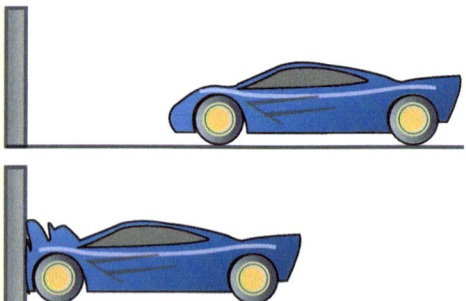

The diagram shows two still frames from the video of a car being crashed at high speed. By making measurements from these frames computer software can be used to analyse the motion of the car. This will allow the acceleration of the car to be calculated and hence the forces acting on occupants of a car in a high speed crash.

1.6.1 Motion of bouncing ball

Motion of bouncing ball   Go online

There is an online activity exploring these relationships.

1.6.2 Motion on a slope

Motion of toy car moving down a slope and hitting a stretched elastic band Go online

This interactivity is available online only.

1.7 Graphical methods for non-uniform acceleration

So far we have assumed uniform (constant) acceleration. In other words, the change in velocity with time has been constant. For non-uniform (varying) acceleration, the change of velocity with time is not constant throughout the motion. So the slope of the velocity-time graph will be changing and it may have curved sections. Therefore, to find the acceleration at a given point in time, you need to draw a tangent to the curve at that point and then calculate its gradient.

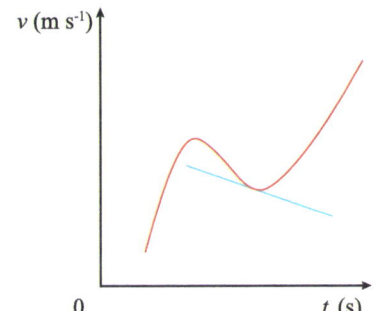

Velocity-time graph Go online

There is an online activity at this stage displaying seven tangents.

Similarly, the velocity at a given point in time can be found from a displacement-time graph by drawing a tangent to the curve at that point and calculating its gradient.

Displacement-time graph Go online

There is an online activity at this stage displaying seven tangents.

If an equation describing the velocity as a function of time is known, then the acceleration at a given time can be found by differentiating the equation and solving for that specific time. Derivatives are just formulae that let us find the slope at any point on a function. In fact, differentiating and finding the gradient of the tangent to the graph are really just the same mathematical process.

You should by now be familiar with the process of finding the displacement from a velocity-time graph by calculating the area under the graph. What you are really doing is integrating. Now let's consider what happens if the acceleration varies and the velocity-time graph is not linear. We will be unable to consider the area of simple shapes. Instead, we will need to calculate the displacement by integrating the velocity function between the limits. This means we must integrate and then substitute in the limits, subtracting the value at the lower limit from the value at the higher limit.

In a similar way, the change in velocity can be determined by finding the definite integral of the acceleration function i.e. integrating with limits.

This can all be summarised as follows

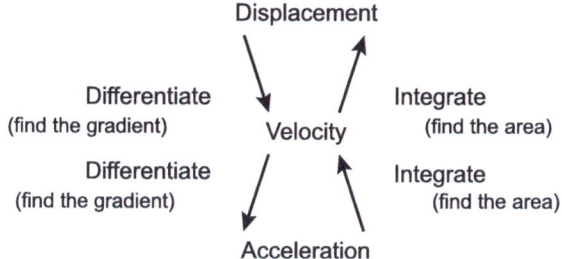

Transformations between graphs of motion   Go online

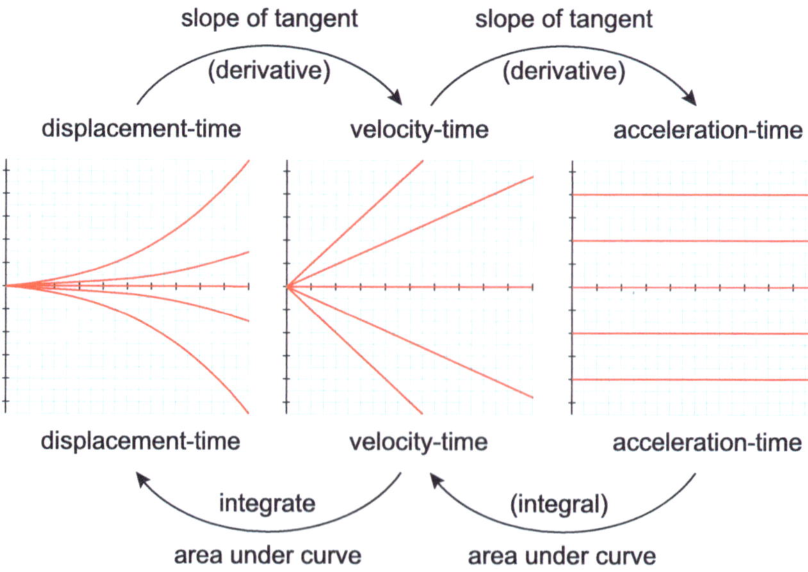

slope of tangent   slope of tangent
(derivative)   (derivative)

displacement-time   velocity-time   acceleration-time

displacement-time   velocity-time   acceleration-time

integrate   (integral)
area under curve   area under curve

There is an animation at this stage displaying the graphs of motion.

1.8 Examples of varying acceleration

Now let's turn our attention to some specific examples of objects moving in a straight line with varying acceleration.

Examples

1.

The velocity in m s^{-1} of a moving particle is given by the equation

$$v = 4t^3 - 5t^2 + 6t$$

where t is the time in seconds.
Find the acceleration at the instant t = 0.5 seconds.

$$v = 4t^3 - 5t^2 + 6t$$
$$a = \frac{dv}{dt} = 12t^2 - 10t + 6$$

At $t = 0.5$ seconds
$$a = (12 \times 0.5^2) - (10 \times 0.5) + 6$$
$$a = 4 \text{ m s}^{-2}$$

....

2.

The acceleration of a space shuttle in m s^{-2} is described by the expression

$$a = (7.0 \times 10^{-7}) t^3 - (3.9 \times 10^{-4}) t^2 + (0.11) t + 0.86$$

where t is the time in seconds.
Determine the change in velocity of the space shuttle throughout the first 100 seconds.

$$\frac{dv}{dt} = a$$
$$\int_u^v dv = \int_0^{100} (7.0 \times 10^{-7}) t^3 - (3.9 \times 10^{-4}) t^2 + (0.11) t + 0.86 \; dt$$
$$[v]_u^v = \left[\frac{7.0 \times 10^{-7} t^4}{4} - \frac{3.9 \times 10^{-4} t^3}{3} + \frac{0.11 t^2}{2} + 0.86 t \right]_0^{100}$$

(Note the constant of integration was omitted because it would be in both parts. It always disappears when the parts are subtracted and so it is normally left out of the working when integrating with limits.)

$$v - u = \left(\frac{7.0 \times 10^{-7}(100)^4}{4} - \frac{3.9 \times 10^{-4}(100)^3}{3} + \frac{0.11(100)^2}{2} + (0.86 \times 100)\right) - (0)$$

$$v = 17.5 - 130 + 550 + 86 - 0$$

$$v = 520 \text{ m s}^{-1}$$

....

3.

The acceleration of a particle in m s⁻² is described by the expression

$$a = 2.5t$$

where t is in seconds. The initial velocity of the particle is 6.0 m s⁻¹. Determine its velocity after 2.0 seconds.

$$\frac{dv}{dt} = a$$

$$\int_6^v dv = \int_0^2 2.5t \; dt$$

$$[v]_6^v = \left[\frac{2.5t^2}{2}\right]_0^2$$

$$v - 6 = \frac{2.5(2)^2}{2} - 0$$

$$v = 11 \text{ m s}^{-1}$$

....

4.

The acceleration of a particle in m s⁻² is described by the equation.

$$a = 5t$$

where t is in seconds. It starts from rest at $t = 0$ seconds. Determine the time taken for the particle to accelerate to 8.0 m s⁻¹.

$$\frac{dv}{dt} = a$$

$$\int_0^8 dv = \int_0^t 5t \; dt$$

$$[v]_0^8 = \left[\frac{5t^2}{2}\right]_0^t$$

$$8 = \frac{5t^2}{2} - 0$$

$$t = 1.8 \text{ s}$$

TOPIC 1. KINEMATIC RELATIONSHIPS

5.

The velocity of a car in m s^{-1} over a period of 8 seconds is described by the equation

$$v = 2.5t - 0.2t^2$$

where t is the time in seconds.
Find the distance travelled in this time interval.

$$v = \frac{ds}{dt}$$
$$\int_0^s ds = \int_0^8 \left(2.5t - 0.2t^2\right) \, dt$$
$$[s]_0^s = \left[\frac{2.5t^2}{2} - \frac{0.2t^3}{3}\right]_0^8$$
$$s = \left[\left(\frac{2.5 \times 8^2}{2} - \frac{0.2 \times 8^3}{3}\right)\right] - [0]$$
$$s = 45.9 \text{ m}$$

6.

When a car brakes to a halt, the displacement of the car from a reference point can be described by the equation

$$s = 5.8t - 2.5t^2$$

where t is the time in seconds.
Sketch a displacement-time graph for the motion.

$$v = \frac{ds}{dt} = \frac{d}{dt}\left(5.8t - 2.5t^2\right)$$
$$v = 5.8 - 5t$$
For $v = 0$ m s^{-1}, $t = \frac{5.8}{5} = 1.16 \text{ s}$

So the car comes to a halt at 1.2 s. The corresponding displacement is as follows.

$$s = (5.8 \times 1.16) - \left(2.5 \times 1.16^2\right)$$
$$s = 3.4 \text{ m}$$

© HERIOT-WATT UNIVERSITY

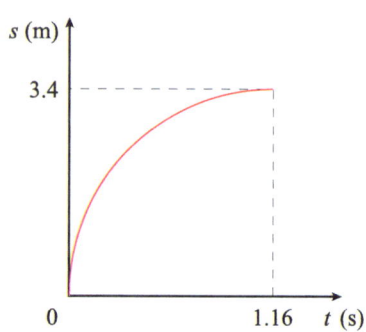

1.9 Extended information

Web links Go online

These web links should serve as an insight to the wealth of information available online and allow you to explore the subject further.

- https://www.youtube.com/watch?v=A5ANjmAzxKE&feature=youtu.be
- https://www.youtube.com/watch?v=WABpcU0mHMU
- http://www.bbc.co.uk/programmes/b009mvj0

1.10 Summary

> **Summary**
>
> You should now be able to:
>
> - use calculus notation to represent velocity as the rate of change of displacement with respect to time;
> - use calculus notation to represent acceleration as the rate of change of velocity with respect to time;
> - use calculus notation to represent acceleration as the second differential of displacement with respect to time;
> - derive the three kinematic relationships from the calculus definitions of acceleration and velocity:
>
> $$v = u + at \qquad s = ut + \tfrac{1}{2}at^2 \qquad v^2 = u^2 + 2as;$$
>
> - apply these equations to describe the motion of a particle with uniform acceleration moving in a straight line;
> - state that the gradient of a graph can be found by differentiation;
> - state that the gradient of a displacement-time graph at a given time is the instantaneous velocity;
> - state that the gradient of a velocity-time graph at a given time is the instantaneous acceleration;
> - state that the area under a graph can be found by integration;
> - state that for a velocity-time graph the displacement can be found by integrating between the limits;
> - state that for an acceleration-time graph the change in velocity can be found by integrating between the limits;
> - use calculus methods to solve problems based on objects moving in a straight line with varying acceleration.

1.11 End of topic 1 test

The following data should be used when required:

Gravitational acceleration on Earth g	9.8 m s^{-2}

Q6: A car accelerates uniformly from rest, travelling 55 m in the first 8.5 s.
Calculate its acceleration.
a = _____ m s $^{-2}$.

....

Q7: The brakes on a car can provide a deceleration of up to 4.9 m s $^{-2}$. The brakes are applied when the car is initially travelling at 20 m s $^{-1}$.
Calculate the minimum stopping distance.
Stopping distance = _____ m

....

Q8: A toy rocket is fired vertically upwards with an initial velocity of 16 m s $^{-1}$.
Calculate how long the rocket stays in the air before it returns to the ground.
t = _____ s

....

Q9: A piano is being raised to the third floor of a building using a rope and pulley. The piano is 15 m above the ground, moving upwards at 0.25 m s $^{-1}$, when the rope snaps.
Calculate how much time elapses before the piano hits the ground.
t = _____ s

....

Q10: The instantaneous acceleration of a body can be found by calculating the _____ of the tangent to the velocity-time graph.

....

Q11: The displacement can be found from a velocity-time graph by determining the area under the graph. This process is equivalent to _____ between limits.
This process is equivalent to _____ between limits.

....

TOPIC 1. KINEMATIC RELATIONSHIPS

Q12: The acceleration of a particle in m s^{-2} is described by the equation $a = 4.0t - 1.2$ where t is the time in seconds.

The particle starts from rest at the time $t = 0$ s.

Find the magnitude of its velocity at the instant $t = 5.0$.

v = _____ m s^{-1}

....

Q13: The acceleration in m s^{-2} of an object can initially be described by the equation $a = 4.20t - 0.600$ where t is in seconds.

The object is already travelling at 5.00 m s^{-1} at $t=0$ s.

Find the magnitude of its velocity after 3.00 s.

v = _____ m s^{-1}

....

Q14: The velocity of a particle moving in a straight line is given by the expression $v = 9.0t^2 - 0.5t$.

At time, $t=0$ s the displacement of the particle is 5.0 m.

Find the displacement of the particle at 2.0 s.

s = _____ m

Unit 1 Topic 2

Angular motion

Contents

- 2.1 Introduction 35
- 2.2 Angular displacement and radians 35
- 2.3 Angular velocity and acceleration 38
- 2.4 Kinematic relationships for angular motion 40
- 2.5 Tangential speed and angular velocity 46
- 2.6 Centripetal acceleration 50
- 2.7 Centripetal force 55
    - 2.7.1 Object moving in a horizontal circle 56
    - 2.7.2 Vertical motion 57
- 2.8 Applications 60
    - 2.8.1 Rollercoaster 60
    - 2.8.2 Conical pendulum 62
    - 2.8.3 Cars cornering 64
    - 2.8.4 Banked tracks 65
    - 2.8.5 Funfair rides 66
- 2.9 Extended information 68
- 2.10 Summary 68
- 2.11 End of topic 2 test 69

Prerequisites

- Kinematic relationships.
- Calculus - differentiation and integration.
- Addition and subtraction of vectors.
- Application of Newton's laws of motion to static and dynamic situations, including the use of free-body diagrams to solve problems.

> **Learning objective**
>
> By the end of this topic you should be able to:
>
> - measure angles and angular displacement in radians, and convert an angle measured in degrees into radians, and vice versa;
> - use angular displacement, angular velocity and angular acceleration to describe motion in a circle;
> - apply the equation $T = \frac{2\pi}{\omega}$ relating the periodic time to the angular velocity;
> - apply the angular kinematic relationships to solving problems of circular motion with uniform angular acceleration;
> - relate the angular velocity to the tangential (linear) speed of a body moving in a circle, and derive the equation $v = r\omega$;
> - apply the expression $a = r\alpha$ relating tangential and angular accelerations;
> - calculate the centripetal acceleration and centripetal force of an object undergoing circular motion;
> - use free-body diagrams to calculate centripetal forces;
> - describe a number of real-life situations in which the centripetal force plays an important role.

2.1 Introduction

In the Kinematics topic we studied the motion of objects travelling in a straight line. Now we will move on to study circular motion, which has a wide range of applications. These include the motion of planets around the Sun, the drum of a washing machine and a car taking a corner. We will consider the angular displacement, velocity and acceleration, and relate these quantities to the linear displacement travelled and the tangential speed and acceleration.

In the second part of this topic we will examine the centripetal force acting on an object. Without this force, an object would move off in a straight line at a tangent to the circle. We will then look at the variables that affect the size of the centripetal force acting on an object moving in a circular path.

2.2 Angular displacement and radians

In the first Topic of the course we investigated motion in a straight line. We are going to apply many of the ideas we met in that Topic to describe the motion of an object moving in a circle. We will see in the next Section that instead of using velocity and acceleration vectors, we will be using angular velocity and angular acceleration. Firstly we will look at angular displacement, which replaces the linear displacement we are used to dealing with. Imagine a disc spinning about a central axis, as shown in Figure 2.1. We can draw a reference line along the radius of the disc. The **angular displacement** after time t is the angle through which this line has swept in time t.

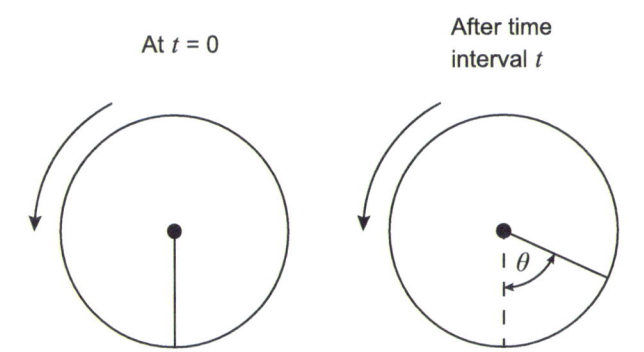

Figure 2.1: Angular displacement

The angular displacement is given the symbol θ, and is measured in **radians** (rad). Throughout this Topic, radians will be used to measure angles and angular displacement. A brief explanation of radian measurement, and how radians and degrees are related, will be given next.

Figure 2.2: Radian measurement

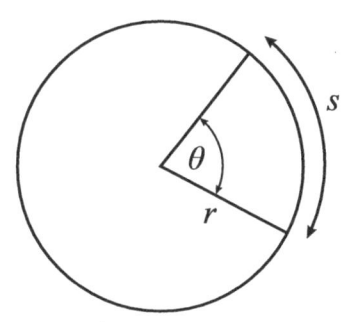

With reference to Figure 2.2, the angle θ, measured in radians, is equal to s/r. Since the radian is defined as being one distance (s) divided by another (r) then strictly speaking it is a dimensionless quantity, however the radian is regarded as a supplementary SI unit.

To compare radians with degrees, consider an angular displacement of one complete circle, equivalent to a rotation of 360°. In this case, the distance s in Figure 2.2 is equal to the circumference of the circle. Hence

$$\theta = \frac{s}{r}$$
$$\therefore \theta = \frac{2\pi r}{r}$$
$$\therefore \theta = 2\pi \text{ rad}$$

(2.1)

So 360° is equivalent to 2π rad, and this relationship can be used to convert from radians to degrees, and vice versa. It is useful to remember that π rad is equivalent to 180° and $\pi/2$ rad is equivalent to 90°. For the sake of neatness and clarity, it is common to leave an angle as a multiple of π rather than as a decimal, so the equivalent of 30° is usually expressed as $\pi/6$ rad rather than 0.524 rad.

Quiz: Radian measurement

Go online

Q1: Convert the angle 145° into radians.

a) 0.395 rad
b) 0.405 rad
c) 2.53 rad
d) 23.1 rad
e) 911 rad

....

Q2: What is the equivalent in degrees (°) to 1.20 radians?

a) 3.76°
b) 7.54°
c) 68.75°
d) 138°
e) 432°

....

Q3: Express 120° in radians.

a) $\pi/120$ rad
b) $\pi/4$ rad
c) $\pi/3$ rad
d) $2\pi/3$ rad
e) $3\pi/2$ rad

....

Q4: An object moves through 3 complete rotations about an axis. What is its total angular displacement?

a) $\pi/6$ rad
b) $\pi/3$ rad
c) 3π rad
d) 6π rad
e) $2\pi^3$ rad

....

Q5: An object moves 5.00 cm around the circumference of a circle of radius 24.0 cm. What is the angular displacement of the object?

a) 0.208 rad
b) 0.208π rad
c) 4.80 rad
d) 4.80π rad
e) 9.60π rad

2.3 Angular velocity and acceleration

Now that we have introduced the concept of angular displacement, it follows that the **angular velocity** ω is equal to the rate of change of angular displacement, just as the linear velocity v is the rate of change of linear displacement s.

$$\omega = \frac{d\theta}{dt} \qquad\qquad v = \frac{ds}{dt}$$

ω is measured in radians per second (rad/s or rad s^{-1}). The average angular velocity over a period of time t is the total angular displacement in time t divided by t.

Example

It takes the Moon 27.3 days to complete one orbit of the Earth. Assuming the Moon travels in a circular orbit at constant angular velocity, what is the angular velocity of the Moon?

The Moon moves through 2π rad in 27.3 days.

Now, 27.3 days is equal to (27.3 × 24) = 655.2 hours
which is equal to (655.2 × 60) = 39312 minutes
which is equal to (39312 × 60) = 2358720 s

So the angular velocity ω is given by

$$\omega = \frac{\text{angular displacement}}{\text{time elapsed}}$$
$$\therefore \theta = \frac{2\pi}{2358720}$$
$$\therefore \theta = 2.66 \times 10^{-6} \text{ rad s}^{-1}$$

You should also be aware of two other useful ways of describing the rate at which a body is moving in circular motion. One is to use the **periodic time** (or **period**) T, which is the time taken for one complete rotation. The other is to express the rate in terms of **revolutions per second**, which is the inverse of the periodic time. In the example above, the periodic time T for the Moon orbiting the Earth is 27.3 days, or 2.36×10^6 s in SI units. The rate of rotation is equal to $1/T = \frac{1}{2.36 \times 10^6} = 4.24 \times 10^{-7}$ revolutions per second.

TOPIC 2. ANGULAR MOTION

The relationship between periodic time and angular velocity is

$$\omega = \frac{2\pi}{T}$$

(2.2)

At National 5 we learned that period and frequency are related by the expression $f = \frac{1}{T}$. Rearranging Equation 2.2 gives us the expression Equation 2.3

$$\omega = 2\pi f$$

(2.3)

Having defined angular displacement θ and angular velocity ω, it should be clear that if ω is changing then we have an **angular acceleration**. The instantaneous angular acceleration α is the rate of change of angular velocity, measured in rad s^{-2}.

$$\alpha = \frac{d\omega}{dt}$$

(2.4)

The average angular acceleration over time t is the total change in angular velocity divided by t.

© HERIOT-WATT UNIVERSITY

UNIT 1. ROTATIONAL MOTION AND ASTROPHYSICS

Orbits of the planets

Following on from the above example, calculate the periodic times and angular velocities of the motion of Mercury, Venus and the Earth around the Sun.

Fill in the gaps in the table below.

Planet	Orbit radius (m)	Period (days)	Period (s)	Angular velocity (rad s^{-1})
Mercury	5.79×10^{10}	88.0		
Venus	1.08×10^{11}	225		
Earth	1.49×10^{11}	365		

2.4 Kinematic relationships for angular motion

In the Kinematics topic we derived the kinematic relationships for linear motion with constant acceleration from the definitions of instantaneous acceleration and velocity.

Similar techniques allow us to derive the equations of circular motion with constant angular acceleration. The derivations are included at the end of the topic for interest, but they are not required.

Key point

Remember that we are only considering motion where angular acceleration α is constant.

The angular motion relationships are the equivalent of the linear relationships. In linear motion we know the expression for final velocity $v = u + at$.

The rotational partner relationship is:

$$\omega = \omega_o + \alpha t \qquad (2.5)$$

Linear displacement for an accelerating object is $s = ut + \frac{1}{2}at^2$.

TOPIC 2. ANGULAR MOTION

Angular displacement is given by:

$$\theta = \omega_o + \frac{1}{2}\alpha t^2$$

(2.6)

Finally, the linear expression $v^2 = u^2 + 2as$ equivalent is:

$$\omega^2 = \omega_o^2 + 2\alpha\theta$$

(2.7)

We now have a set of three kinematic relationships that describe circular motion with constant angular acceleration. If you have trouble remembering them, you should be able to work them out by comparison with their linear equivalents:

Linear motion	Circular motion
$v = u + at$	$\omega = \omega_0 + \alpha t$
$s = ut + \frac{1}{2}at^2$	$\theta = \omega_0 t + \frac{1}{2}\alpha t^2$
$v^2 = u^2 + 2as$	$\omega^2 = \omega_0^2 + 2\alpha\theta$

For interest only

In the Kinematics topic we derived the kinematic relationships for linear motion with constant acceleration from the definitions of instantaneous acceleration and velocity. We can use the same technique to derive the equations of circular motion with constant angular acceleration by starting from the definitions of angular velocity and angular acceleration.

We begin with the definition of instantaneous angular acceleration

$$\alpha = \frac{d\omega}{dt}$$

Remember that we are only considering motion where α is a constant. Integrating the above expression over the time interval from 0 to t gives us

$$\int_{\omega_0}^{\omega} d\omega = \int_{t=0}^{t} \alpha\, dt = \alpha \int_{t=0}^{t} dt$$

Carrying out this integration

$$[\omega]_{\omega_0}^{\omega} = \alpha\,[t]_0^t$$
$$\therefore \omega - \omega_0 = \alpha t$$
$$\therefore \omega = \omega_0 + \alpha t$$

(2.8)

Equation 2.8 gives us the angular velocity after time t in terms of the angular velocity at $t = 0$ and the angular acceleration. Now we can substitute for $\omega = {d\theta}/{dt}$ in this equation

$$\frac{d\theta}{dt} = \omega_0 + \alpha t$$

Integrating this equation over the same time interval

$$\int_{\theta=0}^{\theta} d\theta = \int_{t=0}^{t} (\omega_0 + \alpha t)$$
$$\therefore \theta = \left[\omega_0 t + \frac{1}{2}\alpha t^2\right]_0^t$$
$$\therefore \theta = \omega_0 t + \frac{1}{2}\alpha t^2$$

(2.9)

TOPIC 2. ANGULAR MOTION

Equation 2.9 gives us the angular displacement after time t, again in terms of the angular velocity at $t = 0$ and the angular acceleration. Finally we use Equation 2.8, rearranged as follows

$$t = \frac{\omega - \omega_0}{\alpha}$$

Substituting this expression for t into Equation 2.9 gives us

$$\begin{aligned}\theta &= \omega_0 \left(\frac{\omega - \omega_0}{\alpha}\right) + \frac{1}{2}\alpha\left(\frac{\omega - \omega_0}{\alpha}\right)^2 \\ \therefore \alpha\theta &= \omega_0(\omega - \omega_0) + \frac{1}{2}(\omega - \omega_0)^2 \\ \therefore 2\alpha\theta &= 2\omega_0(\omega - \omega_0) + (\omega - \omega_0)^2 \\ \therefore 2\alpha\theta &= 2\omega_0\omega - 2\omega_0^2 + \omega^2 - 2\omega_0\omega + \omega_0^2 \\ \therefore 2\alpha\theta &= -\omega_0^2 + \omega^2 \\ \therefore \omega^2 &= \omega_0^2 + 2\alpha\theta\end{aligned}$$

(2.10)

We can now use these equations to solve problems involving motion with constant angular acceleration.

Examples

1.

An electric fan has blades that rotate with angular velocity 80 rad s^{-1}. When the fan is switched off, the blades come to rest after 12 s. What is the angular deceleration of the fan blades?

We follow the same procedure as we used to solve problems in linear motion - list the data and select the appropriate kinematic relationship.

Here we are told
$\omega_0 =$ 80 rad s^{-1}
$\omega =$ 0 rad s^{-1}
$t =$ 12 s
$\alpha =$?

With ω_0, ω and t known and α unknown, we use $\omega = \omega_0 + \alpha t$ to find α.

$$\omega = \omega_0 + \alpha t$$
$$\therefore \omega - \omega_0 = \alpha t$$
$$\therefore \alpha = \frac{\omega - \omega_0}{t}$$
$$\therefore \alpha = \frac{0 - 80}{12}$$
$$\therefore \alpha = -6.7 \text{ rad s}^{-2}$$

So the angular acceleration is -6.7 rad s^{-2}, equivalent to a deceleration of 6.7 rad s^{-2}.

....

2.

A wheel is rotating at 35 rad s^{-1}. It undergoes a constant angular deceleration. After 9.5 seconds, the wheel has turned though an angle of 280 radians.
What is the angular deceleration?

$$\theta = \omega_0 t + \frac{1}{2}\alpha t^2$$
$$280 = (35 \times 9.5) + \left(\frac{1}{2} \times \alpha \times 9.5^2\right)$$
$$\alpha = -1.2 \text{ rad s}^{-2}$$

So the angular acceleration is -1.2 rad s^{-2}, equivalent to a deceleration of 1.2 rad s^{-2}.

....

3.

An ice skater spins with an angular velocity of 15 rad s^{-1}. She decelerates to rest over a short period of time. Her angular displacement during this time is 14.1 rad.
Determine the time during which the ice skater decelerates.

$$\omega^2 = \omega_0^2 + 2\alpha\theta$$
$$0^2 = 15^2 + (2\alpha \times 14.1)$$
$$\alpha = -7.98 \text{ rad s}^{-2}$$

$$\omega = \omega_0 + \alpha t$$
$$0 = 15 - 7.98t$$
$$t = 1.9 \text{ s}$$

TOPIC 2. ANGULAR MOTION

Quiz: Angular velocity and angular kinematic relationships

Q6: A compact disc is spinning at a rate of 8.00 revolutions per second. What is the angular velocity of the disc?

a) 0.020 rad s^{-1}
b) 0.785 rad s^{-1}
c) 1.27 rad s^{-1}
d) 25.1 rad s^{-1}
e) 50.3 rad s^{-1}

....

Q7: A turntable rotates through an angle of 1.8π rad in 4.0 s. What is the average angular velocity of the turntable?

a) 0.45π rad s^{-1}
b) 0.90π rad s^{-1}
c) 2.2π rad s^{-1}
d) 4.0π rad s^{-1}
e) 7.2π rad s^{-1}

....

Q8: A disc is spinning about an axis through its centre with constant angular velocity 7.50 rad s^{-1}. What is the angular displacement of the disc in 8.00 s?

a) 0.938 rad
b) 1.07 rad
c) 9.55 rad
d) 60.0 rad
e) 377 rad

....

Q9: At the top of a hill, the wheels of a bicycle are rotating at angular velocity 5.0 rad s^{-1}. When the cyclist reaches the bottom of the hill, the angular velocity has increased to 12 rad s^{-1}. If it took the cyclist 10 s to cycle down the hill, what was the angular acceleration of the wheels?

a) 0.35 rad s^{-2}
b) 0.70 rad s^{-2}
c) 1.4 rad s^{-2}
d) 1.7 rad s^{-2}
e) 5.95 rad s^{-2}

....

Q10: A car parked on a slope begins to slowly roll forward. The angular acceleration of the wheels from rest is 0.20 rad s^{-2}. How long does it take for the wheels to rotate through one complete revolution?

a) 3.5 s
b) 5.0 s
c) 7.9 s
d) 12.6 s
e) 31 s

2.5 Tangential speed and angular velocity

When an object moves in a circle of radius r with angular velocity ω, what is the relationship between ω and v, the velocity of the object measured in m s^{-1}?

We can answer this question by considering the way the radian is defined. Looking back to Equation 2.2, the angular displacement θ, in radians, is given by $\theta = \frac{s}{r}$, where s is the distance swept out.

Since r is constant, we can derive this equation:

$$v = r\omega$$

(2.11)

Equation 2.11 gives us the relationship between the speed of the object (in m s^{-1}) and its angular velocity (in rad s^{-1}).

We shall see in the next topic that this speed is not the same as the linear velocity of the object, since the object is not moving in a straight line, and the velocity describes both the rate and direction at which an object is travelling. (Remember that velocity is a vector quantity.)

Figure 2.3: Tangential speed at several points around a circle

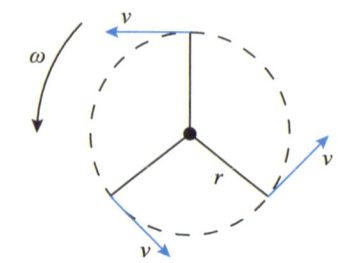

TOPIC 2. ANGULAR MOTION

Figure 2.3 shows that the speed v at any point is the **tangential speed**, and is always perpendicular to the radius of the circle at that point.

Imagine that Figure 2.3 shows a rubber bung on a string being whirled in a circle, what would happen if the string broke?

The bung continues to travel in a straight line in the direction of the linear speed arrow. Why?

With the string broken there is no unbalanced force acting on the bung so it travels at a constant speed in a straight line, that is, it would travel at a tangent to the circle. We will explore this question more fully in the next topic.

Equation 2.11 also shows that there is an important difference between v and ω - that two objects with the same angular velocity can be moving with different tangential speeds. This point is illustrated in the next worked example.

Example

Consider a turntable of radius 0.30 m rotating at constant angular velocity 1.5 rad s^{-1}. Compare the tangential speeds of a point on the circumference of the turntable and a point midway between the centre and the circumference.

The point on the circumference has r = 0.30 m, so

$$v_1 = r_1 \omega$$
$$\therefore v_1 = 0.30 \times 1.5$$
$$\therefore v_1 = 0.45 \text{ ms}^{-1}$$

The point midway between the centre and the circumference is moving with the same angular velocity, but the radius of the motion is only 0.15 m.

$$v_2 = r_2 \omega$$
$$\therefore v_2 = 0.15 \times 1.5$$
$$\therefore v_2 = 0.225 \text{ ms}^{-1}$$

The point on the circumference is moving at twice the speed (in m s^{-1}) of the point with the smaller radius.

This difference in tangential speeds is emphasised in Figure 2.4.

© HERIOT-WATT UNIVERSITY

Figure 2.4: Points moving at the same angular velocity, but with different tangential velocities

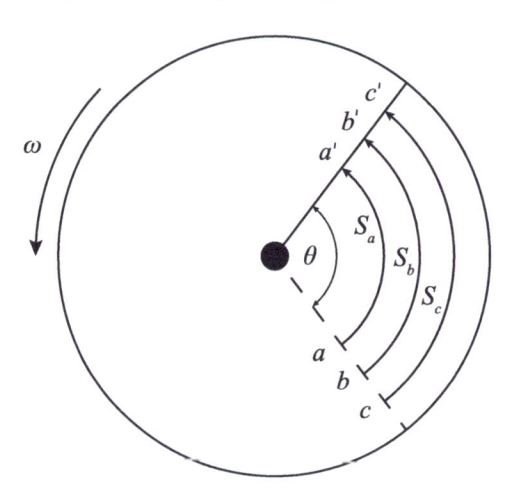

The points a, b and c all lie on the same radius of the circle, which is rotating at angular velocity ω about the centre of the circle. The radius moves through angle θ in time δt. In this same time δt, the points a, b and c move through distances S_a, S_b and S_c respectively, where $S_a < S_b < S_c$. The tangential speeds at each point on the radius are therefore $S_a/\delta t$, $S_b/\delta t$ and $S_c/\delta t$ where $S_a/\delta t < S_b/\delta t < S_c/\delta t$.

Compact disc player — Go online

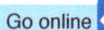

There is an online activity showing how the information on a CD is read by the laser in a CD player.

We can use Equation 2.11 to find an expression relating the angular acceleration to the **tangential acceleration**, the rate at which the speed of an object is changing.

$$v = r\omega$$

We can differentiate this equation with respect to time.

TOPIC 2. ANGULAR MOTION

$$\frac{dv}{dt} = \frac{d}{dt}(r\omega)$$
$$\therefore \frac{dv}{dt} = r\frac{d\omega}{dt}$$
$$\therefore a = r\alpha$$

(2.12)

As with the tangential speed, the tangential acceleration depends on the radius of the motion as well as the rate of change of angular velocity. This acceleration is always perpendicular to the radius of the circle.

Quiz: Angular velocity and tangential speed   Go online

Q11: A mass on the end of a 1.20 m length of rope is being swung round in a circle at an angular velocity of 4.00 rad s^{-1}. What is the tangential speed of the mass?

a) 0.300 m s^{-1}
b) 3.33 m s^{-1}
c) 4.80 m s^{-1}
d) 3.33π m s^{-1}
e) 4.80π m s^{-1}

....

Q12: An object is moving in a circular path with angular velocity 12.5 rad s^{-1}. If the speed of the object is 20.0 m s^{-1}, find the radius of the path.

a) 0.255 m
b) 0.625 m
c) 1.60 m
d) 10.1 m
e) 250 m

....

Q13: What is the speed of an object moving in a circle of radius 1.75 m with periodic time 2.40 s?

a) 0.729 m s^{-1}
b) 1.50 m s^{-1}
c) 2.29 m s^{-1}
d) 4.20 m s^{-1}
e) 4.58 m s^{-1}

....

50 UNIT 1. ROTATIONAL MOTION AND ASTROPHYSICS

Q14: A spinning disc slows down from ω = 4.50 rad s^{-1} to 1.80 rad s^{-1} in 5.00 s. If the radius of the disc is 0.200 m, find the tangential deceleration of a point on the circumference of the disc.

a) 0.108 m s^{-2}
b) 0.370 m s^{-2}
c) 0.540 m s^{-2}
d) 2.70 m s^{-2}
e) 3.39 m s^{-2}

....

Q15: An object is moving in a circle of radius 0.25 m with a constant angular velocity of 6.4 rad s^{-1}. Through what distance does the object travel in 2.0 s?

a) 1.6 m
b) 3.2 m
c) 10.2 m
d) 12.8 m
e) 51.2 m

2.6 Centripetal acceleration

Up until now in our discussion of circular motion, we have concentrated upon the kinematic relationships for circular motion. We will now shift our attention to how Newton's laws of motion apply to circular motion.

We will see that the velocity of an object moving in a circle is constantly changing, hence there is a constant acceleration, requiring a force that acts towards the centre of the circle. This acceleration is quite separate from the tangential acceleration (a) we have met previously.

After discussing Newton's laws we will move on to look at some practical examples of circular motion. We will be paying particular attention to the different forces that are involved, using free-body diagrams to determine the magnitude and direction of these forces.

Consider an object moving in a circle of radius r with constant angular velocity ω. We know that at any point on the circle, the object will have tangential velocity $v = r\omega$. The object moves through an angle $\Delta\theta$ in time Δt.

Figure 2.5 shows the velocity vectors at two points A and B on the circumference of the circle. The tangential velocities v_a and v_b at these points have the same magnitude, but are clearly pointing in different directions. If the angular displacement between A and B is $\Delta\theta$, then the angle between the vectors v_a and v_b is also $\Delta\theta$. Check you can prove this before you proceed.

© HERIOT-WATT UNIVERSITY

TOPIC 2. ANGULAR MOTION

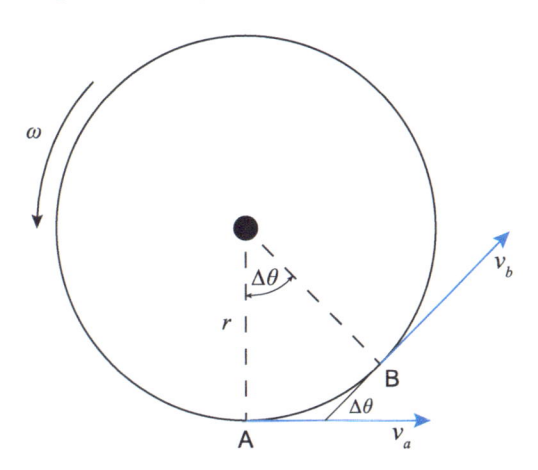

Figure 2.5: Velocity vectors at two points on a circle

The change in velocity Δv is equal to $v_b - v_a$. We can use a 'nose-to-tail' vector diagram to determine Δ_v, as shown in Figure 2.6. Both v_b and v_a have magnitude v, so the vector XY represents v_b and vector YZ represents $-v_a$.

Figure 2.6: Determination of the change in velocity

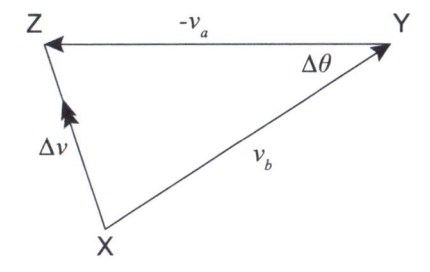

In the limit where Δt is small, $\Delta \theta$ tends to zero. In this case the angle ZXY tends to 90°, and the vector Δv is **perpendicular** to the velocity, so Δv points towards the centre of the circle. The velocity change, and hence the acceleration, is directed towards the **centre of the circle**. To distinguish this acceleration from any tangential acceleration that may occur, we will denote it by the symbol $a\perp$, since it is perpendicular to the velocity vector.

To calculate the magnitude of $a\perp$, we use the equation

$$a_\perp = \frac{\Delta v}{\Delta t}$$

© HERIOT-WATT UNIVERSITY

Since $\Delta\theta$ is small, $\Delta v = v\Delta\theta$, so long as $\Delta\theta$ is measured in radians. The acceleration is therefore

$$a_\perp = \frac{v\Delta\theta}{\Delta t}$$

Taking the limit where Δt approaches zero, this equation becomes

$$a_\perp = \frac{v d\theta}{dt}$$
$$\therefore a_\perp = v\omega$$

Finally we can substitute for $v = r\omega$ to get

$$a_\perp = \frac{v^2}{r} = r\omega^2$$

(2.13)

This acceleration $a_\perp$ is called the **centripetal acceleration**. It is always directed towards the centre of the circle, and it must not be confused with the tangential acceleration a we met previously, which occurs when an orbiting object changes its tangential speed. The centripetal acceleration occurs whenever an object is moving in a circular path, even if its tangential speed is constant. (In some textbooks it is referred to as the radial acceleration, as it is always directed along a radius of the circle.)

Example

Find the centripetal acceleration of an object moving in a circular path of radius 1.20 m with constant tangential speed of 4.00 m s^{-1}.

The centripetal acceleration is calculated using the formula

$$a_\perp = \frac{v^2}{r}$$
$$\therefore a_\perp = \frac{4.00^2}{1.20}$$
$$\therefore a_\perp = 13.3 \text{ m s}^{-2}$$

TOPIC 2. ANGULAR MOTION

Work through the next example to make sure you understand the difference between tangential and centripetal acceleration.

Example

A model aeroplane on a rope 10 m long is circling with angular velocity 1.2 rad s^{-1}. If this speed is increased to 2.0 rad s^{-1} over a 5.0 s period, calculate

1. the angular acceleration;
2. the tangential acceleration;
3. the centripetal acceleration at these two velocities.

1. The data in the question tells us ω_0 = 1.2 rad s^{-1}, ω = 2.0 rad s^{-1}, t = 5.0 s and α is unknown. Using the relationship $\omega = \omega_0 + \alpha t$ we can calculate the angular acceleration α.

$$\omega = \omega_0 + \alpha t$$
$$\therefore \alpha t = \omega - \omega_0$$
$$\therefore \alpha = \frac{\omega - \omega_0}{t}$$
$$\therefore \alpha = \frac{2.0 - 1.2}{5.0}$$
$$\therefore \alpha = 0.16 \text{ rad s}^{-2}$$

2. To calculate the tangential acceleration a, use the relationship between tangential and angular acceleration derived previously.

$$a = r\alpha$$
$$\therefore a = 10 \times 0.16$$
$$\therefore a = 1.6 \text{ m s}^{-2}$$

3. We use the formula $a_\perp = r\omega^2$ to calculate the centripetal acceleration. When ω = 1.2 rad s^{-1}

$$a_\perp = r\omega^2$$
$$\therefore a_\perp = 10 \times 1.2^2$$
$$\therefore a_\perp = 14.4 \text{ m s}^{-2}$$

When ω = 2.0 rad s^{-1}

$$a_\perp = r\omega^2$$
$$\therefore a_\perp = 10 \times 2.0^2$$
$$\therefore a_\perp = 40 \text{ m s}^{-2}$$

© HERIOT-WATT UNIVERSITY

Quiz: Centripetal acceleration

Q16: A disc of radius 0.50 m is rotating with angular velocity 4.1 rad s^{-1}. What is the centripetal acceleration of a point on its circumference?

a) 2.1 m s^{-2}
b) 4.2 m s^{-2}
c) 8.4 m s^{-2}
d) 17 m s^{-2}
e) 34 m s^{-2}

....

Q17: Which *one* of the following statements is true?

a) The centripetal acceleration is always at a tangent to the circle.
b) The centripetal acceleration is always directed towards the centre of the circle.
c) There is only a centripetal acceleration when the tangential speed is changing.
d) The centripetal acceleration does not depend on the angular velocity.
e) The centripetal acceleration does not depend on the radius of the circular motion.

....

Q18: A car takes a corner at 15 m s^{-1}. If the radius of the corner is 40 m, what is the centripetal acceleration of the car?

a) 0.18 m s^{-2}
b) 5.6 m s^{-2}
c) 107 m s^{-2}
d) 600 m s^{-2}
e) 9000 m s^{-2}

....

Q19: A particle moves in a circular path of radius 0.10 m at a constant rate of 6.0 revolutions per second. What is the centripetal acceleration of the particle?

a) 3.6 m s^{-2}
b) 3.8 m s^{-2}
c) 14 m s^{-2}
d) 36 m s^{-2}
e) 140 m s^{-2}

....

Q20: An object is moving in a circular path with speed v m s^{-1}. If the radius of the path doubles whilst the speed v remains constant, what happens to the centripetal acceleration?

a) The centripetal acceleration halves in value.
b) The centripetal acceleration doubles in value.
c) The new centripetal acceleration equals the square of the original value.
d) The new centripetal acceleration equals the square root of the original value.
e) The centripetal acceleration remains constant.

2.7 Centripetal force

Newton's second law of motion tells us that if an object is undergoing acceleration, then a net force must be acting on the object in the direction of the acceleration. Since we have a centripetal acceleration acting towards the centre of the circle, there must be a **centripetal force** acting in that direction.

Newton's law can be summed up by the equation

$$F = ma$$

If the centripetal acceleration is given by $a_\perp = v^2/r = r\omega^2$, then the centripetal force acting on a body of mass m moving in a circle of radius r is

$$F = \frac{mv^2}{r} = mr\omega^2$$

(2.14)

Figure 2.7: Centripetal force

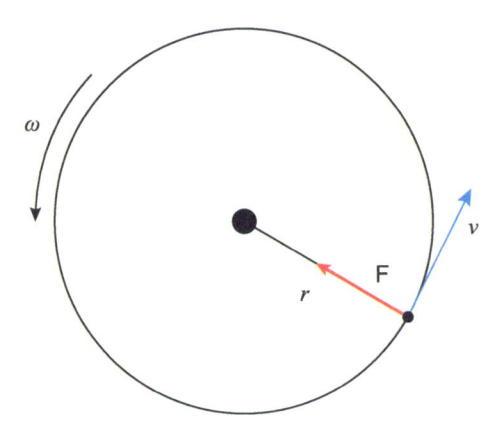

This is the force which must act on a body to make it move in a circular path. If this force is suddenly removed, the body will move in a straight line at a tangent to the circle with speed v, since there will be no force acting to change the velocity of the body.

Example

Compare the centripetal forces required for a 2.0 kg mass moving in a circle of radius 40 cm if the velocity is:

1. 3.0 m s^{-1};
2. 6.0 m s^{-1}.

1. Using Equation 2.14, the force required is

$$F = \frac{mv^2}{r}$$
$$\therefore F = \frac{2.0 \times 3.0^2}{0.40}$$
$$\therefore F = 45 \text{ N}$$

2. In this case

$$F = \frac{mv^2}{r}$$
$$\therefore F = \frac{2.0 \times 6.0^2}{0.40}$$
$$\therefore F = 180 \text{ N}$$

So doubling the velocity means that the centripetal force required increases by a factor of four, since $F \propto v^2$.

2.7.1 Object moving in a horizontal circle

Consider the case shown in Figure 2.8(a) of an object moving at constant angular velocity ω in a horizontal circle, such as a mass being whirled overhead on a string.

Figure 2.8: (a) An object moving in a horizontal circle; (b) horizontal force acting on the object

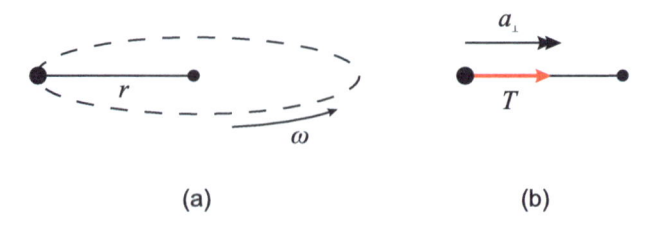

In the horizontal direction, Figure 2.8(b) shows there is an acceleration $a_\perp$ acting towards the centre of the circle. The only force acting in the horizontal direction is the tension in the string, so this tension must provide the centripetal force $mr\omega^2$. So in this case

TOPIC 2. ANGULAR MOTION

$$T = mr\omega^2$$

(2.15)

2.7.2 Vertical motion

Let us now consider the same object being whirled in a vertical circle. Figure 2.9 shows the object at three points on the circle, with the forces acting at each point.

Figure 2.9: Object undergoing circular motion in a vertical plane

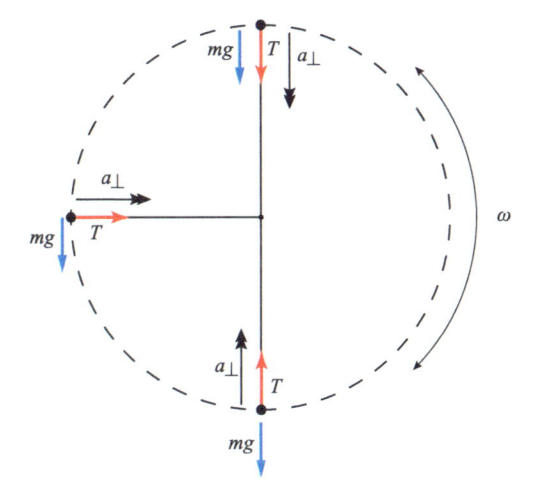

If we consider the forces acting when the object is at the top of the circle, we find that both the weight and the tension in the string are acting downwards. Thus the resultant force acting towards the centre of the circle is

$$T_{top} + mg = mr\omega^2$$

This resultant force must provide the centripetal force to keep the object moving in its circular path. Hence

$$T_{top} = mr\omega^2 - mg$$

(2.16)

© HERIOT-WATT UNIVERSITY

On the other hand, when the object reaches the lowest point on the circle, the tension and the weight are acting in opposite directions, so the resultant force acting towards the centre of the circle is

$$T_{bottom} - mg = mr\omega^2$$

Again, this force supplies the centripetal force so in this case

$$T_{bottom} = mr\omega^2 + mg$$

(2.17)

Equation 2.16 and Equation 2.17 give us the minimum and maximum values for the tension in the string.

When the string is horizontal, there is no component of the weight acting towards the centre of the circle, so the tension in the string provides all the centripetal force,

$$T_{horiz.} = mr\omega^2$$

(2.18)

Motion in a vertical circle

A man has tied a 1.2 kg mass to a piece of rope 0.80 m long, which he is twirling round in a vertical circle, so that the rope remains taut. He then starts to slow down the speed of the mass.

1. At what point of the circle is the rope likely to go slack?
2. What is the speed (in m s^{-1}) at which the rope goes slack?

TOPIC 2. ANGULAR MOTION

Quiz: Horizontal and vertical motion

Q21: What is the centripetal force acting on a 2.5 kg mass moving in a circle of radius 4.0 m when its speed is 8.0 m s^{-1}?

a) 5.0 N
b) 6.3 N
c) 40 N
d) 102 N
e) 640 N

....

Q22: A centripetal force of 25 N causes a 1.4 kg mass to move in a circular path of radius 0.50 m. What is the angular velocity of the mass?

a) 1.8 rad s^{-1}
b) 3.0 rad s^{-1}
c) 6.0 rad s^{-1}
d) 8.9 rad s^{-1}
e) 36 rad s^{-1}

....

Q23: Consider an object of fixed mass m moving in a circle with a constant tangential speed v. Which *one* of the following statements is true?

a) Since the tangential speed is constant, there is no centripetal force.
b) The centripetal force increases if the radius increases.
c) The centripetal acceleration increases if the centripetal force decreases.
d) If the radius of the circle is increased, the centripetal force decreases.
e) There is no centripetal acceleration.

....

Q24: A 1.20 kg mass on a 2.00 m length of string is being whirled in a horizontal circle. If the maximum tension in the string is 125 N, what is the maximum possible speed of the mass?

a) 8.66 m s^{-1}
b) 14.4 m s^{-1}
c) 17.3 m s^{-1}
d) 20.4 m s^{-1}
e) 208 m s^{-1}

....

Q25: A mass on a string is being whirled around in a vertical circle. At which point in its motion is the string most likely to snap?

a) When the mass is at the bottom of the circle.
b) When the mass is at the top of the circle.
c) When the string is horizontal.
d) When the string is at 45° to the horizontal.
e) There is an equal chance of the string snapping everywhere around the circle.

© HERIOT-WATT UNIVERSITY

2.8 Applications

The centripetal acceleration and centripetal force of an object moving in a circular path depend on the speed or angular velocity of the object and the radius of the circle. Without this force, the object would move off in a straight line at a tangent to the circle.

We will now calculate the centripetal force and acceleration in some practical applications. In each case a free-body diagram will be used to calculate the tensions and other forces that supply the centripetal force. Before carrying out the self-assessment test for this topic you should make sure you understand the forces involved in each of the examples. Try to think of other situations involving circular motion, and the forces that provide the centripetal force.

2.8.1 Rollercoaster

When a person rides on a rollercoaster that follows a loop the loop track, they often feel very "light" at the top of the loop. However, their weight does not change throughout the motion. In fact, it is the normal reaction force applied to them by the seat which alters. It is this which causes the strange sensation.

Assume the rollercoaster is moving at a constant speed, then the centripetal force required to keep them moving in a circle will also be constant. However, when they are at the top of the loop, the centripetal force is provided by both their weight and the normal reaction force. So the normal reaction force is very small and the person feels "light".

Rollercoaster Go online

There is an animation online displaying the forces acting on a person throughout a rollercoaster ride.

At the bottom

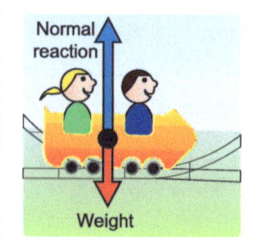

$$\frac{mv^2}{r} = Normal\ reaction\ -\ Weight$$

$$Normal\ reaction\ =\ \frac{mv^2}{r} + Weight$$

At the side

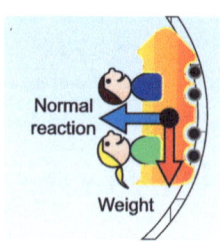

$$\frac{mv^2}{r} = Normal\ reaction$$

At the top

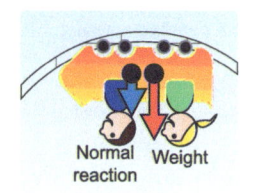

$$\frac{mv^2}{r} = Normal\ reaction + Weight$$

$$Normal\ reaction = \frac{mv^2}{r} - Weight$$

Infact, if the speed is sufficiently large, then the normal reaction force applied to the person at the top of the loop will become zero and the person will feel weightless.
At this point

$$\frac{mv^2}{r} = mg$$

Simplifying and rearranging, this becomes

$$v^2 = gr$$

Astronauts train in aircrafts that follow a circular path like this. At the top of a circular manoeuvre, they feel weightless. This allows them to become accustomed to experiencing the sensation of zero gravity. Since this was found to have an unsettling effect on their stomachs, such aircrafts were nicknamed "vomit comets".

2.8.2 Conical pendulum

The next situation we will study is the **conical pendulum** - a pendulum of length l whose bob moves in a circle of radius r at a constant height. Figure 2.10 shows such a pendulum, with a free-body diagram of all the forces acting on the bob.

Figure 2.10: (a) Conical pendulum moving in a horizontal circle (b) free-body diagram showing the forces acting on the bob

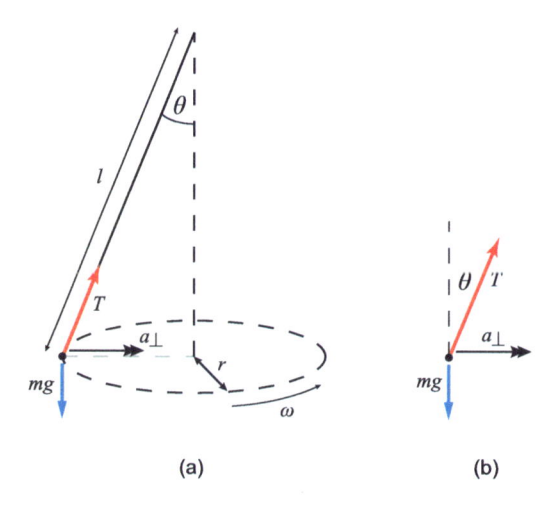

The string of the pendulum makes an angle θ with the vertical, such that $\sin\theta = {}^r\!/\!_l$. To calculate this angle, we use the free-body diagram in Figure 2.10(b) and resolve vertically and horizontally.

Vertically, the system is in equilibrium, so

$$T\cos\theta = mg$$

Horizontally, we have an acceleration towards the centre of the circle, so there must be a centripetal force which is provided by the horizontal component of the tension, $T\sin\theta$, so

$$T\sin\theta = mr\omega^2$$
$$\therefore T\sin\theta = ml\sin\theta\,\omega^2$$
$$\therefore T = ml\omega^2$$

We can substitute for $T = ml\omega^2$ in the first equation, which gives us

TOPIC 2. ANGULAR MOTION

$$ml\omega^2 \cos\theta = mg$$
$$\therefore l\omega^2 \cos\theta = g$$
$$\therefore \cos\theta = \frac{g}{l\omega^2}$$

(2.19)

We have an inverse square dependence of $\cos\theta$ on the angular velocity. If ω increases, $\cos\theta$ decreases and hence θ increases. The faster the bob is moving, the closer to horizontal the string becomes.

Example

Consider a conical pendulum of length 1.0 m. Compare the angle the string makes with the vertical when the pendulum completes exactly 1 revolution and 2 revolutions per second ($\omega = 2\pi$ rad s^{-1} and $\omega = 4\pi$ rad s^{-1}).

In the first instance, when $\omega = 2\pi$ rad s^{-1}

$$\cos\theta = \frac{g}{l\omega^2}$$
$$\therefore \cos\theta = \frac{9.8}{1.0 \times (2\pi)^2}$$
$$\therefore \cos\theta = \frac{9.8}{4\pi^2}$$
$$\therefore \cos\theta = 0.248$$
$$\therefore \theta = 76°$$

When $\omega = 4\pi$ rad s^{-1}

$$\cos\theta = \frac{g}{l\omega^2}$$
$$\therefore \cos\theta = \frac{9.8}{1.0 \times (4\pi)^2}$$
$$\therefore \cos\theta = \frac{9.8}{16\pi^2}$$
$$\therefore \cos\theta = 0.062$$
$$\therefore \theta = 86°$$

Clearly the angle θ approaches 90° as the speed of the pendulum bob gets higher and higher.

© HERIOT-WATT UNIVERSITY

| Conical pendulum | Go online |

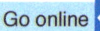

This simulation shows a practical example of a conical pendulum.

2.8.3 Cars cornering

When a car takes a corner, the frictional force between the car wheels and the road provides the centripetal force. If there is insufficient friction the car will skid. If we know the size of the frictional force, we can calculate the maximum speed at which a car can safely negotiate a corner.

Figure 2.11: (a) A car taking a corner; (b) the forces acting on the car

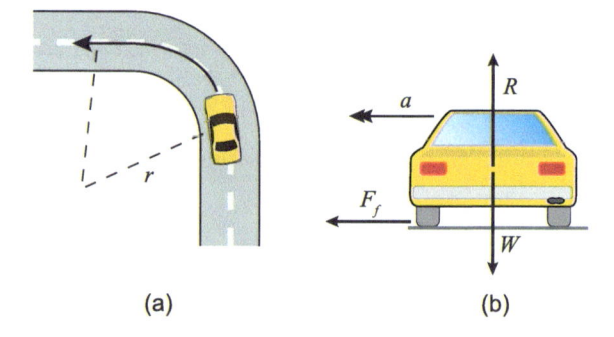

(a)   (b)

Figure 2.11(a) shows a car of mass m taking a corner. The radius of the bend is r. A free-body diagram of the forces acting on the car is shown in Figure 2.11(b). The centripetal force is provided by the frictional force F_f. Resolving vertically, using Newton's third law of motion, the normal reaction force R is equal to the weight $W = mg$. The frictional force provides the centripetal force acting on the car. If $F_{f\,\text{max}}$ is the maximum value of the frictional force, then the maximum speed at which the car can take the corner is given by

$$F_{f\,\text{max}} = \frac{mv_{\text{max}}^2}{r}$$

(2.20)

It is worth noting that the maximum value of the frictional force becomes extremely small when the road is icy, therefore reducing the maximum speed at which the car can take the corner.

2.8.4 Banked tracks

We saw that the safe cornering of a car on a flat road depends on whether the tyres and the road surface can provide sufficient friction. However, if a vehicle is on a banked track, then it can in fact still successfully take a corner when there is no frictional force present. Banked tracks are sloped at an angle. This means the horizontal component of the normal reaction force provides the required centripetal force. The vertical component of the normal reaction force balances the weight.

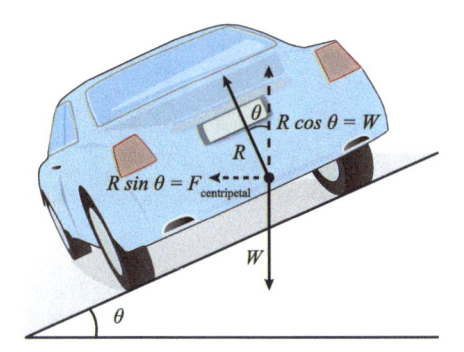

So for a track sloped at an angle θ to the horizontal

$$R \sin \theta = \frac{mv^2}{r}$$
$$R \cos \theta = mg$$

Dividing the top equation by the bottom gives

$$\tan \theta = \frac{v^2}{gr}$$

So

$$v = \sqrt{gr \tan \theta}$$

Looking at the equation above, we can see that a steeper bank allows the corner to be taken at a greater speed.

Banked corners   Go online

There is an animation online looking at a bobsleigh taking a banked corner on ice.

A velodrome is a bicycle racing track that consists of two 180° circular bends connected by two straight sections. It is another example of a banked track.

2.8.5 Funfair rides

A popular amusement park ride is called the sticky wall or the rotor. This involves a drum that spins around a vertical axis. People stand with their backs to the inside wall and the drum spins. Once it is rotating with a sufficiently large angular velocity, the floor drops away. Everyone has the sensation that they are "stuck" or "pinned" to the wall, but what is really happening is the drum wall is exerting a normal reaction force. This provides the centripetal force to keep them moving in a circular path. They do not fall down since the friction acting upwards balances their weight.

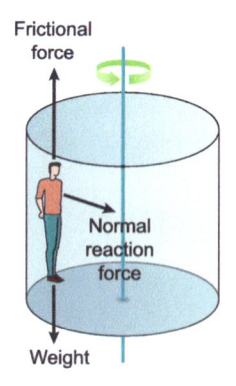

Quiz: Conical pendulum and cornering

Q26: The string of a conical pendulum makes an angle θ with the vertical, and the tension in its string is T. What is the centripetal force acting on the bob, in terms of T and θ?

a) $T \times \theta$
b) $T \times \cos\theta$
c) $T/\cos\theta$
d) $T \times \sin\theta$
e) $T/\sin\theta$

....

Q27: Consider a conical pendulum of length 50 cm. What angle does the string make with the vertical when the angular speed of the pendulum is 6.0 rad s^{-1}?

a) 0.55°
b) 33°
c) 57°
d) 87°
e) 89°

....

Q28: The conical pendulum is rotating at constant angular velocity. What is the period of the pendulum?

a) $2\pi \frac{g}{l\cos\theta}$
b) $2\pi \sqrt{g/l\cos\theta}$
c) $\sqrt{l\cos\theta/g}$
d) $2\pi \frac{l\cos\theta}{g}$
e) $2\pi \sqrt{l\cos\theta/g}$

....

Q29: For a car cornering on a flat (unbanked) road, the

a) centripetal force is provided by the frictional force.
b) centripetal force is provided by the normal reaction force.
c) centripetal force is provided by the weight.
d) normal reaction force is equal to the frictional force.
e) weight is equal to the frictional force.

....

Q30: A car can take a corner on a banked road at a higher speed than on a horizontal road because

a) the weight of the car increases.
b) less centripetal force is required.
c) the radius of the corner decreases.
d) a component of the normal reaction force contributes to the centripetal force.
e) the mass of the car decreases.

2.9 Extended information

Web links — Go online

These web links should serve as an insight to the wealth of information available online and allow you to explore the subject further.

- https://www.youtube.com/watch?v=w2J6eC43_mQ
- https://www.youtube.com/watch?v=K3ouHnEenoo
- https://www.youtube.com/watch?v=g3GC3F7zrok
- https://www.youtube.com/watch?v=31MMw3Eazqw
- https://www.youtube.com/watch?v=x4PoepVaLDQ

2.10 Summary

Summary

You should now be able to:

- measure angles and angular displacement in radians, and convert an angle measured in degrees into radians, and vice versa;
- use angular displacement, angular velocity and angular acceleration to describe motion in a circle;
- apply the equation $T = \frac{2\pi}{\omega}$ relating the periodic time to the angular velocity;
- apply the angular kinematic relationships to solving problems of circular motion with uniform angular acceleration;
- relate the angular velocity to the tangential (linear) speed of a body moving in a circle, and derive the equation $v = r\omega$;
- apply the expression $a = r\alpha$ relating tangential and angular accelerations;
- calculate the centripetal acceleration and centripetal force of an object undergoing circular motion;
- use free-body diagrams to calculate centripetal forces;
- describe a number of real-life situations in which the centripetal force plays an important role.

2.11 End of topic 2 test

Q31: How many radians are equivalent to 66°?

....

Q32: A flywheel is rotating with constant angular velocity 21 rad s^{-1}.
Calculate the time taken for the flywheel to complete exactly 6 revolutions.
Time taken = _____ s

Q33: A moon orbiting a distant planet has a period of 24 days.
Calculate the angular velocity of the moon, in rad s^{-1}.
Angular velocity = _____ rad s^{-1}

Q34: An electric fan is switched from a "Low" setting (angular velocity 4.9 rad s^{-1}) to a "High" setting (angular velocity 15 rad s^{-1}).
The angular acceleration is 6.6 rad s^{-2}.
Calculate the time taken for the fan to reach the higher angular velocity.
t = _____ s

Q35: As the brakes are applied to a bicycle wheel, the angular velocity of the wheel changes from 23.50 rad s^{-1} to 5.00 rad s^{-1}.
The wheel turns through exactly 10 revolutions whilst the brakes are being applied.
Calculate the magnitude of the angular deceleration of the wheel.
Angular deceleration = _____ rad s^{-2}

Q36: Passengers in a Ferris Wheel sit in cars 11 m from the centre of the wheel. The wheel takes 100 s to complete one revolution, travelling with uniform angular velocity.
Calculate the speed of a passenger car.
Speed = _____ m s^{-1}

Q37: A cyclist travelling around a circular track of radius 32 m increases his speed from 13 m s^{-1} to 25 m s^{-1} in 5.0 s.
Calculate the average angular acceleration of the cyclist.
Average angular acceleration = _____ rad s^{-2}

UNIT 1. ROTATIONAL MOTION AND ASTROPHYSICS

Q38: A cyclist is riding in a circle of radius 17.0 m at an average speed of 9.5 m s^{-1}.

Calculate the cyclist's centripetal acceleration.

Centripetal acceleration = _____ m s^{-2}

....

Q39: A disc is rotating at constant angular velocity about its centre. At the edge of the disc, the centripetal acceleration is 6.9 m s^{-2}, whilst at a point 6.5 cm from the centre of the disc, the centripetal acceleration is 5.2 m s^{-2}.

Calculate the radius of the disc.

Radius = _____ cm

....

Q40: A 1.25 kg mass is moving in a circular path of radius 1.82 m, with periodic time 1.95 s.

Calculate the centripetal force acting on the mass.

Centripetal force = _____ N

....

Q41: A model aeroplane (mass 0.80 kg) can fly at up to 28 m s^{-1} in a horizontal circle on a guideline which has a breaking tension of 100 N.

Calculate the minimum radius in which the plane could be flown if it is travelling at its maximum speed.

Minimum radius = _____ m

....

Q42: A conical pendulum consists of a string of length 1.1 m and a bob of mass 0.34 kg. The string makes an angle of 25° with the vertical.

Calculate the tension in the string and the angular velocity of the pendulum.

1. Tension = _____ N
2. Angular velocity = _____ rad s^{-1}

....

Q43: A motorcyclist is approaching a hump-backed bridge, as shown in the diagram. The bridge forms part of a circle of radius r = 20.0m. The combined mass of the motorcycle and rider is 165 kg.

Calculate the maximum speed at which the rider could travel without leaving the road at the top of the bridge.

© HERIOT-WATT UNIVERSITY

Maximum speed = _____ m s^{-1}

....

Q44: The static frictional force between a car of mass 1990 kg and a dry road has a maximum value of 1.71×10^4 N.

Calculate the maximum speed at which the car could take an unbanked bend of radius 40.0 m without skidding.

Maximum speed = _____ m s^{-1}

Unit 1 Topic 3

Rotational dynamics

Contents

3.1 Introduction	75
3.2 Torque and moment	75
3.3 Newton's laws applied to rotational dynamics	78
3.3.1 Torques in equilibrium	78
3.3.2 Angular acceleration and moment of inertia	82
3.4 Calculating the moment of inertia	83
3.4.1 Combinations of rotating bodies	87
3.5 Extended information	89
3.6 Summary	89
3.7 End of topic 3 test	90

Prerequisites

- Applying Newton's laws of motion.
- Angular velocity and acceleration.
- Understanding of the principle of conservation of linear momentum.

> **Learning objective**
>
> By the end of this topic you should be able to:
>
> - state that the moment (torque) of a force is the tendency of a force to cause or change rotational motion of an object;
> - state and apply the equation $T = Fr$;
> - state that an unbalanced torque acting on an object produces an angular acceleration;
> - state that the moment of inertia of an object is a measure of its resistance to angular acceleration about a given axis;
> - explain that the moment of inertia of an object about an axis depends on the mass of the object, and the distribution of the mass about the axis;
> - state that the moment of inertia I of a point mass of mass m at a distance r from a fixed axis is given by the equation $I = mr^2$;
> - use the relevant equation to calculate the moment of inertia for a point mass and the following shapes - rod about centre, rod about end, disc about centre, sphere about centre;
> - calculate the moment of inertia of a combination of rotating bodies;
> - state and apply the equation $T = I\alpha$.

TOPIC 3. ROTATIONAL DYNAMICS

3.1 Introduction

In the previous topic we derived kinematic relationships for describing circular motion which compared directly to the linear kinematic relationships. In this topic, we will now perform a similar exercise for other aspects of rotational dynamics. We will begin by looking at the turning effect (torque) of a force being used to make an object rotate about an axis. This will lead us to the concept of moment of inertia.

3.2 Torque and moment

We will begin by studying the turning effect of a force, and the example we shall look at is a force being used to push a door open. The axis of rotation is the door hinge. What happens if you push the door in the same place with different forces?

Figure 3.1: (a) Small force, and (b) large force used to push open a door

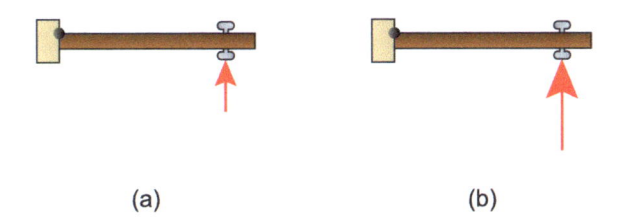

(a)   (b)

Since $F = ma$, the door opens more quickly the larger the force you use, something you can easily check for yourself! The turning effect increases when the force is increased.

Now think about what happens if you apply the same force near the hinge or near the edge of the door.

Figure 3.2: Same force being applied (a) far from the axis, and (b) close to the axis

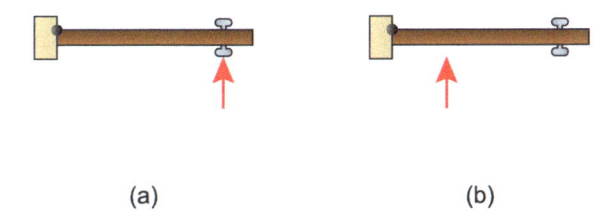

(a)   (b)

The force has a greater turning effect when it is applied near the edge of the door - it is harder to open the door when you push it near the hinge. So the turning effect also depends on how far from the axis of rotation the force is being applied.

© HERIOT-WATT UNIVERSITY

The turning effect of a force is called the **torque** T or **moment**, and is defined by

$$T = Fr$$

(3.1)

In this equation F is the size of the force and r is the distance between the axis and the point where the force is being applied. Since the torque is producing a circular motion, r is the radius of the motion. T has the units N m. Looking at Figure 3.3 below, we can see that Equation 3.1 does not quite tell the full story. This equation is only valid when the force is applied at right angles to the radius.

Figure 3.3: Same force applied (a) perpendicular to the door, and (b) at an angle to the door

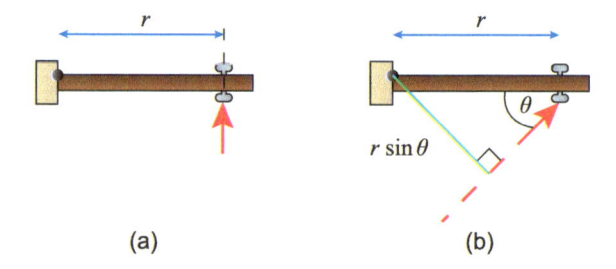

The dashed line in Figure 3.3 represents the **line of action** of the force. When calculating the torque, the important distance is not where the force is applied, but the **perpendicular distance from the line of action to the axis of rotation**. If F makes an angle θ to the line r joining the axis to the point where the force is applied, then the torque is given by

$$T = Fr \sin \theta$$

(3.2)

Example

A gardener picks up her wheelbarrow to wheel it along by applying a perpendicular force of 50 N to the handles. If the handle grips are 1.2 m from the axis of the wheel, what is the torque that she applies?

The situation is shown schematically in Figure 3.4

TOPIC 3. ROTATIONAL DYNAMICS

Figure 3.4: Torque applied to the wheelbarrow

Since a perpendicular force is applied, we can use Equation 3.1

$$T = Fr$$
$$\therefore T = 50 \times 1.2$$
$$\therefore T = 60 \, \text{N m}$$

Quiz: Torques

Go online

Q1: Which of the following sets of dimensions could be used to measure a torque?

a) kg m
b) kg m^2
c) N kg m^2
d) N m^{-1}
e) N m

....

Q2: What is the torque when a 35 N force is applied at a distance 1.4 m from an axis, perpendicular to the radius?

a) 0.04 N m
b) 25 N m
c) 35 N m
d) 49 N m
e) 69 N m

....

Q3: A 16.0 N force produces a 60.0 N m torque when applied at right angles to the radius of rotation. How far from the axis is the force applied?

a) 3.75 m
b) 26.7 m
c) 48.0 m
d) 60.0 m
e) 960 m

....

Q4: A force of 12.0 N is applied 0.750 m from the axis to produce a torque. The force is applied at an angle of 70° to the radius of motion. What is the size of the torque?

a) 0.156 N m
b) 3.08 N m
c) 8.46 N m
d) 9.00 N m
e) 630 N m

....

Q5: A torque of 48.0 N m is produced when a force is applied 0.500 m from an axis. If the force makes an angle of 60° with the radius, what is the magnitude of the force?

a) 20.8 N
b) 83.1 N
c) 96.0 N
d) 111 N
e) 192 N

3.3 Newton's laws applied to rotational dynamics

We are used to the idea of forces in equilibrium producing no acceleration, and a resultant force producing an acceleration in the direction of the force. A similar situation exists when we are applying torques instead of forces. A system of torques in equilibrium will produce no net turning effect, whilst a resultant torque produces an angular acceleration about the axis.

3.3.1 Torques in equilibrium

To consider the equilibrium situation, let us return to the problem of the wheelbarrow. We will assume the barrow stays horizontal, and we will remove its support so that the gardener has to apply a perpendicular force to keep the barrow horizontal. Suppose the barrow is loaded with 20 kg of rubble, and the centre of mass of this rubble is 50 cm from the wheel axis. If we can neglect the mass of the barrow, what force is necessary to keep it horizontal?

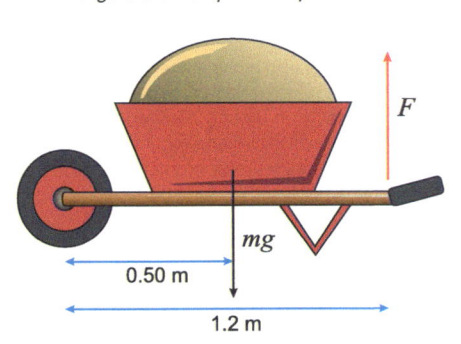

Figure 3.5: Torques in equilibrium

For the above system to be in equilibrium, the clockwise torque (due to the mass of the rubble) must equal the anti-clockwise torque (due to the gardener applying a force). Therefore

$$T_{a-c} = T_c$$
$$\therefore F \times 1.2 = mg \times 0.50$$
$$\therefore F \times 1.2 = 20 \times 9.8 \times 0.50$$
$$\therefore F \times 1.2 = 98$$
$$\therefore F = 82\,\text{N}$$

The problem is similar to a Newton's first law problem, and the method of solving it is the same: draw a free-body diagram to show all the forces acting. The difference is that we are no longer considering point objects, so we must also include where the forces are acting on the diagram. In the above problem, we have been told to ignore the weight of the barrow. In a situation where this cannot be ignored, the weight acts through the centre-of-mass of the object, and the distance from the axis to the centre-of-mass must be known to calculate the torque. The next example illustrates how to solve more complicated problems.

Example

A serving table of mass 2.5 kg and length 80 cm is hinged to a wall, and is supported by a chain which makes an angle of 50° with the horizontal table top. The chain is attached to the edge of the table furthest from the wall (r_c = 80 cm). The table is uniform, so its centre-of-mass is r_t = 40 cm from the wall. A full serving dish of mass 1.0 kg is placed on the table r_d = 60 cm from the wall. Calculate the tension in the chain.

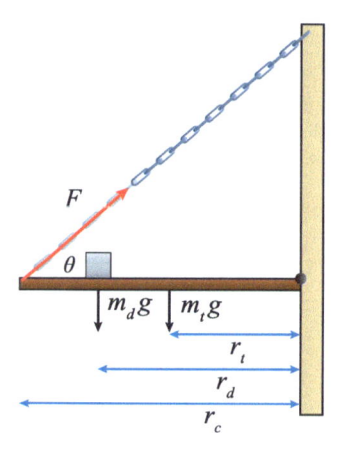

Figure 3.6: Free-body diagram showing the forces acting on the table

In equilibrium, the anti-clockwise torques must balance the clockwise torques. This means the torques due to the two masses are balanced by the torque due to the tension in the chain.

$$F \times \sin\theta \times r_c = (m_d \times g \times r_d) + (m_t \times g \times r_t)$$
$$\therefore F \times 0.766 \times 0.80 = (1.0 \times 9.8 \times 0.60) + (2.5 \times 9.8 \times 0.40)$$
$$\therefore 0.6128 F = 5.88 + 9.8$$
$$\therefore 0.6128 F = 15.68$$
$$\therefore F = 25.6 \, N$$

If you try this calculation again with the dish at a different position, you will find that the tension in the chain depends on whereabouts on the table the dish is positioned. If the dish is placed on the edge of the table, (r_d = 80 cm), then the tension F = 28.8 N, whereas placing the dish at r_d = 20 cm from the wall reduces F to 19.2 N. Clearly, when we are considering torques, we are not just interested in the mass of an object or objects, but how that mass is distributed relative to the axis.

TOPIC 3. ROTATIONAL DYNAMICS

Torque and static equilibrium

In questions 1 to 4, try to balance the beam by making sure that the moment (sum of force x distance) is the same on each side of the fulcrum (pivot). The length of the beam is 8.0 m and the fulcrum is at the centre of the beam.

Q6:

Load A = 450N, sits 4 m from the fulcrum.

Where should B = 600N sit?

To the left or right?

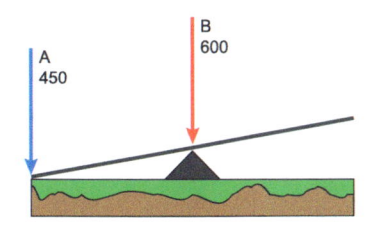

....

Q7:

Load A now moves to 3 m from the fulcrum.

Load B distance = _____ ?

To the left or right?

....

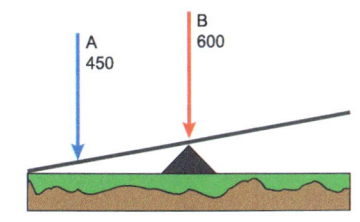

Q8:

Load C = 480N is added on A's side, sitting 2 m from the fulcrum.

Load B distance = _____ ?

To the left or right?

....

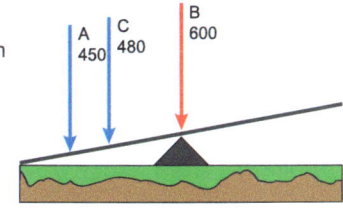

Q9:

Load D = 360N is added on B's side, 4 m from fulcrum.

Load B distance = _____ ?

To the left or right?

© HERIOT-WATT UNIVERSITY

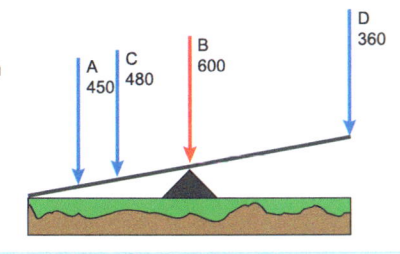

3.3.2 Angular acceleration and moment of inertia

Let us now consider a Newton's second law-type problem. We are used to a force producing an acceleration, and the relationship $F = ma$. We will now see if there is an equivalent expression involving torques and angular acceleration. The problem we will look at is that of an object moving in a horizontal circle of fixed radius r at an angular velocity ω.

Figure 3.7: (a) Object moving around a circle; (b) a tangential force applied to the object

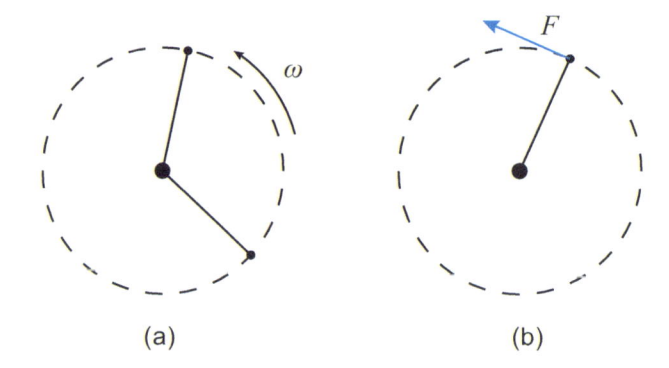

If we apply an anticlockwise force F to the object at a tangent to the circle, as shown in Figure 3.7(b), the object will accelerate. We can use Newton's second law

$$F = ma$$

We can express this acceleration as an angular acceleration using the expression $a = r\alpha$. Hence

$$F = mr\alpha$$

Since the force is being applied at a distance r from the axis of rotation, perpendicular to the radius, we can substitute for F using $T = Fr$

$$\frac{T}{r} = mr\alpha$$
$$\therefore T = mr^2\alpha$$
$$\therefore T = I\alpha$$

(3.3)

TOPIC 3. ROTATIONAL DYNAMICS

We have used the symbol I to represent mr^2 to give us an expression in Equation 3.3 which looks similar to $F = ma$. The quantity I is called the **moment of inertia**, and for a point mass m rotating with radius r about an axis.

$$I = mr^2$$

(3.4)

I has the units kg m^2. Throughout this topic we will be using I in our calculations, as it takes into account not only the mass of an object, but also how far the mass is from the axis of rotation.

There are some important properties about the moment of inertia that you must make sure you understand. First, we cannot state a fixed value of I for any object. The value of I depends not only on the mass of the object, but the position of the axis. The axis **must** be specified when giving the moment of inertia of an object. Although there are similarities between mass and moment of inertia, there is a major difference in that the mass of a body is constant but the moment of inertia changes for different axes of rotation.

Secondly, the equation $I = mr^2$ is only valid for a point mass with a known position. For a collection of masses $m_1, m_2, m_3...m_i$, the total moment of inertia is

$$I_{\text{total}} = m_1 r_1^2 + m_2 r_2^2 + m_3 r_3^2 + ... = \sum_i m_i r_i^2$$

(3.5)

3.4 Calculating the moment of inertia

You should be aware that a satellite can be considered as a point mass, since it is so small compared to the radius of its orbit. This means that the moment of inertia of a satellite can be found from the equation $I = mr^2$.

It is also worth noting that the mass of a thin hoop is all located at the same distance r from its central axis. So, a thin hoop's moment of inertia can also be found using the equation $I = mr^2$.

Finding the moment of inertia for other rotating bodies is considerably more complicated. Using a calculus approach, the body is divided into infinitely small sections. I is calculated for each section and then an integration is carried out to find the total moment of inertia. Such a calculation is beyond the scope of this course. However, you will be expected to employ the following equations for a number of standard shapes. You will need to remember to use the same formula for a hoop as for a point mass, but the other equations should be provided in the exam.

© HERIOT-WATT UNIVERSITY

Table 3.1: The moment of inertia

Shape	Axes of rotation	Equation for moment of inertia
Hoop about centre		$I = mr^2$
Rod about centre		$I = \dfrac{1}{12}ml^2$
Rod about end		$I = \dfrac{1}{3}ml^2$
Disc about centre		$I = \dfrac{1}{2}mr^2$

Shape	Axes of rotation	Equation for moment of inertia
Sphere about centre		$I = \dfrac{2}{5}mr^2$

Moment of inertia — Go online

There is animated version available online showing different axes of rotation.

TOPIC 3. ROTATIONAL DYNAMICS

Use the relevant information from Table 3.1 for following examples.

Examples

1.

A baton of length 80 cm and mass 120 g is rotating about one end. Calculate its moment of inertia.

$$I = \frac{1}{3} m l^2$$
$$I = \frac{1}{3} \times 0.12 \times 0.8^2$$
$$I = 0.0256 \text{ kg m}^2$$

....

2.

A baton of length 80 cm and mass 120 g is rotating about its centre. Calculate its moment of inertia.

$$I = \frac{1}{12} m l^2$$
$$I = \frac{1}{12} \times 0.12 \times 0.8^2$$
$$I = 0.0064 \text{ kg m}^2$$

....

© HERIOT-WATT UNIVERSITY

3.

A flat disk of pizza dough of mass 800 g and radius 25 cm is spun in a circle about its middle.
Calculate its moment of inertia.

$$I = \frac{1}{2}mr^2$$
$$I = \frac{1}{2} \times 0.8 \times 0.25^2$$
$$I = 0.025 \text{ kg m}^2$$

....

4.

A hula hoop has an axis of rotation as shown. Its mass is 2.0 kg and radius is 60 cm.
Find its moment of inertia.

$$I = mr^2$$
$$I = 2 \times 0.6^2$$
$$I = 0.72 \text{ kg m}^2$$

....

5.

A football of mass 249 g is made to rotate about its centre. The diameter of the ball is 15.0 cm.
Find the ball's moment of inertia about this axis of rotation.

$$I = \frac{2}{5}mr^2$$
$$I = \frac{2}{5} \times 0.249 \times 0.075^2$$
$$I = 5.60 \times 10^{-4} \text{ kg m}^2$$

3.4.1 Combinations of rotating bodies

Let us now consider systems involving more than one rotating body.

Example

A Catherine wheel consists of a light rod 0.64 m in length, pivoted about its centre. At either end of the rod a 0.50 kg firework is attached, each of which provides a force of 25 N perpendicular to the rod when lit. Calculate the angular acceleration of the Catherine wheel when the fireworks are lit.

The angular acceleration is calculated using the equation $T = I\alpha$, so we need to calculate the moment of inertia of the Catherine wheel and the total torque provided by the two fireworks.

Figure 3.8: Catherine wheel diagram

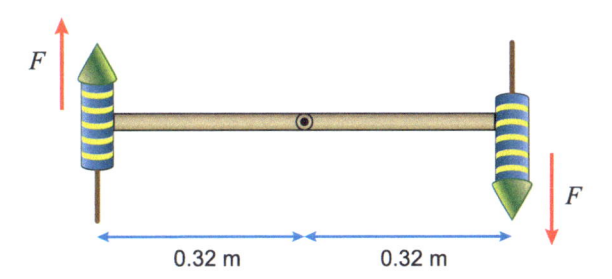

The total moment of inertia is

$$I = \sum mr^2$$
$$\therefore I = (0.50 \times 0.32^2) + (0.50 \times 0.32^2)$$
$$\therefore I = 0.1024 \text{ kg m}^2$$

The total torque is the sum of the torques provided by each firework

$$T = \sum Fr$$
$$\therefore T = (25 \times 0.32) + (25 \times 0.32)$$
$$\therefore T = 16 \text{ N m}$$

The angular acceleration can now be calculated

$$\alpha = \frac{T}{I}$$
$$\therefore \alpha = \frac{16}{0.1024}$$
$$\therefore \alpha = 160 \text{ rad s}^{-2}$$

Combinations of rotating bodies

Different masses are placed on a rotating platform to study the moment of inertia of a system. Investigate how the same moment of inertia can be obtained with different combinations of mass and position.

We will be considering a platform of negligible mass, with an axis through its centre. The radius of the platform is 50 cm.

Q10: What is the moment of inertia when a 1.0 kg mass is placed 40 cm from the axis as shown?

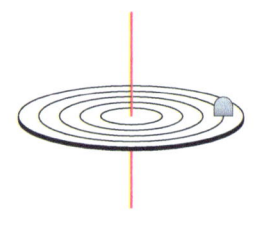

....

Q11: If we remove the 1.0 kg mass, where should a 4.0 kg mass be placed on the platform to produce the same moment of inertia?

....

Q12: Suppose both masses are placed on the platform, with the 4.0 kg mass 30 cm from the centre and the 1.0 kg mass 20 cm from the centre. What is the moment of inertia now?

....

Q13: Calculate the moment of inertia of the following system shown: a 1.0 kg mass 10 cm from the centre, a 2.5 kg mass 20 cm from the centre and a 2.0 kg mass 40 cm from the centre.

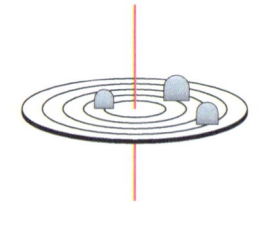

....

Q14: Finally, consider the masses in previous question. If they were combined to form a single object of mass 5.5 kg, where should it be placed to give the same moment of inertia as obtained in question 4?

3.5 Extended information

Web links — Go online

These web links should serve as an insight to the wealth of information available online and allow you to explore the subject further.

- https://www.youtube.com/watch?v=m9weJfoW5J0
- https://www.youtube.com/watch?v=cB8GNQuyMPc
- http://www.ck12.org/physics/Moment-of-Inertia/lecture/Moment-of-Inertia-Demonstrated-with-Rulers/
- https://www.youtube.com/watch?v=CHQOctEvtTY
- http://hyperphysics.phy-astr.gsu.edu/hbase/mi.html

3.6 Summary

Summary

You should now be able to:

- state that the moment (torque) of a force is the tendency of a force to cause or change rotational motion of an object;
- state and apply the equation $T = Fr$;
- state that an unbalanced torque acting on an object produces an angular acceleration;
- state that the moment of inertia of an object is a measure of its resistance to angular acceleration about a given axis;
- explain that the moment of inertia of an object about an axis depends on the mass of the object, and the distribution of the mass about the axis;
- state that the moment of inertia I of a point mass of mass m at a distance r from a fixed axis is given by the equation $I = mr^2$;
- use the relevant equation to calculate the moment of inertia for a point mass and the following shapes - rod about centre, rod about end, disc about centre, sphere about centre;
- calculate the moment of inertia of a combination of rotating bodies;
- state and apply the equation $T = I\alpha$.

3.7 End of topic 3 test

End of topic 3 test Go online

The following data should be used when required:

Gravitational acceleration on Earth g	9.8 m s^{-2}
Moment of inertia of a point mass	$I = mr^2$
Moment of inertia of a rod about its centre	$I = \frac{1}{12}ml^2$
Moment of inertia of a rod about its end	$I = \frac{1}{3}ml^2$
Moment of inertia of a disc about its centre	$I = \frac{1}{2}mr^2$
Moment of inertia of a sphere about its centre	$I = \frac{2}{5}mr^2$

Q15: A spanner, length 35 cm, is being used to try to undo a nut stuck on a bolt. A force of 71 N is applied at the end of the spanner.

Calculate the moment which is applied.

Give your answer to *at least* 1 decimal place.

Moment = _____ m

....

Q16: A torque of 65 Nm is applied to a solid flywheel of moment of inertia 17 kg m^2.

Calculate the resulting angular acceleration.

Give your answer to at least 1 decimal place.

Angular acceleration = _____ rad s^{-2}

....

Q17: Three objects are arranged on a circular turntable of negligible mass, which is able to rotate about an axis through its centre.

Mass A (0.46 kg) is 20 cm from the axis.

TOPIC 3. ROTATIONAL DYNAMICS

Mass B (0.61 kg) is 40 cm from the axis.

Mass C (0.28 kg) is 70 cm from the axis.

Calculate the total moment of inertia of the loaded turntable about the central axis.

I = _____ kg m^2

....

Q18: An object of mass 2.9 kg is moving in a circle of radius 1.2 m. Calculate the moment of inertia of the mass about the centre of the circle.

Give your answer rounded to *at least* 1 decimal places.

I = _____ kg m^2

....

Q19: Masses A (1.5 kg) and B (2 kg) are placed on a platform of negligible mass, which can rotate about an axis XY.

Mass A is positioned 10 cm from XY.

Mass B sits 23 cm from XY.

Calculate the total moment of inertia about the axis XY.

Give your answer rounded to *at least* 2 decimal places.

I = _____ kg m^2

....

Q20: The moment of inertia of an object is a measure of its resistance to _____ acceleration about a given axis.

The moment of inertia of an object about an axis depends on the _____ of the object, and the distribution of the _____ about the axis.

....

Q21: A solid cylinder of mass 0.3 kg rolls down a ramp. The diameter of the cylinder is 4.0 cm.

Find the moment of inertia of the cylinder.

I = _____ kg m^2

....

Q22: The moment of inertia of a CD about a perpendicular axis through its centre is 2.70 × 10^{-5} kg m^2. The CD has a diameter of 12 cm.

Find its mass in grams.

Give your answer rounded to 3 significant figures.

m = _____ g

....

Q23: A ball bearing of mass 29.0 g rolls down a slope. The moment of inertia of the ball bearing is 1.05 × 10^{-6} kg m^2.

Find its radius in cm.

Give your answer rounded to 3 significant figures.

r = _____ cm

© HERIOT-WATT UNIVERSITY

Unit 1 Topic 4

Angular momentum

Contents

- 4.1 Introduction ... 94
- 4.2 Angular momentum ... 94
- 4.3 Conservation of angular momentum ... 94
  - 4.3.1 Angular momentum in sports ... 95
  - 4.3.2 Satellite spin ... 96
- 4.4 Rotational kinetic energy ... 97
- 4.5 Comparing linear and angular motion ... 100
- 4.6 Extended information ... 101
- 4.7 Summary ... 102
- 4.8 End of topic 4 test ... 102

Prerequisites

- Understand the relationship between torque, moment of inertia and angular acceleration.
- Understand the principle of conservation of linear momentum.

Learning objective

By the end of this topic you should be able to:

- state that the angular momentum L of a rigid body is given by the equation $L = I\omega$;
- use the equation $L = mvr = mr^2\omega$ for a point mass;
- state that in the absence of external torques, angular momentum is conserved;
- state the expression $E_K = \frac{1}{2}I\omega^2$ for the rotational kinetic energy of a rigid body, and carry out calculations using this relationship.

4.1 Introduction

In the last topic, we studied the moment of inertia of shapes about different axes of rotation. We will now find the equations for angular momentum and rotational kinetic energy. Again, these compare directly to the linear equations with which you should be familiar.

4.2 Angular momentum

The **angular momentum** L of a body rotating with moment of inertia I and angular velocity ω is defined as

$$L = I\omega$$

(4.1)

L is measured in units of kg m² s⁻¹. Equation 4.1 is analogous to the expression $p = mv$ for linear momentum, and the same conservation of momentum principle applies. In this case, we can state that the angular momentum of a rotating rigid body is conserved unless an external torque acts on the body.

If we want to describe the angular momentum at a point, then we can substitute for $I = mr^2$ and $\omega = v/r$ in Equation 4.1

$$L = I\omega \qquad\qquad L = mr^2 \times v/r$$
$$\therefore L = mr^2\omega \qquad\qquad \therefore L = mvr$$

4.3 Conservation of angular momentum

The conservation of angular momentum has some interesting consequences. Since angular momentum depends on the moment of inertia, then it depends on the distribution of mass about the axis. If the moment of inertia increases, the angular velocity must decrease to keep the angular momentum constant, and vice versa. This effect can be seen if you spin on a swivel chair. Flinging out your arms and legs moves more of your body mass away from the axis of rotation. Your moment of inertia increases, so your angular velocity must decrease. (This works even better if you have a heavy book in each hand!) If you then hunch up on the chair, your moment of inertia decreases, and hence your angular velocity increases.

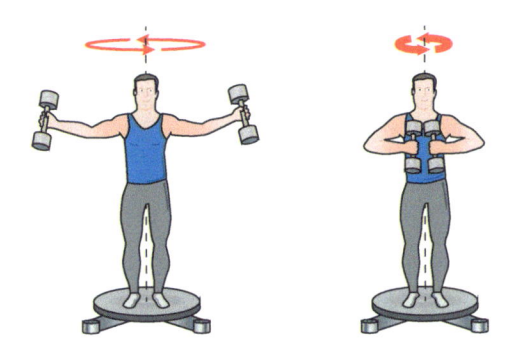

4.3.1 Angular momentum in sports

Gymnasts, snowboarders, divers and acrobats all use the same principle. As they move from a stretched to a tucked position, their moment of inertia decreases and this makes their angular velocity increase.

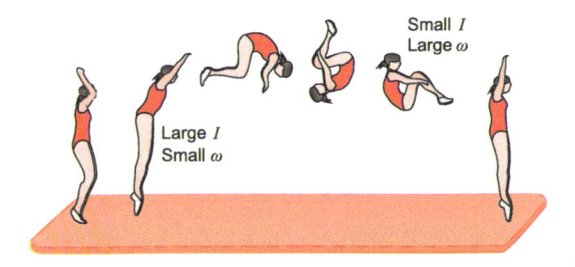

Example

An ice skater is spinning with his arms extended. His angular velocity is 7.0 rad s^{-1}. What is his angular velocity if he pulls his arms in to his sides? The moment of inertia of the skater is 4.7 kg m^2 when his arms are extended, and 1.8 kg m^2 when they are by his sides.

The angular momentum of the skater must be conserved, so the initial angular momentum must equal the final angular momentum.

$$L_i = L_f$$
$$\therefore I_i \omega_i = I_f \omega_f$$
$$\therefore 4.7 \times 7.0 = 1.8 \times \omega_f$$
$$\therefore \omega_f = \frac{4.7 \times 7.0}{1.8}$$
$$\therefore \omega_f = 18 \text{ rad s}^{-1}$$

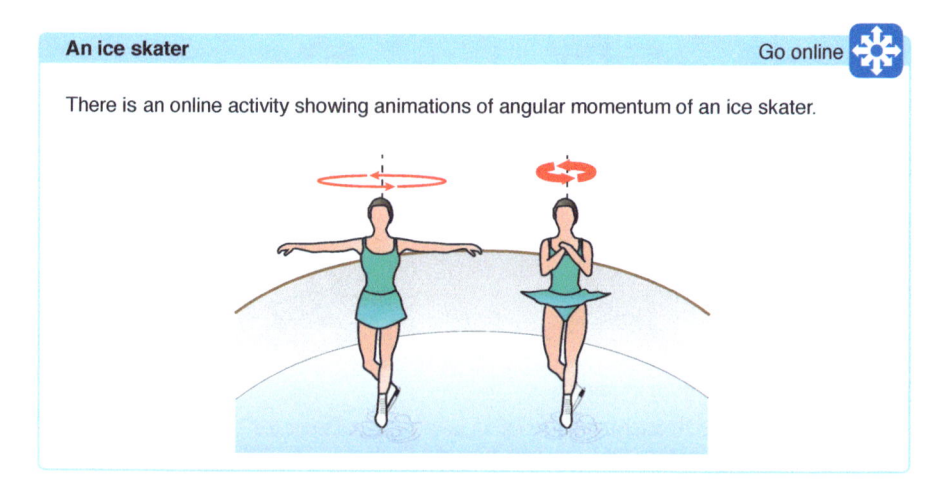

> **An ice skater** Go online
>
> There is an online activity showing animations of angular momentum of an ice skater.

4.3.2 Satellite spin

A special device is used to reduce the spin of satellites. It involves two cables wrapped around the satellite, with masses attached to the other end. When activated, the masses are released and move away from the satellite like yoyos, extending to the end of their cables. The moment of inertia therefore increases. In the absence of an external torque, angular momentum must be conserved. So the cord length can be chosen to reduce the angular velocity of the satellite to the desired value. At this point, the masses are then detached.

The Mars exploration rover's angular velocity was reduced in this way.

TOPIC 4. ANGULAR MOMENTUM

Quiz: Conservation of angular momentum   Go online

Q1: A solid object with moment of inertia 2.40 kg m² is rotating at 3.00 rad s⁻¹. What is its angular momentum?

a) 0.800 kg m² s⁻¹
b) 1.25 kg m² s⁻¹
c) 1.92 kg m² s⁻¹
d) 7.20 kg m² s⁻¹
e) 17.3 kg m² s⁻¹

....

Q2: A disc of mass m and radius r spinning about an axis through its centre has moment of inertia $I = \frac{1}{2}mr^2$. What is the angular momentum of a disc (m = 0.400 kg, r = 0.25 m) spinning at 20 rad s⁻¹?

a) 0.05 kg m² s⁻¹
b) 0.25 kg m² s⁻¹
c) 5.0 kg m² s⁻¹
d) 8.0 kg m² s⁻¹
e) 250 kg m² s⁻¹

....

Q3: A horizontal turntable is rotating on a frictionless mount at constant angular velocity. What happens if a lump of clay is dropped onto the turntable, sticking to it?

a) The turntable would slow down.
b) The turntable would speed up.
c) The turntable would continue rotating at the same speed.
d) The lump of clay would rotate in the opposite direction to the turntable.
e) The turntable would stop rotating but the lump of clay would continue to rotate.

4.4 Rotational kinetic energy

There are also expressions for calculating work done and kinetic energy in rotational motion, that are again analogous to expressions used to describe linear motion. You should be familiar with the expression *work done = force × displacement*. In the case of rotational work, the equivalent expression is

$$\text{work done} = T \times \theta$$

(4.2)

In Equation 4.2, T is the torque and θ is the angular displacement. This equation tells us the work done when a torque T applied to an object produces an angular displacement θ of the object.

A spinning body has kinetic energy. The **rotational kinetic energy** of a body is given by

$$\text{Rotational } E_K = \frac{1}{2}I\omega^2$$

(4.3)

Note that this energy is due to the rotation about an axis. If we consider a wheel spinning about its axis, then this is the total kinetic energy. What if the wheel is rolling along the ground? In this case the wheel has both rotational kinetic energy, due to its turning motion, and translational kinetic energy due to its moving along the ground. The total kinetic energy in this case is

$$\text{Total } E_K = \text{Rotational } E_K + \text{Translational } E_K$$
$$\therefore \text{Total } E_K = \frac{1}{2}I\omega^2 + \frac{1}{2}mv^2$$

Note that the translational speed is the same as the tangential speed at the circumference of the wheel, so long as the wheel does not slip. If you cannot see why, imagine the wheel rotating through one complete circle. A point on the circumference moves through a distance $2\pi r$ in one revolution. The wheel must have rolled the same distance, and it has done so in the same period of time. The translational speed must therefore be equal to the tangential speed at the circumference.

Example

A solid sphere of radius r and mass m has moment of inertia $I = \frac{2}{5}mr^2$ about any axis though its centre. Calculate the translational and rotational kinetic energies of a sphere of mass 0.40 kg and radius 20 cm rolling (without slipping) along a horizontal surface with translational speed 5.0 m s^{-1}.

The translational kinetic energy is

$$\text{Translational } E_K = \frac{1}{2}mv^2$$
$$\therefore \text{Translational } E_K = \frac{1}{2} \times 0.40 \times 5.0^2$$
$$\therefore \text{Translational } E_K = 5.0 \text{ J}$$

TOPIC 4. ANGULAR MOMENTUM

To find the rotational kinetic energy, we need to find the moment of inertia I and angular velocity ω of the sphere. If the sphere has a translational speed of v m s^{-1}, then a point on the circumference of the sphere must also have speed v. The angular velocity is therefore equal to v/r, which in this case is

$$\omega = \frac{v}{r} = \frac{5.0}{0.20} = 25 \text{ rad s}^{-1}$$

The rotational kinetic energy is

$$\text{Rotational } E_K = \frac{1}{2}I\omega^2$$
$$\therefore \text{Rotational } E_K = \frac{1}{2} \times \frac{2}{5}mr^2 \times \omega^2$$
$$\therefore \text{Rotational } E_K = \frac{1}{5} \times 0.40 \times 0.20^2 \times 25^2$$
$$\therefore \text{Rotational } E_K = 2.0 \text{ J}$$

Flywheels

Flywheels are spinning wheels or discs with a fixed axle so that rotation is only about one axis. They can be used to store energy in machines, such as push and go toy cars. Some experimental vehicles use flywheels to charge up their batteries. Your teacher may ask you to carry out a practical experiment to find the moment of inertia of a flywheel.

Total energy of a rolling body Go online

At this stage there is an online activity allowing you to plot out the energy changes that occur for different objects rolling down a slope..

Quiz: Angular momentum and rotational kinetic energy Go online

Q4: A square sheet of mass m and side length l has a moment of inertia $I = \frac{1}{3}ml^2$ when rotating about one edge. What is the rotational kinetic energy of a square sheet of mass 1.0 kg and side 0.50 m rotating at 0.60 rad s^{-1} about one edge?

a) 9.0×10^{-4} J
b) 4.2×10^{-3} J
c) 0.015 J
d) 0.030 J
e) 0.18 J

Q5: A rotating platform is spinning at a rate of 1.50 revolutions per second. If the moment of inertia of the platform is 4.20 kg m², what is the rotational kinetic energy of the platform?

a) 4.73 J
b) 19.8 J
c) 39.6 J
d) 47.3 J
e) 187 J

4.5 Comparing linear and angular motion

Using the moment of inertia allows us to describe the angular motion of rigid bodies in the same way that we describe linear motion of point objects. The same conservation rules apply, and the following tables should clarify the different quantities and relationships:

Table 4.1: Kinematic relationships

Linear motion	Angular motion
$v = u + at$	$\omega = \omega_0 + \alpha t$
$v^2 = u^2 + 2as$	$\omega^2 = \omega_0^2 + 2\alpha\theta$
$s = ut + \frac{1}{2}at^2$	$\theta = \omega_0 t + \frac{1}{2}\alpha t^2$

Table 4.2: Linear and angular motion

Linear motion	Angular motion
Displacement s	Angular displacement θ
Force F	Torque T
Velocity v	Angular velocity ω
Acceleration a	Angular acceleration α

Table 4.3: Newton's second law and kinetic energy

Linear motion	Angular motion
Newton's second law $F = ma$	Newton's second law $T = I\alpha$
Momentum $p = mv$	Angular momentum $L = I\omega$
Work done $W = Fs$	Work done $W = T\theta$
Translational kinetic energy Translational $E_K = \dfrac{1}{2}mv^2$	Rotational kinetic energy Rotational $E_K = \dfrac{1}{2}I\omega^2$

The moment of inertia of a rigid body depends on the mass of the body, and how that mass is distributed about the axis of rotation. The axis must be specified when a moment of inertia is being calculated. For a collection of point masses, the total moment of inertia is the sum of the individual moments.

4.6 Extended information

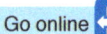

Web links   Go online

These web links should serve as an insight to the wealth of information available online and allow you to explore the subject further.

- https://www.youtube.com/watch?v=dtVs4SGZYvM
- http://www.nsf.gov/news/special_reports/olympics/figureskating.jsp
- http://www.our-space.org/materials/states-of-matter/angular-momentum
- https://www.youtube.com/watch?v=8I4ii1xEeG0
- https://www.youtube.com/watch?v=ty9QSiVC2g0
- http://hyperphysics.phy-astr.gsu.edu/hbase/top.html
- http://www.animations.physics.unsw.edu.au/jw/foucault_pendulum.html

4.7 Summary

Summary

You should now be able to:

- state that the angular momentum L of a rigid body is given by the equation $L = I\omega$;
- use the equation $L = mvr = mr^2\omega$ for a point mass;
- state that in the absence of external torques, angular momentum is conserved;
- state the expression $E_K = \frac{1}{2}I\omega^2$ for the rotational kinetic energy of a rigid body, and carry out calculations using this relationship.

4.8 End of topic 4 test

End of topic 4 test — Go online

The following data should be used when required:

Gravitational acceleration on Earth g	9.8 m s^{-2}
Moment of inertia of a point mass	$I = mr^2$
Moment of inertia of a rod about its centre	$I = \frac{1}{12}ml^2$
Moment of inertia of a rod about its end	$I = \frac{1}{3}ml^2$
Moment of inertia of a disc about its centre	$I = \frac{1}{2}mr^2$
Moment of inertia of a sphere about its centre	$I = \frac{2}{5}mr^2$

Q6: A solid sphere, radius 20.0 cm and mass 1.36 kg, is rotating about an axis through its centre with angular velocity 4.65 rad s^{-1}.

Calculate the rotational kinetic energy of the sphere.

Rotational E$_K$ = _____ J

....

Q7: A flat horizontal disc of moment of inertia 1.2 kg m^2 is rotating at 4.5 rad s^{-1} about a vertical axis through its centre.

A 0.13 kg mass is dropped onto the disc, landing without slipping 1.7 m from the centre.

Calculate the new angular velocity of the disc.

ω = _____ rad s^{-1}

....

TOPIC 4. ANGULAR MOMENTUM

Q8: A solid cylinder of radius 0.76 m and mass 7.9 kg is at rest. A 4 N m torque is applied to the cylinder about an axis through its centre.

Calculate the angular velocity of the cylinder after the torque has been applied for 2.0 s.

$\omega = $ _____ rad s^{-1}

....

Q9: A solid cylinder of mass 3.5 kg and radius 1.2 m is rolling without slipping along a horizontal road.

The translational speed of the cylinder is 5.4 m s^{-1}.

1. Calculate the angular velocity of the cylinder.
   $\omega = $ _____ rad s^{-1}
2. Calculate the total kinetic energy of the cylinder.
   Total $E_K = $ _____ J

Unit 1 Topic 5

Gravitation

Contents

5.1	Introduction	107
5.2	Newton's law of gravitation	107
	5.2.1 The Cavendish-Boys experiments to determine G	109
	5.2.2 Maskelyne's experiment	110
5.3	Weight	111
5.4	Gravitational fields	114
5.5	Gravitational potential and potential energy	119
	5.5.1 Gravitational potential	119
	5.5.2 Gravitational potential energy	121
5.6	Escape velocity	124
5.7	Extended information	128
5.8	Summary	128
5.9	End of topic 5 test	129

Prerequisites

- Newton's laws of motion.
- Circular motion - centripetal force, periodic time.

> **Learning objective**
>
> By the end of this topic you should be able to:
>
> - state and apply the equation $F\frac{GMm}{r^2}$ to calculate the gravitational force between two objects;
> - calculate the weight of an object using Newton's Universal Law of Gravitation;
> - state what is meant by a gravitational field, and calculate the gravitational field strength at a point in the field;
> - calculate the value of the acceleration due to gravity at a point in a gravitational field, given the local conditions;
> - sketch the field lines around a planet and a planet-moon system;
> - state the expression $V = \frac{-GM}{r}$ and use it to calculate the gravitational potential V at a point in a gravitational field;
> - calculate the gravitational potential energy of a mass in a gravitational field and calculate the change in the potential energy when a mass is moved between points in the field;
> - define escape velocity as the minimum velocity required to allow a mass to escape a gravitational field;
> - derive the expression $v = \sqrt{\frac{2GM}{r}}$ and use it to calculate the escape velocity.

5.1 Introduction

You studied Newton's Universal Law of Gravitation at Higher. In this topic we will look at how this law can be used to establish a relationship between gravitational field strength and the height above the Earth. Throughout this topic, some approximations will be made with regard to planets and their orbits. It will be assumed that the Sun, the Moon, the Earth and other planets are all spherical objects and that all orbits are circular. We will see how the centripetal force on a satellite due to the gravitational attraction determines the speed and period of the satellite.

The second half of the topic explores gravitational fields. We will look at the potential energy of a body in a gravitational field, so that the work done in moving a body around in a gravitational field can be calculated. Finally, we will consider the **escape velocity** of a rocket, which is how fast it needs to be travelling when it takes off in order to escape a body's gravitational field. This will help us understand black holes in a future topic.

5.2 Newton's law of gravitation

Working in the 17th century, Sir Isaac Newton discovered the **Universal Law of Gravitation**. He used his own observations, along with those of Johannes Kepler, who had formulated a set of laws that described the motion of the planets around the Sun. Kepler's laws will be studied in the next topic.

Newton's Law of Gravitation states that there is a force of attraction between any two objects in the universe. The size of the force is proportional to the product of the masses of the two objects, and inversely proportional to the square of the distance between them. This law can be summed up in the equation

$$F = \frac{Gm_1m_2}{r^2}$$

(5.1)

In Equation 5.1, m_1 and m_2 are the masses of the two objects, and r is the distance between them. The constant of proportionality is G, the Universal constant of gravitation. The value of G is 6.67×10^{-11} m^3 kg^{-1} s^{-2}. A simple example will show us the order of magnitude of this force.

Example

Consider two point masses, each 2.00 kg, placed 1.20 m apart on a table top. Calculate the magnitude of the gravitational force between the two masses.

Using Newton's Law of Gravitation

$$F = \frac{Gm_1m_2}{r^2}$$
$$\therefore F = \frac{6.67 \times 10^{-11} \times 2.00 \times 2.00}{1.20^2}$$
$$\therefore F = \frac{2.668 \times 10^{-10}}{1.44}$$
$$\therefore F = 1.85 \times 10^{-10} \text{ N}$$

The gravitational force between these two masses is only 1.85×10^{-10} N. This is an extremely small force, one which is not going to be noticeable in everyday life.

We rarely notice the gravitational force that exists between everyday objects as it is such a small force. You do not have to fight against gravity every time you walk past a large building, for example, as the gravitational force that the building exerts on you is too small to notice. Because the constant G in Newton's Law of Gravitation is so small, the gravitational force between everyday objects is usually negligible. The force only really becomes important when we are dealing with extremely large masses such as planets.

The gravitational force is always attractive, and always acts in the direction of the straight line joining the two objects. According to Newton's third law of motion, the gravitational force is exerted on **both** objects. As the Earth exerts a gravitational force on you, so you exert an equal force on the Earth.

Most of the work we will be doing on gravitation concerns the forces acting between planets and stars. So far we have only considered point objects, so do we need to adapt Newton's Law of Gravitation when we are dealing with larger bodies? The answer is no - for spherical objects (or more accurately, objects with a spherically symmetric mass distribution), the gravitational interaction is exactly the same as it would be if all the mass was concentrated at the centre of the sphere. Remember, we will assume in all our calculations that the planets and stars we are dealing with are spherical.

Example

The Earth has a radius of 6.4×10^6 m and a mass 6.0×10^{24} kg. What is the gravitational force due to the Earth acting on a woman of mass 60.0 kg standing on the surface of the Earth?

The solution is obtained by calculating the force between two point objects placed 6.4×10^6 m apart, if we treat the Earth as a uniform sphere. Hence

$$F = \frac{Gm_1m_2}{r^2}$$
$$\therefore F = \frac{6.67 \times 10^{-11} \times 6.0 \times 10^{24} \times 60.0}{(6.4 \times 10^6)^2}$$
$$F = \frac{2.4012 \times 10^{16}}{4.096 \times 10^{13}}$$
$$F = 590 \, \text{N}$$

This answer has been rounded to two significant figures since the mass of the Earth was only stated to two significant figures.

The gravitational force acting on the woman is 590 N, and this force is directed towards the centre of the Earth.

If you calculate the force due to gravity acting on the woman by another method, using $F = m \times g$, you should also get the answer 590 N, when rounded to two significant figures.

5.2.1 The Cavendish-Boys experiments to determine G

An accurate measurement of G can be carried out using the Cavendish-Boys method. Cavendish first performed this experiment in the late 18th century and the accurate determination was performed nearly 100 years later by Boys. The experiment uses a **torsion balance** to measure the gravitational force between lead spheres.

Figure 5.1: Cavendish Boys-experiment, viewed from (a) the side, and (b) the top

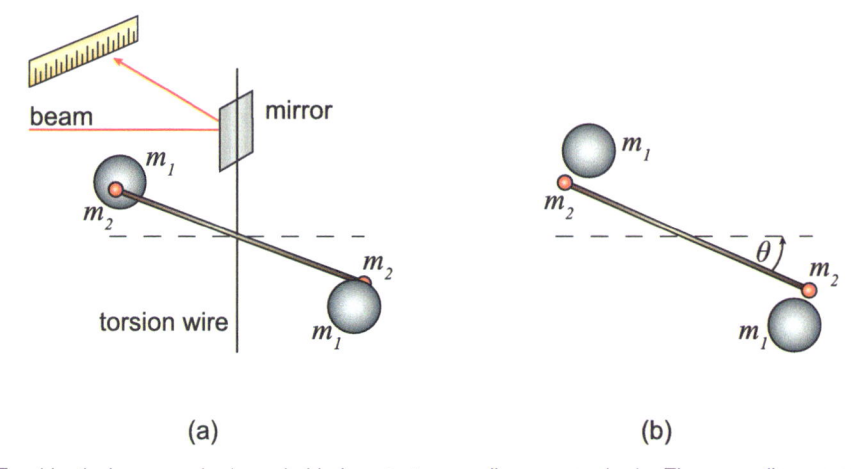

Two identical masses (m_1) are held close to two smaller masses (m_2). These smaller masses are attached to the ends of a light rod of length l suspended from a torsion wire in a draft-free chamber. By reflecting a narrow beam of light from a mirror attached to the torsion wire, the angular deflection of the wire can be measured as the masses m_1 are brought close to the masses m_2. In equilibrium, the rotation caused by the gravitational force between m_1 and m_2 is balanced by the restoring torque (turning force) in the wire. The equipment must be calibrated first to determine a quantity called the torsional constant c of the wire. The restoring torque is then equal to $c \times \theta$ when the angular displacement of the rod is θ rad (see Figure 5.1(b)). Hence in equilibrium,

Torque due to gravitational force = Restoring torque

$$\therefore G\frac{m_1 m_2}{r^2} l = c\theta$$

(5.2)

5.2.2 Maskelyne's experiment

One of the consequences of Newton's Universal Law of Gravitation is that on level ground a pendulum ought to hang vertically, since it is attracted towards the centre of the Earth. However, if a mountain is nearby, then its gravitational attraction will pull the pendulum slightly off vertical. For this reason, in the late 18th century, an astronomer called Maskelyne used a pendulum and a Scottish mountain called Schiehallion to estimate the mass of the Earth. He chose to use this Scottish mountain since it was in an isolated location and its regular shape allowed its mass to be estimated.

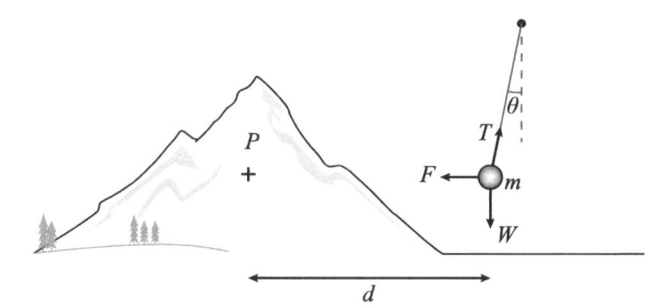

Two gravitational forces act on the pendulum bob of mass m. They are the gravitational attraction of Schiehallion and the bob's weight. They can be found from

$$F = \frac{GM_{\text{Schiehallion}} m}{d^2}$$

and

$$W = \frac{GM_{\text{Earth}} m}{r_E^2}$$

where d is the distance between the mountain and the bob. As usual, r_E is the radius of the Earth.

TOPIC 5. GRAVITATION

Maskelyne knew that the vertical component of the tension T must balance the bob's weight. He knew the horizontal component of the tension T must balance the force exerted by Schiehallion. Therefore, the angle θ of the pendulum to the vertical could be expressed by the equation

$$\tan \theta = \frac{\frac{GM_{\text{Schiehallion}} m}{d^2}}{\frac{GM_{\text{Earth}} m}{r_E^2}}$$

This meant that, by measuring the angle of the pendulum to the vertical, Maskelyne was then able to solve for the mass of the Earth.

5.3 Weight

The **weight** W of an object of mass m can be defined as the gravitational force exerted on it by the Earth.

$$W = F_{\text{grav}} = \frac{GM_E m}{r_E^2}$$

(5.3)

In Equation 5.3, M_E and r_E are the mass and radius of the Earth.

The acceleration due to gravity, g, is found from Newton's second law of motion

$$F = ma$$
$$\therefore W = mg$$

We can substitute for F in this equation

$$mg = \frac{GM_E m}{r_E^2}$$
$$\therefore g = \frac{GM_E}{r_E^2}$$

(5.4)

© HERIOT-WATT UNIVERSITY

So the acceleration of an object due to gravity close to the Earth's surface does not depend on the mass of the object. In the absence of friction, all objects fall with the same acceleration.

Equation 5.4 is a specific equation for calculating g on the Earth's surface. In general, at a distance r from the centre of a body (a star or a planet, say) of mass M, the value of g is given by

$$g = \frac{GM}{r^2}$$

(5.5)

The mass of an object is constant; it is an intrinsic property of that object. The weight of an object tells us the magnitude of the gravitational force acting upon it, so it is not a constant.

Example

Compare the values of g on the surfaces of the Earth (M_E = 6.0 × 10^{24} kg, r_E = 6.4 × 10^6 m) and the Moon (M_M = 7.3 × 10^{22} kg, r_M = 1.7 × 10^6 m).

To solve this problem, use Equation 5.5 with the appropriate values

$$g = \frac{GM_E}{r_E^2}$$
$$\therefore g = \frac{6.67 \times 10^{-11} \times 6.0 \times 10^{24}}{(6.4 \times 10^6)^2}$$
$$\therefore g = 9.8 \text{ m s}^{-2}$$

$$g = \frac{GM_M}{r_M^2}$$
$$\therefore g = \frac{6.67 \times 10^{-11} \times 7.3 \times 10^{22}}{(1.7 \times 10^6)^2}$$
$$\therefore g = 1.7 \text{ m s}^{-2}$$

This gives a value for g on the surface of the Earth of 9.8 m s^{-2}, compared to a value on the surface of the Moon of 1.7 m s^{-2}. The value for g on the surface of the Moon is usually quoted as 1.6 m s^{-2}, though it varies over its entire surface by about 0.03 ms^{-2}. The value calculated here differs since the mass and radius of the Moon were only quoted to two significant figures in the question.

TOPIC 5. GRAVITATION

Quiz: Gravitational force

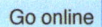

Useful data:

Universal constant of gravitation G	6.67×10^{-11} N m² kg⁻²
Mass of the Moon M_M	7.3×10^{22} kg
Radius of the Moon r_M	1.7×10^6 m
Mass of Venus M_V	4.87×10^{24} kg
Radius of Venus r_V	6.05×10^6 m

Q1: Two snooker balls, each of mass 0.25 kg, are at rest on a snooker table with their centres 0.20 m apart. What is the magnitude of the gravitational force that exists between them?

a) 2.1×10^{-11} N
b) 4.3×10^{-11} N
c) 8.3×10^{-11} N
d) 1.0×10^{-10} N
e) 4.2×10^{-10} N

....

Q2: The Sun exerts a gravitational force F_S on the Earth. The Earth exerts a gravitational force F_E on the Sun. Which one of these statements about the magnitudes of F_S and F_E is true?

a) $F_S = F_E$
b) $F_S < F_E$
c) $F_S > F_E$
d) $F_S/F_E = \text{mass}_S/\text{mass}_E$
e) $F_S/F_E = (\text{mass}_S)^2/(\text{mass}_E)^2$

....

Q3: What is the weight of a 5.00 kg mass placed on the surface of the Moon?

a) 0.12 N
b) 1.4 N
c) 1.6 N
d) 8.4 N
e) 49 N

....

© HERIOT-WATT UNIVERSITY

Q4: An object is taken from sea level to the top of Mount Everest. Which one of the following statements is true?

a) Its mass remains constant but its weight increases.
b) Its mass remains constant but its weight decreases.
c) Its weight remains constant but its mass increases.
d) Its weight remains constant but its mass decreases.
e) Neither its mass nor its weight alter.

....

Q5: What is the value of the acceleration due to gravity on the surface of the planet Venus?

a) 0.887 m s^{-2}
b) 5.37 m s^{-2}
c) 6.67 m s^{-2}
d) 8.29 m s^{-2}
e) 8.87 m s^{-2}

5.4 Gravitational fields

The region of space around an object A, in which A exerts a gravitational force on another object B placed in that region, is called the **gravitational field** of A. The concept of a field is used in many situations in physics, such as the electric field surrounding a charged particle, or the magnetic field around a bar magnet.

The **gravitational field strength** g at a point in a gravitational field is defined as the gravitational force acting on a unit mass placed at that point in the field. At a distance r from a point object of mass M, the gravitational field strength g is given by

$$g = \frac{GM}{r^2}$$

(5.6)

The units of g are N kg^{-1}. These units are equivalent to m s^{-2}, so the gravitational field strength at a point in a gravitational field is equal to the acceleration due to gravity at that point, and Equation 5.6 is identical to Equation 5.5. This is why the same symbol g is used for both the gravitational field strength and the acceleration due to gravity.

TOPIC 5. GRAVITATION

Example

The Earth orbits the Sun with a mean radius of 1.5×10^{11} m. What is the gravitational field strength on Earth due to the Sun, given that the mass of the Sun is 2.0×10^{30} kg.

In the Sun's gravitational field, the field strength is given by

$$g = \frac{GM_S}{r^2}$$

So, at the location of the Earth, the field strength is

$$g = \frac{GM_S}{r^2}$$
$$\therefore g = \frac{6.67 \times 10^{-11} \times 2.0 \times 10^{30}}{\left(1.5 \times 10^{11}\right)^2}$$
$$\therefore g = 5.9 \times 10^{-3} \text{ N kg}^{-1}$$

The gravitational field around a body is often shown diagramatically by drawing field lines. Figure 5.2 shows the gravitational field lines around a point object. The pattern of the field lines is symmetrical in three dimensions. This pattern would be exactly the same outside an object with a spherically symmetric mass distribution. The lines are symmetrical about the centre of the object and show the direction of the force exerted on any mass placed in the field. The closer together the field lines are, the greater the field strength.

Figure 5.2: Gravitational field lines around a point mass

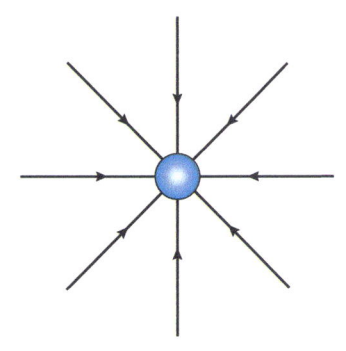

The gravitational field is an example of a **conservative field**. This is a field in which the work done in moving from one point to another in the field is independent of the path taken.

The field lines representing the gravitational field caused by two or more objects can also be sketched. The gravitational field due to two point masses is shown in Figure 5.3.

© HERIOT-WATT UNIVERSITY

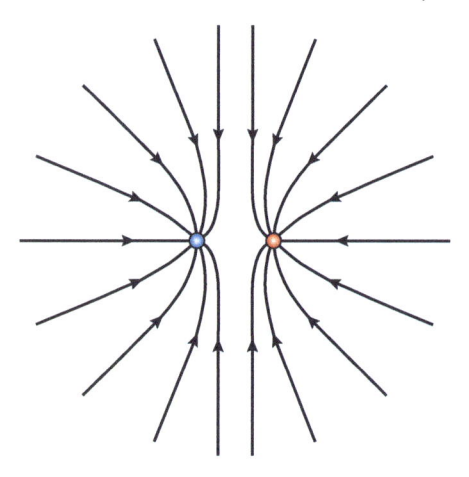

Figure 5.3: Gravitational field lines around two identical point masses

To calculate the gravitational field strength at any point in the field due to two point masses or spheres, we take the vector sum of the two individual fields.

Example

The mean distance between the centre of the Earth and the centre of the Moon is 3.84×10^8 m. Given that the mass of the Earth is 6.0×10^{24} kg and the mass of the Moon is 7.3×10^{22} kg, at what distance from the centre of the Earth is the point where the total gravitational field strength due to the Earth and the Moon is zero?

A sketch is useful in solving this sort of problem, as it can clearly show the direction of the field vectors.

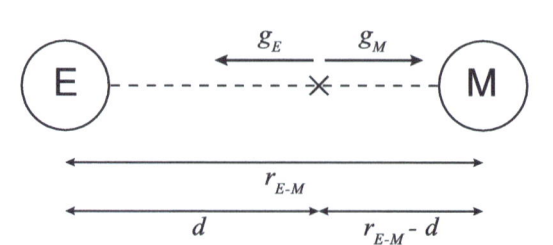

Figure 5.4: Earth - Moon system

At position X, the point where the total gravitational field strength is zero, the gravitational field strength g_E of the Earth is balanced by g_M, the gravitational field strength of the Moon. Position X is a distance d from the centre of the Earth. To find the distance d, we must solve the equation $g_E = g_M$

TOPIC 5. GRAVITATION

$$g_E = g_M$$
$$\therefore \frac{GM_E}{d^2} = \frac{GM_M}{(r_{E-M} - d)^2}$$
$$\therefore \frac{M_E}{d^2} = \frac{M_M}{(r_{E-M} - d)^2}$$
$$\therefore \frac{\sqrt{M_E}}{d} = \frac{\sqrt{M_M}}{(r_{E-M} - d)}$$
$$\therefore \sqrt{M_E} \times (r_{E-M} - d) = \sqrt{M_M} \times d$$
$$\therefore \sqrt{M_E} \times r_{E-M} = d \times \left(\sqrt{M_M} + \sqrt{M_E}\right)$$
$$\therefore 2.449 \times 10^{12} \times 3.84 \times 10^8 = d \times \left(2.702 \times 10^{11} + 2.449 \times 10^{12}\right)$$
$$\therefore d = \frac{9.404 \times 10^{20}}{2.719 \times 10^{12}}$$
$$\therefore d = 3.5 \times 10^8 \text{ m}$$

The combined gravitational field strength is zero at a point 3.5×10^8 m from the centre of the Earth. This point is very much closer to the Moon than the Earth because the Earth is much more massive than the Moon.

Gravitational fields   Go online

At this stage there is an online activity where you can see the pattern of the gravitational field lines around a single object or a collection of objects.

The gravitational field lines around an isolated object are plotted. A second object can be added to the system, and the field lines are re-plotted if the separation between the objects or their masses are changed.

With this simulation you can see the pattern of the gravitational field lines around a single object or a collection of objects. Note that masses are always added at the same position, so if you do not move the masses they will all be located at the same point.

The Earth - Moon system

The last interactivity allowed us to explore the gravitational field line pattern around two masses. The gravitational field line pattern for a planet-moon system is therefore as follows.

© HERIOT-WATT UNIVERSITY

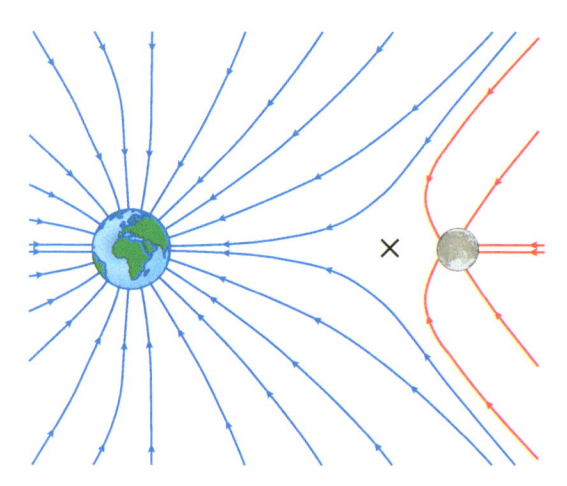

Note the gravitational field lines meet the surface of the Earth and the Moon at 90 ° to their surface. The point at which the gravitational field strength is zero has been labelled X. We saw earlier that this is closer to the Moon than the Earth because the Earth has a much larger mass.

Quiz: Gravitational fields

Go online

Useful data:

Universal constant of gravitation G	6.67×10^{-11} m^3 kg^{-1} s^{-2}

Q6: The planet Jupiter has a mass of 1.90×10^{27} kg and a radius of 6.91×10^7 m. What is the gravitational field strength on the surface of Jupiter?

a) 0.040 N kg^{-1}
b) 1.83 N kg^{-1}
c) 9.81 N kg^{-1}
d) 26.5 N kg^{-1}
e) 168 N kg^{-1}

....

Q7: Which one of the following units is equivalent to the units used to express gravitational field strength?

a) m s^{-2}
b) N m s^{-2}
c) kg m^{-2}
d) kg N^{-2}
e) N s^{-2}

TOPIC 5. GRAVITATION

Q8: The gravitational field strength at a distance 5.0×10^5 m from the centre of a planet is 8.0 N kg^{-1}. At what distance from the centre of the planet is the field strength equal to 4.0 N kg^{-1}?

a) 7.1×10^5 m
b) 9.0×10^5 m
c) 1.0×10^6 m
d) 2.0×10^6 m
e) 2.5×10^{11} m

Q9: P is a point mass (mass $9m$) and Q is a point mass (mass m). P and Q are separated by 4.0 m. At what point on the line joining P to Q is the net gravitational field strength zero?

a) 0.44 m from P
b) 1.33 m from P
c) 3.0 m from P
d) 3.6 m from P
e) 3.95 m from P

Q10: Astronomical observations tell us that the radius of the planet Neptune is 2.48×10^7 m, and the gravitational field strength at its surface is 11.2 N kg^{-1}. What is the mass of Neptune?

a) 3.66×10^3 kg
b) 4.59×10^4 kg
c) 4.16×10^{18} kg
d) 1.03×10^{26} kg
e) 1.16×10^{27} kg

5.5 Gravitational potential and potential energy

We are now going to cover Gravitational potential and Gravitational potential energy.

5.5.1 Gravitational potential

The concept of gravitational potential energy should already be familiar to you from Newtonian mechanics. If a mass m is raised through a height h, it gains potential energy mgh. If the mass is then allowed to fall back down, this potential energy is converted to kinetic energy as it is accelerated downwards.

When an object moves through large distances in a gravitational field, we can no longer use this simple expression for the change in potential energy, as the value of the gravitational field strength g is not constant. The gravitational potential is used to describe how the potential energy of an object changes with its position in a gravitational field, and how much work is done in moving an object

© HERIOT-WATT UNIVERSITY

within the field.

The **gravitational potential** V at a point in a gravitational field is defined as the work done by external forces in moving a unit mass from infinity to that point. Suppose we are considering the field around a mass m, and moving a unit mass from infinity to a point a distance r from m. We cannot use the simple expression *work done = force × distance* to calculate the work done against the gravitational force, as the gravitational force acting on the unit mass increases as it moves closer to m. Instead, we have to use a calculus approach, in which we consider the small amount of work dV done in moving a unit mass a distance dr in the field. Integrating over the range from ∞ to r

$$V = \int_{\infty}^{r} F \, dr$$

Using Newton's law of gravitation for the force acting on the body

$$V = \int_{\infty}^{r} \frac{Gm_1 m_2}{r^2} \, dr$$

In this expression, the values of m_1 and m_2 are M and 1 kg.

$$V = \int_{\infty}^{r} \frac{GM}{r^2} \, dr$$

Performing this integration

$$V = GM \int_{\infty}^{r} \frac{1}{r^2} \, dr$$
$$\therefore V = GM \left[\frac{-1}{r} \right]_{\infty}^{r}$$
$$\therefore V = GM \left(-\frac{1}{r} - \left(-\frac{1}{\infty} \right) \right)$$
$$\therefore V = GM \left(\frac{1}{\infty} - \frac{1}{r} \right)$$
$$\therefore V = -\frac{GM}{r}$$

(5.7)

The gravitational potential V is measured in J kg^{-1}. Remember that there is a minus sign in Equation 5.7. The zero of gravitational potential is at an infinite distance from M. The potential becomes lower and lower the closer we get to M, so it must be a negative number. The gravitational potential at a distance r from the Earth is plotted in Figure 5.5.

TOPIC 5. GRAVITATION

Figure 5.5: Gravitational potential around the Earth

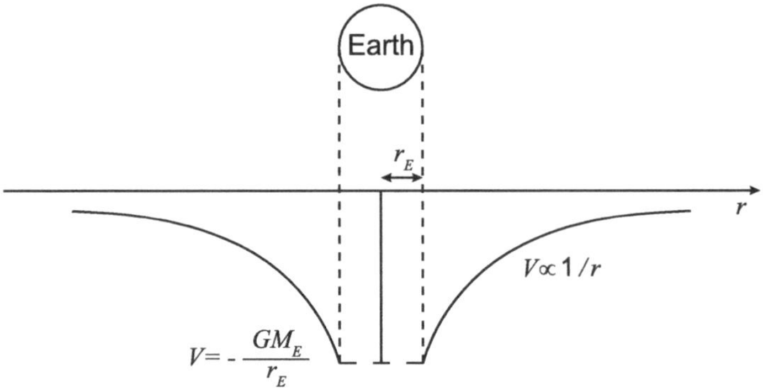

Example

The planet Pluto orbits at a mean distance of 5.92×10^{12} m from the Sun. What is the gravitational potential due to the Sun's gravitational field at this distance?

We will use Equation 5.7, remembering that M here represents the mass of the Sun, since we are calculating the potential due to the Sun's gravitational field. The mass of the Sun is 2.0×10^{30} kg, so

$$V = -\frac{GM}{r}$$
$$\therefore V = -\frac{6.67 \times 10^{-11} \times 2.0 \times 10^{30}}{5.92 \times 10^{12}}$$
$$\therefore V = -2.3 \times 10^{7} \text{ J kg}^{-1}$$

The gravitational potential due to the Sun's gravitational field is -2.3×10^{7} J kg^{-1}.

5.5.2 Gravitational potential energy

Equation 5.7 tells us the gravitational potential at a point in a gravitational field per unit mass. To find the gravitational potential energy of an object of mass m_2 placed at that point in the field around a mass m_1, we simply multiply by m_2.

$$E_P = V \times m_2 = -\frac{Gm_1m_2}{r}$$

(5.8)

© HERIOT-WATT UNIVERSITY

We can now calculate the work done in moving an object in a gravitational field using this equation. The work done is equal to the change in potential energy of the object. When carrying out these calculations, it is important to take care to get the sign correct.

> **Example**
>
> A rocket ship, mass 4.00×10^5 kg, is travelling away from the Moon. The ship's rockets are fired when the ship is at a distance of 3.00×10^6 m from the centre of the Moon. If the mass of the Moon is 7.3×10^{22} kg, how much work is done by the rockets in moving the ship to a distance 3.20×10^6 m from the Moon's centre?
>
> The rocket ship has an initial gravitational potential energy of
>
> $$E_{P1} = -\frac{GM_M m_s}{r}$$
> $$\therefore E_{P1} = -\frac{6.67 \times 10^{-11} \times 7.3 \times 10^{22} \times 4.00 \times 10^5}{3.00 \times 10^6}$$
> $$\therefore E_{P1} = -6.492 \times 10^{11} \text{ J}$$
>
> The final value of the potential energy is
>
> $$E_{P2} = -\frac{GM_M m_s}{r}$$
> $$\therefore E_{P2} = -\frac{6.67 \times 10^{-11} \times 7.3 \times 10^{22} \times 4.00 \times 10^5}{3.20 \times 10^6}$$
> $$\therefore E_{P2} = -6.086 \times 10^{11} \text{ J}$$
>
> The change in potential energy ΔE_P is
>
> $$\Delta E_P = E_{P2} - E_{P1}$$
> $$\therefore \Delta E_P = -6.086 \times 10^{11} - (-6.492 \times 10^{11})$$
> $$\therefore \Delta E_P = 4.1 \times 10^{10} \text{ J}$$
>
> The potential energy of the rocket ship has increased by 4.1×10^{10} J, so the amount of work done by the rockets must also be equal to 4.1×10^{10} J.

The gravitational field is a **conservative field**, which means that the work done in moving a mass between two points in the field is independent of the path taken. In the above example, all that we are concerned with are the initial and final locations in the gravitational field. We do not need any details about the path taken between these two locations. (An example of doing non-conservative work would be the work done against friction in sliding a heavy mass between two points. Obviously we do less work if we slide the mass directly from one point to the other rather than taking a longer route.)

Quiz: Gravitational potential

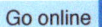

Useful data:

Universal constant of gravitation G	6.67×10^{-11} m^3 kg^{-1} s^{-2}
Mass of the Earth M_E	6.0×10^{24} kg
Radius of the Earth r_E	6.4×10^6 m
Mass of the Moon M_M	7.3×10^{22} kg
Radius of the Moon r_M	1.7×10^6 m

Q11: Consider the gravitational potential at a point A in the Earth's gravitational field. The value of the gravitational potential depends on

a) the mass of an object placed at A.
b) the speed of an object passing through A.
c) the distance of A from the centre of the Earth.
d) the density of an object placed at A.
e) the mass of the Earth only.

....

Q12: What is the gravitational potential on the surface of the Moon, due to the Moon's gravitational field?

a) -1.6 J kg^{-1}
b) -2.4×10^3 J kg^{-1}
c) -2.9×10^6 J kg^{-1}
d) -4.2×10^{17} J kg^{-1}
e) -2.9×10^{28} J kg^{-1}

....

Q13: A satellite orbits the Earth at an altitude of 3.00×10^5 m *above the Earth's surface*. What is the gravitational potential at this altitude?

a) -9.8 J kg^{-1}
b) -4.4×10^3 J kg^{-1}
c) -6.0×10^7 J kg^{-1}
d) -1.3×10^9 J kg^{-1}
e) -5.3×10^{24} J kg^{-1}

....

Q14: What is the gravitational potential energy of a satellite of mass 800 kg, moving in the Earth's gravitational field with orbit radius 6.60×10^6 m from the centre of the Earth?

a) -7.3×10^3 J
b) -6.1×10^7 J
c) -2.5×10^{10} J
d) -4.9×10^{10} J
e) -4.9×10^{32} J

....

Q15: The gravitational potential energy of a satellite orbiting the Earth changes from -5.0×10^9 J to -7.0×10^9 J. Which *one* of the following statements could be true?

a) The satellite has moved closer to the Earth.
b) The satellite has moved further away from the Earth.
c) The mass of the satellite has decreased only.
d) The orbit and mass have stayed the same, but the satellite is moving faster.
e) The orbit and mass have stayed the same, but the satellite is moving slower.

5.6 Escape velocity

At the beginning of the section on satellite motion, you should have carried out an interactive activity in which a projectile was launched into orbit with different speeds. If the speed was sufficiently great, the projectile did not complete an orbit, but escaped from the Earth's gravitational field. We will now calculate what minimum speed is required for a projectile to escape from the Earth's gravitational field, and then generalise it to all other gravitational fields.

To escape from the gravitational field, an object must have sufficient kinetic energy. We know that at an infinite distance from Earth, the potential energy of a body will be zero. We can also work out the (negative) potential energy of the body when placed at the Earth's surface. If the body has sufficient kinetic energy to raise its total energy above zero J, then it can escape from the gravitational field.

The gravitational potential energy of an object such as a rocket of mass m at a point on the Earth's surface can be calculated using Equation 5.8.

$$E_P = -\frac{GM_E m}{r_E}$$

To escape the Earth's gravitational field, the work done by the rocket must equal the potential difference between infinity and the point on the Earth's surface.

TOPIC 5. GRAVITATION

$$\Delta E_P = \frac{-GM_Em}{\infty} - \frac{-GM_Em}{r_E}$$
$$\therefore \Delta E_P = 0 + \frac{GM_Em}{r_E}$$
$$\therefore \Delta E_P = \frac{GM_Em}{r_E}$$

The rocket must have an initial kinetic energy at least equal to this, so that its velocity does not drop to zero before it has escaped from the field. We can use the equation for kinetic energy to calculate the initial velocity of the rocket.

$$\frac{1}{2}mv^2 = \frac{GM_Em}{r_E}$$
$$\therefore v^2 = \frac{2GM_E}{r_E}$$
$$\therefore v = \sqrt{\frac{2GM_E}{r_E}}$$

(5.9)

This velocity is the minimum velocity required. Now we will use the expression for the acceleration due to gravity g, equivalent to the gravitational field strength at the Earth's surface, which is given by the equation

$$g = \frac{GM_E}{r_E^2}$$
$$\therefore gr_E = \frac{GM_E}{r_E}$$

Substituting this expression into the equation for the escape velocity gives us

$$v = \sqrt{\frac{2GM_E}{r_E}}$$
$$\therefore v = \sqrt{2gr_E}$$

(5.10)

The escape velocity for a rocket fired from Earth is given by Equation 5.10. Putting in the values of $g = 9.8$ m s^{-2} and $r_E = 6.4 \times 10^6$ m, the escape velocity has a value

$$v = \sqrt{2gr_E}$$
$$\therefore v = \sqrt{2 \times 9.8 \times 6.4 \times 10^6}$$
$$\therefore v = 1.1 \times 10^4 \, \text{m s}^{-1} \text{ or } 11 \, \text{km s}^{-1}$$

You should note that the escape velocity does not depend on the mass of the rocket - the escape velocity is the same for any object launched from the Earth's surface. In general, the escape velocity from the gravitational field around a body of mass m, starting from a point r from the centre of the field, is given by the following equation

$$v = \sqrt{\frac{2GM}{r}}$$

(5.11)

From Equation 5.11, we can see that the escape velocity is greater for a planet with a higher mass. This causes interesting differences between the planets' atmospheres. The Earth's atmosphere is mainly nitrogen and oxygen, with a very low incidence of helium. In contrast, larger planets such as Jupiter have plenty of helium. Molecules such as helium have a very low mass. This meant that when the Earth was forming, the helium molecules were able to reach velocities in excess of the Earth's escape velocity and they managed to escape into space. Meanwhile, Mercury has an extremely thin atmosphere made up of atoms blasted off its surface by the solar wind. Mercury's atmosphere is constantly escaping and being replaced, since its high temperature means that its gas molecules can travel faster than the comparatively low escape velocity.

Example

What is the escape velocity for a lunar probe taking off from the surface of the Moon? (M_M = 7.3×10^{22} kg, r_M = 1.7×10^6 m)

Using Equation 5.11 with the values given in the question, the escape velocity from the Moon is

$$v = \sqrt{\frac{2GM_M}{r_M}}$$
$$v = \sqrt{\frac{2 \times 6.67 \times 10^{-11} \times 7.3 \times 10^{22}}{1.7 \times 10^6}}$$
$$v = \sqrt{5.728 \times 10^6}$$
$$v = 2.4 \times 10^3 \, \text{m s}^{-1}$$

The escape velocity from the Moon's gravitational field is 2.4×10^3 m s^{-1}, or 2.4 km s^{-1}.

Quiz: Escape velocity

Useful data:

Universal constant of gravitation G	6.67×10^{-11} m^3 kg^{-1} s^{-2}
Mass of the Earth M_E	6.0×10^{24} kg
Radius of the Earth r_E	6.4×10^6 m

Q16: The escape velocity of an object taking off from Earth is the minimum velocity required to

a) place the object in a geostationary orbit.
b) place the object in a non-geostationary orbit.
c) reach the point where the combined field of the Earth and the Moon is zero.
d) escape from the Earth's atmosphere.
e) escape from the Earth's gravitational field.

....

Q17: The escape velocity of an object from the Earth depends on

a) the masses of the Earth and the object.
b) the mass and radius of the Earth.
c) the mass and density of the object.
d) the mass of the object and the radius of the Earth.
e) the mass and radius of the Earth, and the mass of the object.

....

Q18: What would be the escape velocity of a Martian spacecraft of mass 900 kg taking-off from the surface of Mars, if the mass and radius of Mars are 6.42×10^{23} kg and 3.40×10^6 m?

a) 2510 m s^{-1}
b) 3550 m s^{-1}
c) 5020 m s^{-1}
d) 1.06×10^5 m s^{-1}
e) 1.51×10^5 m s^{-1}

5.7 Extended information

Web links   Go online

These web links should serve as an insight to the wealth of information available online and allow you to explore the subject further.

- https://www.youtube.com/watch?v=Uz4tjl0cla0
- https://www.youtube.com/watch?v=2PdiUoKa9Nw
- http://vimeo.com/16031284
- http://www.countingthoughts.com/schiehallionclip.php
- http://www.esa.int/spaceinvideos/Videos/2011/03/GOCE_Geoid
- https://www.youtube.com/watch?v=YS6IqnvnOak
- https://www.youtube.com/watch?v=gftT3wHJGtg

5.8 Summary

Summary

You should now be able to:

- state and apply the equation $F \frac{GMm}{r^2}$ to calculate the gravitational force between two objects;
- calculate the weight of an object using Newton's Universal Law of Gravitation;
- state what is meant by a gravitational field, and calculate the gravitational field strength at a point in the field;
- calculate the value of the acceleration due to gravity at a point in a gravitational field, given the local conditions;
- sketch the field lines around a planet and a planet-moon system;
- state the expression $V = \frac{-GM}{r}$ and use it to calculate the gravitational potential V at a point in a gravitational field;
- calculate the gravitational potential energy of a mass in a gravitational field and calculate the change in the potential energy when a mass is moved between points in the field;
- define escape velocity as the minimum velocity required to allow a mass to escape a gravitational field;
- derive the expression $v = \sqrt{\frac{2GM}{r}}$ and use it to calculate the escape velocity.

5.9 End of topic 5 test

End of topic 5 test   Go online

The following data should be used when required:

Universal constant of gravitation G	6.67×10^{-11} m^3 kg^{-1} s^{-2}
Mass of the Earth M_E	6.0×10^{24} kg
Mass of the Moon M_M	7.3×10^{22} kg
Mass of the Sun M_S	2.0×10^{30} kg
Radius of the Earth r_E	6.4×10^6 m
Radius of the Moon r_M	1.7×10^6 m

Q19: A distant planet has mass 5.45×10^{25} kg. A moon, mass 3.04×10^{22} kg, orbits this planet with an orbit radius of 7.16×10^8 m.

Calculate the size of the gravitational force that exists between the moon and the planet.

F = _____ N

....

Q20: Two identical solid spheres each have mass 0.853 kg and diameter 0.245 m.

Find the gravitational force between them when they are touching.

F = _____ N

....

Q21: The mass of planet Neptune is 1.03×10^{26} kg and its radius is 2.48×10^7 m.

Calculate the weight of a 6.64 kg mass on the surface of Neptune.

Weight on Neptune = _____ N

....

Q22: The value of the acceleration due to gravity is not constant, decreasing with height above the surface of the Earth.

What is the value of the acceleration due to gravity in the ionosphere at a height 3.03×10^5 m above the Earth's surface?

g = _____ m s^{-2}

....

Q23: On the surface of the Earth, a particular object has a weight of 22.0 N.

Calculate its weight on the surface of the Moon.

_____ N

....

UNIT 1. ROTATIONAL MOTION AND ASTROPHYSICS

Q24: A planet in a distant galaxy has mass 6.67×10^{25} kg and radius 4.02×10^7 m.
Calculate the value of the gravitational field strength on the surface of this planet.
_____ N kg^{-1}

....

Q25: The gravitational field strength at a distance 2.13×10^6 m from the centre of a planet is 6.05 N kg^{-1}.
Calculate the field strength at a distance 8.52×10^6 m from the centre of the planet.
Note: It is possible to solve this problem without having to calculate the mass of the planet.
_____ N kg^{-1}

....

Q26: In a distant solar system, a planet (mass 2.02×10^{28} kg) is orbiting a star (mass 5.41×10^{30} kg) with an orbit radius of 4.44×10^{11} m.
Calculate the magnitude of the net gravitational field strength midway between the planet and the star.
_____ N kg^{-1}

....

Q27: Calculate the gravitational potential at a distance 1.25×10^7 m from the centre of the Earth, due to the Earth's gravitational field.
V = _____ J kg^{-1}

....

Q28: A spaceship, mass 5.4×10^4 kg is travelling through the solar system. At one point in its journey, the spaceship passes near the planet Jupiter taking photographs at a distance 9.5×10^7 m from the centre of Jupiter.
If the mass of Jupiter is 1.9×10^{27} kg, calculate the potential energy of the spaceship at this point.
E_p = _____ J

....

Q29: The gravitational potential at a distance 2.46×10^7 m from the centre of a planet of radius 8.55×10^6 m is -5.24×10^7 J kg^{-1}.
Calculate the gravitational potential at a distance 4.92×10^7 m (twice the original distance) from the centre of the planet.
Note: It is possible to solve this problem without having to calculate the mass of the planet.
V = _____

....

Q30: A meteorological satellite is orbiting the Earth. The mass of the satellite is 5.25×10^3 kg and it orbits at a height 1.35×10^5 m above the Earth's surface.
Calculate the gravitational potential energy of the satellite.
E_p = _____ J

....

© HERIOT-WATT UNIVERSITY

TOPIC 5. GRAVITATION

Q31: Scientists wish to launch a satellite (mass 2.8×10^4 kg) which will orbit the Earth once every 5400 seconds.

At what height above the Earth's surface should the satellite be placed?

_____ m

....

Q32: The planet Zaarg has mass 7.54×10^{25} kg and radius 2.88×10^7 m.

Calculate the escape velocity of a rocket ship launched from the planet Zaarg.

_____ m s^{-1}

....

Q33: Astronomers observing a distant solar system have noticed a planet orbiting a star with a period 6.82×10^7 s. The distance from the planet to the star is 2.95×10^{11} m.

Calculate the mass of the star.

_____ kg

Unit 1 Topic 6

General relativity and spacetime

Contents

6.1	Introduction	135
6.2	Comparison of special and general relativity	135
6.3	The equivalence principle	136
6.4	Understanding the consequences	138
6.5	Show me the evidence	140
6.6	Spacetime diagrams	142
6.7	Worldlines	144
6.8	The curvature of spacetime	146
6.9	Black holes	151
6.10	Extended information	154
6.11	Summary	155
6.12	End of topic 6 test	156

Prerequisites
- Special relativity (Higher).
- Gravitation.

134 UNIT 1. ROTATIONAL MOTION AND ASTROPHYSICS

Learning objective

By the end of this topic you should be able to:

- state that:
  - an inertial frame of reference is one that is stationary or has a constant velocity;
  - a non-inertial frame of reference is one that is accelerating;
- state that Einstein's theory of special relativity is appropriate for inertial frames of reference;
- state that Einstein's theory of general relativity is appropriate for non-inertial frames of reference;
- describe Einstein's equivalence principle in terms of an accelerated frame of reference being equivalent to a reference frame at rest in a gravitational field;
- state that when an object is in freefall, the downwards acceleration exactly cancels out the effects of being in a gravitational field;
- explain:
  - some consequences of the equivalence principle, such as that clocks in a weaker gravitational field run faster than those in a stronger gravitational field;
  - some of the pieces of evidence for general relativity;
- state that a free falling object or light follows a geodesic in spacetime;
- explain that spacetime represents the three dimensions of space and one dimension of spacetime;
- interpret spacetime diagrams for stationary objects, those moving at a constant speed and accelerating objects;
- state that the greater the gradient of a worldline, the smaller the velocity;
- explain that curved lines on spacetime diagrams correspond to non-inertial (accelerating) frames of reference i.e. accelerations are represented by worldlines of changing gradient;
- explain that general relativity allows understanding that mass curves spacetime and that gravity is causes by the curvature of spacetime;
- state that the event horizon is the boundary of a black hole and that no matter or radiation can escape from within the event horizon.
- state that the density of a black hole is so great that the escape velocity at the event horizon is equal to the speed of light;
- explain that to a distant observer, time appears to be frozen at the event horizon of a black hole;
- explain what is meant by the Schwarzschild radius of a black hole;
- solve problems using the equation for the Schwarzschild radius of a black hole $r_{Schwarzschild} = \frac{2GM}{c^2}$;

© HERIOT-WATT UNIVERSITY

6.1 Introduction

Einstein's special theory of relativity only applies when an object is moving at a constant speed in a straight line. It does not let us consider what happens when something turns or changes speed. For this we need Einstein's general theory of relativity.

In this topic we will compare the two theories and explore how **general relativity** allows us to more fully understand gravity. We will then turn our attention to the effect of mass on **spacetime** and the curious behaviour of black holes.

6.2 Comparison of special and general relativity

You studied the special theory of relativity at Higher. You may remember that in this theory Einstein made the following two points:

1. The speed of light is absolute. It is always the same for all observers irrespective of their relative velocities.
2. The laws of Physics are the same for all observers inside their frame of reference.

These observations lead to some conclusions that at first seemed strange but were confirmed by observation, such as time dilation and length contraction. However, the theory did not allow us to consider what happens when a frame of reference is accelerating. For this we will need general relativity.

In other words, **special relativity** only considered inertial frames of reference. That is ones that are stationary or have a constant velocity. General relativity will allow us to study a **non-inertial frame of reference**, which is one that is accelerating.

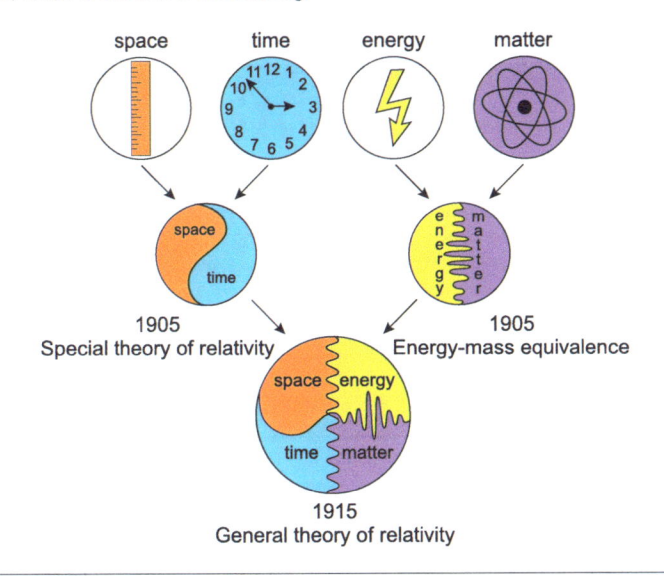

6.3 The equivalence principle

Einstein used the following thought experiment to show how he was extending relativity to include accelerating objects. Consider two observers in identical spaceships, one at rest on the Earth and the other accelerating at 9.8 ms^{-2} in deep space, far away from any astronomical bodies. Einstein said it would be impossible to distinguish between the two situations and that an experiment would produce the same results in both spaceships. For instance, if both observers stand on a newton balance, then provided they have the same mass as each other, the readings would be identical in the two scenarios.

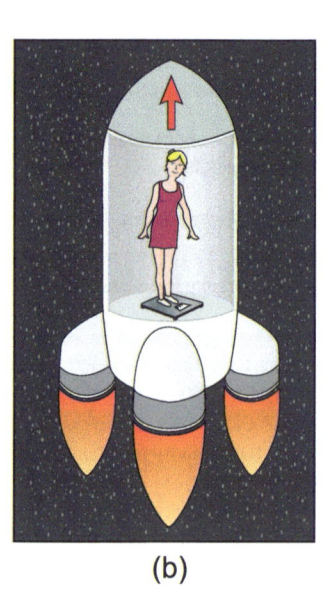

(a) At rest in the Earth's gravitational field. (b) In deep space accelerating upwards at 9.8 m s^{-2}.

Furthermore, a ball dropped in one spaceship would fall to the floor in the same way as in the other spaceship. When the ball is released in the spaceship on Earth, the gravitational field will cause it to accelerate downwards towards the floor at 9.8 ms^{-2}. When the ball is released in the spaceship in outer space, it becomes a free object. That is, there is no unbalanced force acting upon the ball. However, since the spaceship continues to accelerate upwards, the floor of the spaceship will accelerate upwards to meet the ball. So, from the observer's viewpoint, the ball appears to accelerate downwards towards the floor at 9.8 ms^{-2}.

TOPIC 6. GENERAL RELATIVITY AND SPACETIME

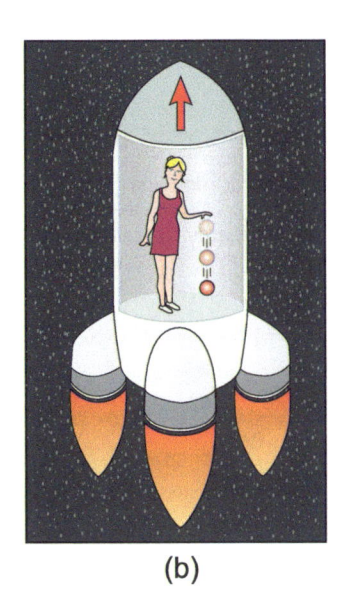

(a)   (b)

(a) At rest in the Earth's gravitational field. (b) In deep space accelerating upwards at 9.8 m s^{-2}.

Einstein's **equivalence principle** summarises this. It states that it is impossible to tell the difference between a uniform gravitational field and a frame of reference that has a constant acceleration. In other words, gravity is equivalent to acceleration.

Einstein also pondered what would happen if a person dropped a ball whilst they were falling off the side of a building. He realised that a person who accelerates downwards along with the ball will not be able to detect the effects of gravity on the ball. Indeed, both the ball and the person would effectively be weightless in this scenario. They would be equivalent to a person motionless in a spaceship in deep space. He identified that a force is experienced when accelerating or when in a gravitational field, but no force is felt when an object is in freefall, since the downwards acceleration exactly cancels out the effect of being in a gravitational field.

So, all of these thought experiments made Einstein realise that the force of gravity is just the acceleration that you feel as you move through spacetime (see section 6.8).

The equivalence principle   Go online

There is an animation available online showing the effects of gravity.

© HERIOT-WATT UNIVERSITY

6.4 Understanding the consequences

Bending light

Consider a laser beam sent across a spaceship that is accelerating upwards. To an observer who watches the spaceship from the ground, the light moves in a straight line. However, the spaceship will move slightly upwards in the time it takes the light to travel across it. Therefore the light will strike a point lower on the spaceship wall than it would if the spaceship did not move. This means that, to an observer inside the spaceship, the light actually appears to bend.

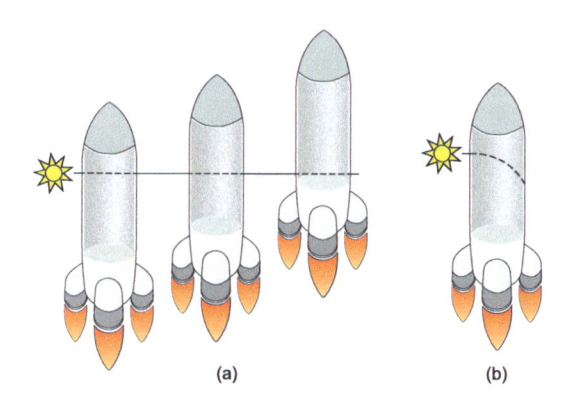

(a) View from the ground. (b) View inside the spaceship.

In the last section we explored the **equivalence principle**, which states that an accelerating frame acts in the same way as a frame in a gravitational field. Therefore, we can conclude that gravitational fields must also bend light. Since we are very familiar with light being bent by a lens, we call this effect gravitational lensing.

The Equivalence Principle - the bending of light due to acceleration and gravity.

Gravitational time dilation

Once more, consider a spaceship which is accelerating upwards. A source of light emits pulses at regular time intervals from the bottom of the spaceship. By the time a pulse reaches an observer at the top of the spaceship, the spaceship will have moved away from the position it occupied when the pulse was emitted. Since the spaceship is accelerating, the distance each consecutive light pulse must travel to reach the observer will be increasing. That means the time for pulses to reach the observer will also be increasing. So the observer at the top of the spaceship will conclude that the clock at the bottom of the spaceship is running more slowly. We already know from the equivalence principle that a spaceship accelerating upwards is equivalent to a spaceship at rest in a gravitational field.

The clock at the rear runs more slowly.

So for a clock in a gravitational field, time runs more slowly than in the absence of a field.

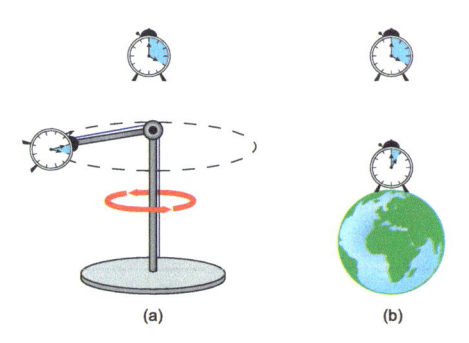

(a) Accelerating clock in a centrifuge. (b) Clock in a gravitational field - the lower the slower.

6.5 Show me the evidence

Gravitational red shift

Consider a source of light at the top of an accelerating spaceship. Due to the Doppler Effect, an observer at the bottom of the spaceship will observe a higher frequency than the source emitted. The **equivalence principle** means that a beam of light travelling in the same direction as a gravitational field should shift towards the higher frequency blue end of the spectrum. Furthermore, a beam of light travelling upwards from the Earth's surface should be shifted towards the lower frequency red end of the spectrum. More generally, any electromagnetic wave originating from a source that is in a gravitational field is reduced in frequency when observed in a region of a weaker gravitational field. We call this effect gravitational red shift. It was confirmed experimentally for the first time by measuring the relative redshift of two sources situated at the top and bottom of a tower at Harvard University.

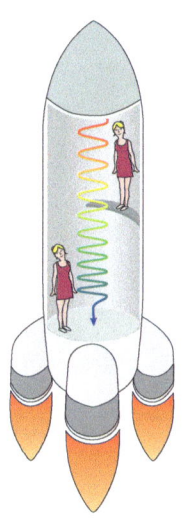

This effect can also be explained in terms of the photons. If a photon moves towards the Earth, the photon's energy increases as it travels to a position of smaller gravitational potential energy. Therefore, from $E = hf$, the photon's frequency ought to decrease.

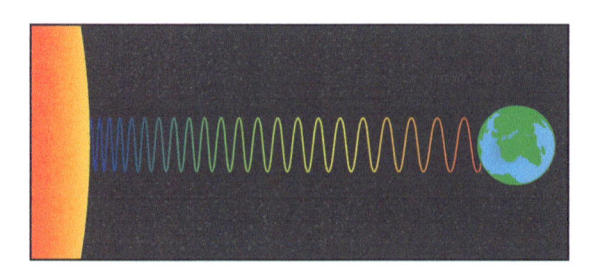

GPS

When Einstein first produced his general theory of relativity, its predictions were far removed from everyday experience and were therefore difficult to test experimentally. However, most people are nowadays accustomed to relying on the theory to keep their GPS (Global Positioning System) receivers working accurately.

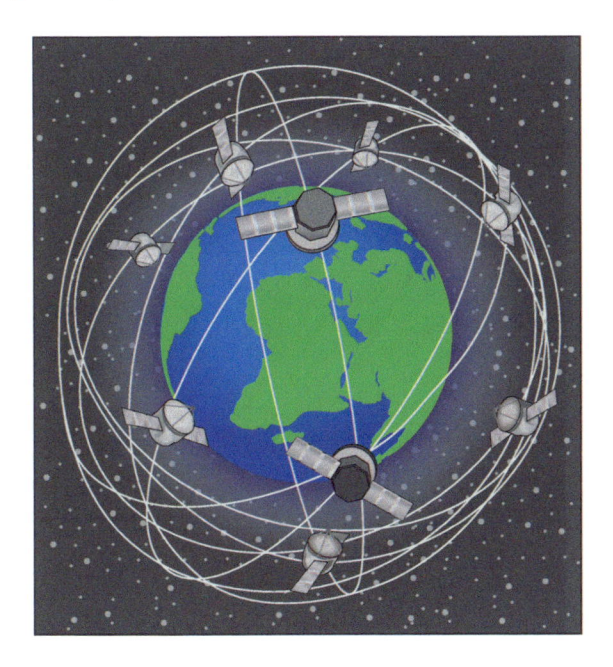

The clocks in GPS satellites need to be adjusted to take account of relativity.

Since the Earth's gravitational field is weaker at altitude, a clock in a GPS satellite will run fast relative to a clock on the ground. Adjustments need to be made to take this into account or the clocks will become out of sync within minutes. Corrections also need to be made to allow for the special relativistic effect of time dilation brought about by the large velocity of the satellite. Ignoring Einstein's theories would result in a discrepancy of around 10 km per day. So within a week you would be confusing Edinburgh with Glasgow.

6.6 Spacetime diagrams

In the last topic we saw that space and time are linked and that gravity is a property of both space and time. So it is helpful to consider spacetime, which is a coordinate system that involves the three dimensions of space (x, y and z), along with time. The motion of a particle can then be plotted as a series of points in this system.

Consider an event (a particular place at a particular time). Let's call it E. Imagine light moving out from a source at E in an expanding spherical shell. To simplify matters, let's consider only two dimensional space for just now. That is, let's consider x and y, but ignore z. Then the light would look like an expanding circle. Now imagine snapshots are taken at regular intervals and stacked on top of each other so that the horizontal plane represents how far the light has moved and the vertical axis represents time. A cone shape would be formed.

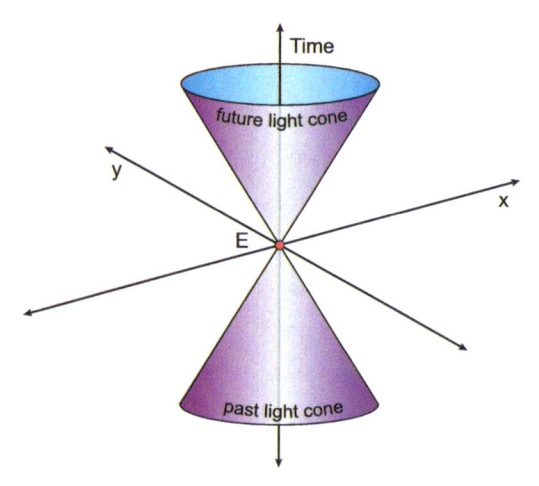

We constructed this lightcone by choosing an event in spacetime and imagining all the possible paths that light could take in moving through this event. However, we could have considered any event in spacetime and its corresponding lightcone. So spacetime is actually filled with light cones; there is one for every event.

Knowing the light cone structure of spacetime allows us to see which events can have an effect upon others. The top half of the light cone represents the future and it is the all events that have not yet happened that could be affected by event E. The bottom half is the past and it is all the events that could have contributed to event E. A region outside the light cone would be inaccessible, since something would need to travel faster than light to reach it. Remember from special relativity that the speed of light is believed to be the greatest possible speed.

To understand this, let us consider Alpha Centauri, our next nearest star after the Sun. It is 4.4 light years from Earth. Now, if you consider yourself at this present moment in time as an event, then Alpha Centauri in three years' time would lie outside your light cone. In other words, you cannot reach it without travelling faster than the speed of light. Therefore, it is not accessible. However, Alpha Centauri in thirty years' time does lie inside your light cone. This means that theoretically you could travel there, though we do not currently have the technology to reach Alpha Centauri within a human's lifetime.

TOPIC 6. GENERAL RELATIVITY AND SPACETIME 143

The path of an object through spacetime is called a worldline. An example is shown in red below. Note that worldlines cannot go beyond the light cone, as to do so would require an object to travel at a speed greater than the speed of light.

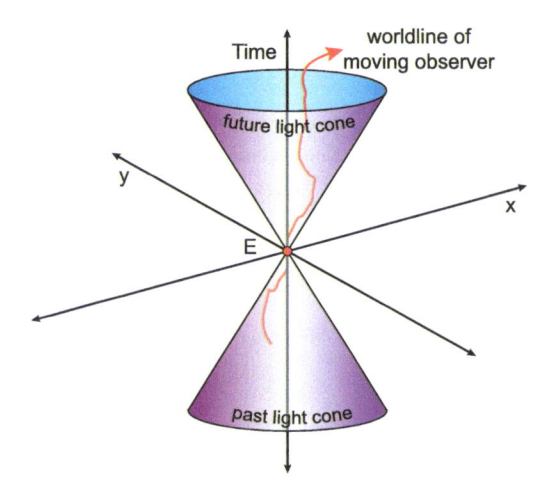

In reality, the light cone would actually be four-dimensional, 3 dimensions for space and one for time. However, the concept is easier to visualise with the number of spatial dimensions reduced. Indeed, the diagrams are most often simplified to show time and only one spatial dimension as shown below.

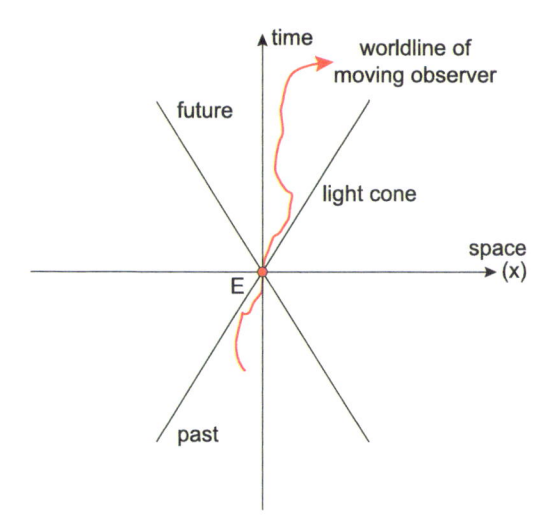

© HERIOT-WATT UNIVERSITY

Again, the different regions of the spacetime diagram have the following significance:

- The diagonal lines above correspond to a speed equal to the speed of light in vacuum i.e. $v = c$.
- The regions within the diagonals are where the speed is less than c.
- The regions outside the diagonals cannot affect or be affected by the event at E, since messages cannot travel faster than c.
- The present corresponds to where the time is zero.
- The region below the x-axis is the past.

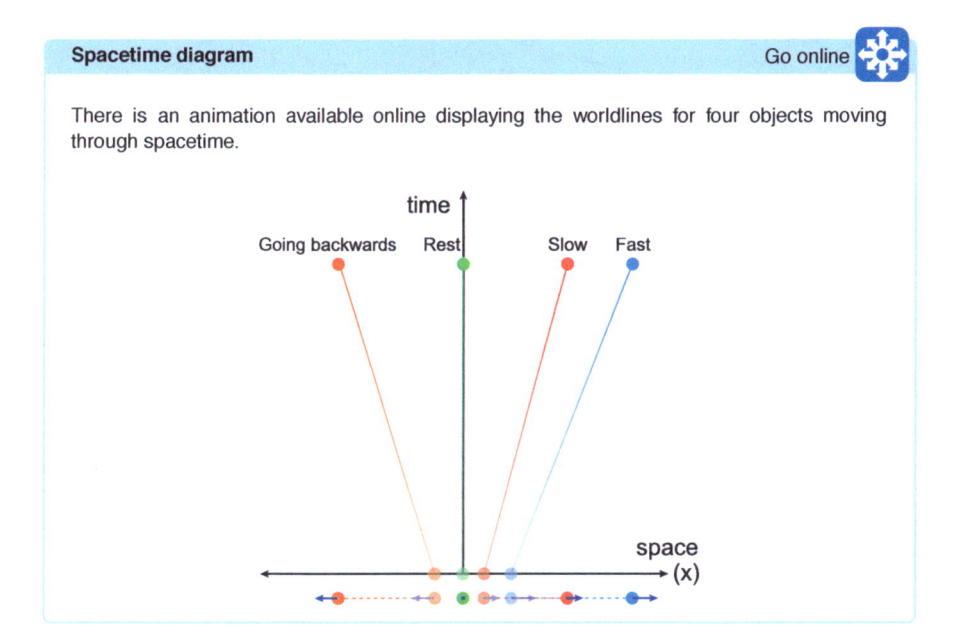

Spacetime diagram Go online

There is an animation available online displaying the worldlines for four objects moving through spacetime.

6.7 Worldlines

As you found out in the last section, a line on a spacetime diagram which maps a particle's spatial location at every instant in time is called a **worldline**. Typically only events in a one dimensional world are considered as this is a lot simpler to understand! Each point of a worldline is an event that can be labelled with the time and the spatial position of the object at that particular time. Note that unlike distance-time graphs, the spatial position will be displayed on the x-axis and the time on the y-axis.

A stationary object's position does not change with time and so its spacetime diagram would be as follows.

TOPIC 6. GENERAL RELATIVITY AND SPACETIME

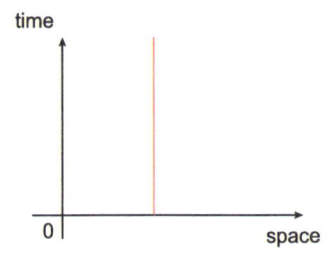

An object moving at constant speed could be shown by either of the following worldlines.

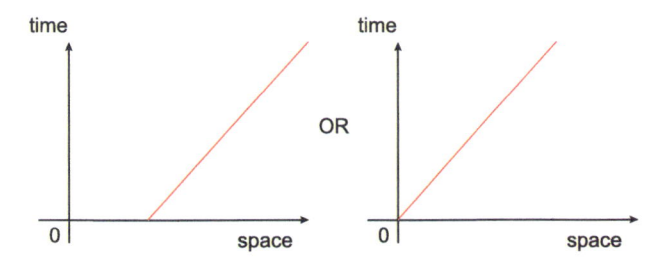

The greater the gradient of a worldline, the smaller the velocity.

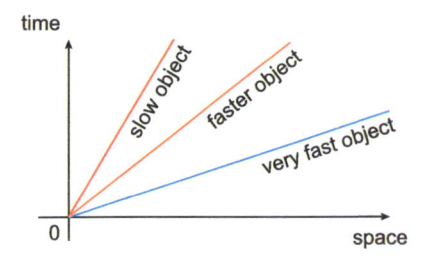

Accelerating objects are shown on spacetime diagrams as having world lines of changing gradient. So curved lines on spacetime diagrams correspond to non-inertial frames of reference i.e. accelerating frames of reference.

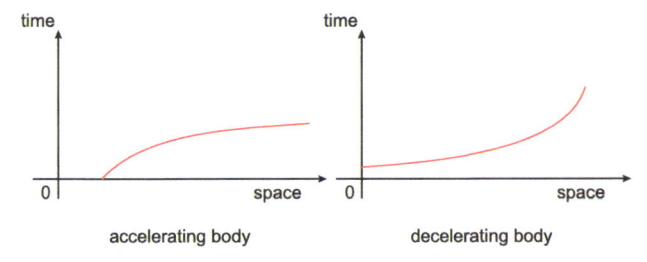

accelerating body

decelerating body

Simultaneous events are ones that occur at the same time. Therefore simultaneous events are shown as a flat line on a spacetime diagram, a line of constant time.

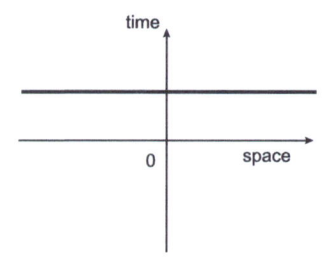

6.8 The curvature of spacetime

We saw earlier that gravitational lensing is the process whereby a large mass can make light bend. In this situation, the light is infact still travelling in a straight line. It is just that the mass has actually warped spacetime into being curved. Light always travels the shortest path between two points in spacetime and this path is called a **geodesic path**.

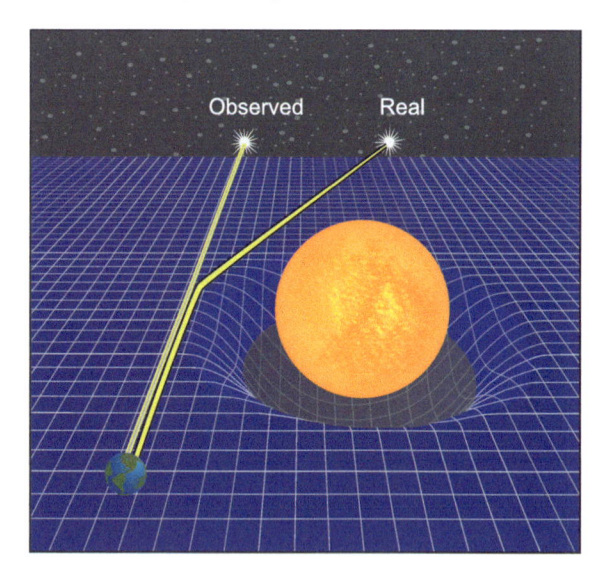

Warped spacetime

Einstein proposed that a large mass curves and stretches the spacetime around it, rather like a bowling ball creating a dimple as it distorts a rubber sheet. This then affects the motion of any other body that enters its spacetime since it will now need to follow the curvature to follow a straight line. So what is called gravity is really the result of the mass of an object creating a curvature in spacetime. Just as an object with a greater mass causes a greater distortion in a rubber sheet, an astronomical body with a larger mass causes a greater distortion and curvature in the fabric of spacetime. So gravity feels strongest where spacetime is most curved, and it vanishes where

TOPIC 6. GENERAL RELATIVITY AND SPACETIME

spacetime is flat.

The same explanation can be used to account for the motion of an object in freefall, such as a satellite, where the only force acting on it is gravity. The gravitational field of the central body makes spacetime curved so that the satellite's geodesic in spacetime is now curved, rather than straight.

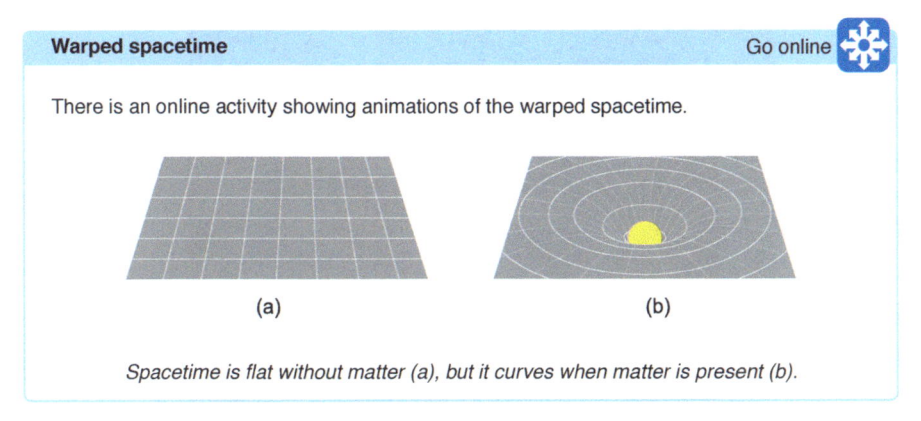

Warped spacetime   Go online

There is an online activity showing animations of the warped spacetime.

(a)   (b)

Spacetime is flat without matter (a), but it curves when matter is present (b).

In short, matter tells spacetime how to curve, and curved spacetime tells matter how to move.

The gravitational force of attraction between the Sun and a planet decreases with distance from the Sun. The closer to the Sun, the greater the degree of curvature in spacetime. The further from the Sun, the less the curvature in spacetime.

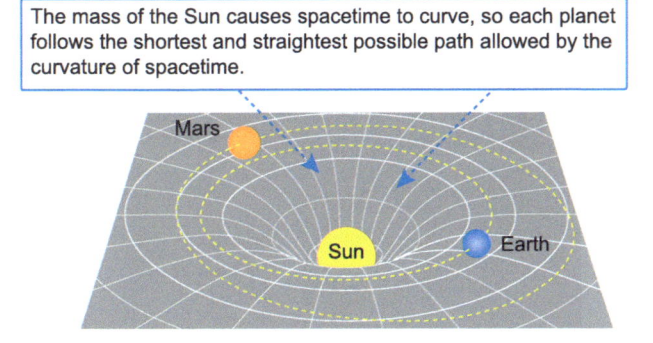

The mass of the Sun causes spacetime to curve, so each planet follows the shortest and straightest possible path allowed by the curvature of spacetime.

Gravity Probe B was launched in 2004 to measure the curvature of spacetime. Tiny deviations in the orientation of spinning gyroscopes allowed astronomers to measure the amount by which the Earth warps the local spacetime in which it resides.

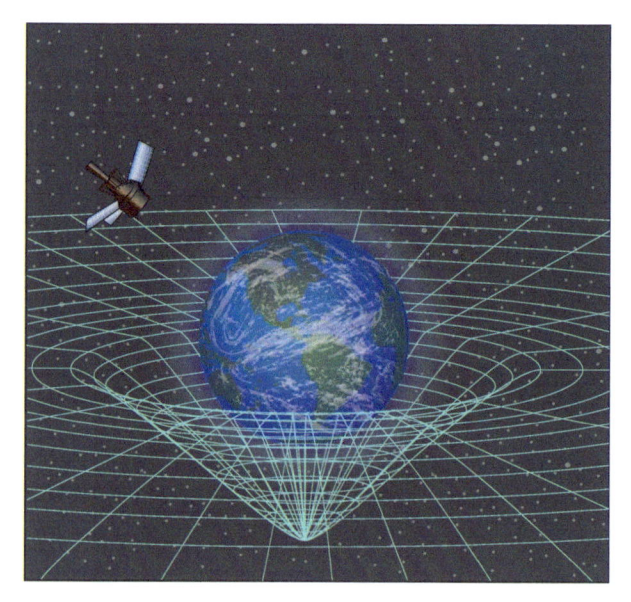

The orbit of Mercury

Einstein's general theory of relativity has now been demonstrated experimentally in various ways. However, one of the earliest pieces of evidence to support it was the motion of Mercury. Newtonian Physics allowed astronomers to accurately describe the orbit of the planets in our solar system, but they remained mystified by the behaviour of Mercury. They understood that the perturbing (changing) effect of the other planets made Mercury's elliptical orbit **precess** around the Sun, like a spinning top. However, Newtonian Physics could not account for the extent to which Mercury precessed. Einstein managed to successfully explain Mercury's behaviour by outlining that the mass of the Sun was creating a curvature in spacetime. In other words, the Sun was warping spacetime and Mercury was simply following the resulting curvature in the fabric of spacetime.

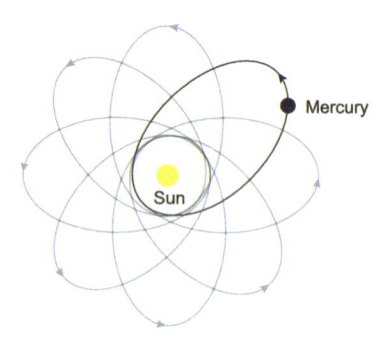

Observing gravitational lensing

General relativity predicts that an astronomical body with a very large mass ought to bend light. The astronomer Arthur Eddington was the first to provide evidence to confirm this. He made measurements to show that the gravitational field of the Sun deflects light from its straight path towards the Earth. Furthermore, **gravitational lensing** by various astronomical bodies has now been verified numerous times with data from radio telescopes.

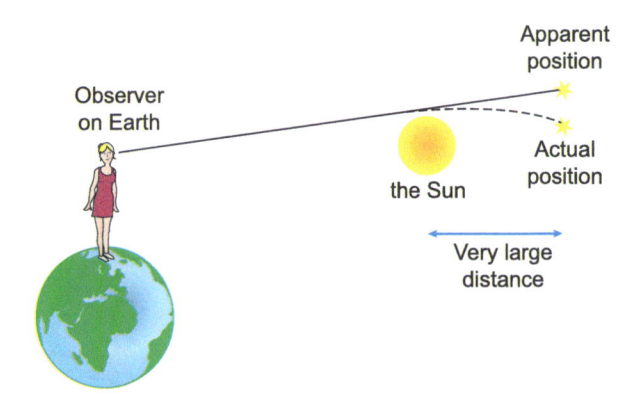

An interesting example of this effect is known as an Einstein ring. If the light from a distant galaxy is made to bend by the gravitational field of another galaxy that is directly behind it, the light can be focused into a visible ring.

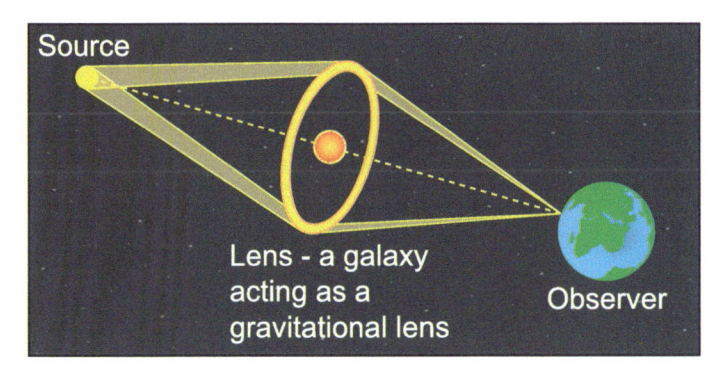

Gravitational waves - Ripples in spacetime

Einstein's general theory of relativity also predicts the existence of **gravitational waves**, which are ripples in spacetime generated during extremely violent astrophysical events, such as the collision of two black holes.

Gravitational waves are so weak that they are very hard to detect, but special interferometers such as LIGO (Laser Interferometer Gravitational Wave Observatory) use precisely calibrated laser beams to search for them. A passing gravitational wave ought to slightly distort spacetime and cause a noticeable shift in the interference pattern created by the lasers.

Image provided by Caltech/MIT/LIGO Laboratory

Astronomers' observations are currently limited to objects which emit electromagnetic radiation, but vast parts of the Universe are obscured by dark clouds. Since gravitational waves can pass through unhindered, their detection would allow astronomers to greatly increase their knowledge. Furthermore, the Big Bang is believed to have created a flood of gravitational waves which still fill the Universe today. So the detection of gravitational waves would also allow astronomers to gain a better understanding of the creation, development and fate of the Universe. It is not surprising some people call them Einstein's messengers.

6.9 Black holes

We will see in Topic 7 that when a star of exceptionally large mass reaches the end of its life, gravitational compression will cause it to collapse to a very small radius, producing an incredibly dense body called a **black hole**. Due to their extraordinary density, black holes exert extremely strong gravitational fields. Or rather, as general relativity would describe it, black holes severely distort spacetime.

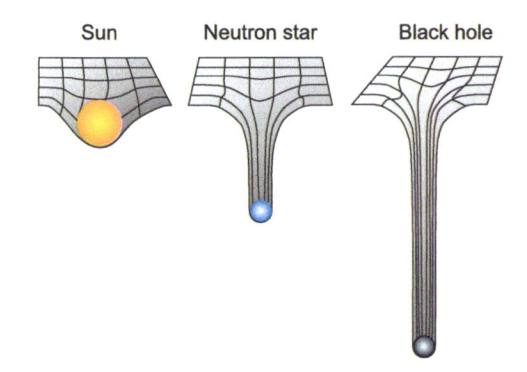

A black hole's extremely large density means it severely distorts the surrounding spacetime.

The spacetime around a black hole is stretched to a point of infinite density known as a **singularity**. This is a single point to which all mass would collapse.

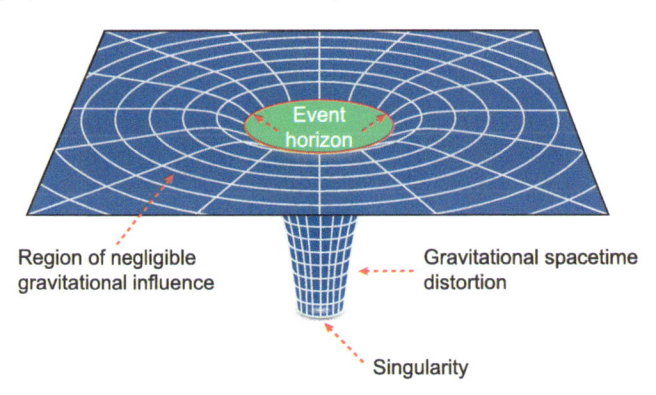

Black hole regions

Up to a certain distance away (**Schwarzschild radius**), the gravitational field around a black hole is so high that nothing can escape, not even light. Another way to look at it is that for distances less than the Schwarzschild radius, the escape velocity is greater than the speed of light. And since nothing can travel faster than the speed of light, nothing can escape, not even photons. This means that a black hole's name is certainly a very apt description. A black hole is effectively cut off from the rest of the Universe. However, a black hole's presence can still be deduced by detecting the stream of X-rays produced as matter falls into it.

The Schwarzschild radius and the event horizon

The Schwarzschild radius is the distance from the centre of a black hole at which not even light can escape. At the Schwarzschild radius the escape velocity is equal to the speed of light.

In Topic 5 we saw that the escape velocity for an object in a gravitational field can be found from:

$$v = \sqrt{\frac{2GM}{r}}$$

So, replacing v with the speed of light, c, gives the equation for the Schwarzschild radius:

$$c = \sqrt{\frac{2GM}{r_{Schwarzschild}}}$$

Squaring both sides results in:

$$c^2 = \frac{2GM}{r_{Schwarzschild}}$$

Rearranging for $r_{Schwarzschild}$ gives:

$$r_{Schwarzschild} = \frac{2GM}{c^2}$$

The Schwarzschild radius is also called the gravitational radius. It effectively forms a boundary called the **event horizon**. No matter or radiation can escape from within the event horizon. If you a looking at the event horizon from a distance, time would appear to stand still. It is possible to go from outside the event horizon to inside, but once you pass it, there is no going back!

TOPIC 6. GENERAL RELATIVITY AND SPACETIME

Launching a clock into a black hole   Go online

There is an online activity showing an animation of a clock launched into a black hole.

Examples

1. A star of mass 7.96×10^{32} kg collapses to form a black hole. Calculate the Schwarzschild radius.

$$r_{Schwarzschild} = \frac{2GM}{c^2}$$
$$r_{Schwarzschild} = \frac{2 \times 6.67 \times 10^{-11} \times 7.96 \times 10^{32}}{(3 \times 10^8)^2}$$
$$r_{Schwarzschild} = 1.18 \times 10^6 \, m$$

....

2. Calculate the mass of a black hole with a Schwarzschild radius of 1.76×10^3 km.

$$r_{Schwarzschild} = \frac{2GM}{c^2}$$
$$1.76 \times 10^6 = \frac{2 \times 6.67 \times 10^{-11} \times M}{(3 \times 10^8)^2}$$
$$M = 1.19 \times 10^{33} \, \text{kg}$$

....

3. A black hole has a mass of 1.50×10^{34} kg. Find the distance of the event horizon from its centre.

$$r_{Schwarzschild} = \frac{2GM}{c^2}$$
$$r_{Schwarzschild} = \frac{2 \times 6.67 \times 10^{-11} \times 1.50 \times 10^{34}}{(3 \times 10^8)^2}$$
$$r_{Schwarzschild} = 2.22 \times 10^7 \, m$$

....

4. Calculate the mass of a black hole with a Schwarzschild radius of 2.98×10^{29} km.

$$r_{Schwarzschild} = \frac{2GM}{c^2}$$
$$2.98 \times 10^{32} = \frac{2 \times 6.67 \times 10^{-11} \times M}{(3 \times 10^8)^2}$$
$$M = 2.01 \times 10^{59} \, \text{kg}$$

© HERIOT-WATT UNIVERSITY

6.10 Extended information

Web links — Go online

These web links should serve as an insight to the wealth of information available online and allow you to explore the subject further.

- https://www.youtube.com/watch?v=lXG-yoUsVS8
- http://www.bbc.co.uk/science/space/universe/questions_and_ideas/general_relativity#p009p1c1
- http://spiff.rit.edu/classes/phys230/lectures/planets/Lens_Nav.swf
- http://www.nsf.gov/news/mmg/mmg_disp.jsp?med_id=58443&from=vid.htm
- http://www.youtube.com/watch?v=0rocNtnD-yI
- https://www.youtube.com/watch?v=dUnbqBxUhf8
- https://www.youtube.com/watch?v=MO_Q_f1WgQI
- https://www.youtube.com/watch?v=7By2Fox_5ic
- https://www.youtube.com/watch?v=HHYECUcfC1Y
- http://highered.mheducation.com/olcweb/cgi/pluginpop.cgi?it=swf::800::600::/sites/dl/free/007299181x/78778/Escape_Nav.swf::Escape%20Velocity%20Interactive
- http://hubblesite.org/explore_astronomy/black_holes/encyclopedia.html
- http://www.bbc.co.uk/science/space/universe/sights/black_holes/#p00frjln
- http://www.youtube.com/watch?v=pL7R6jwKPo0
- http://www.youtube.com/watch?v=VP5KsGEsl-w
- https://www.youtube.com/watch?v=cJvYobIkVRo

6.11 Summary

> **Summary**
>
> You should now be able to:
>
> - state that:
>   - an inertial frame of reference is one that is stationary or has a constant velocity;
>   - a non-inertial frame of reference is one that is accelerating;
> - state that Einstein's theory of special relativity is appropriate for inertial frames of reference;
> - state that Einstein's theory of general relativity is appropriate for non-inertial frames of reference;
> - describe Einstein's equivalence principle in terms of an accelerated frame of reference being equivalent to a reference frame at rest in a gravitational field;
> - state that when an object is in freefall, the downwards acceleration exactly cancels out the effects of being in a gravitational field;
> - explain:
>   - some consequences of the equivalence principle, such as that clocks in a weaker gravitational field run faster than those in a stronger gravitational field;
>   - some of the pieces of evidence for general relativity;
> - state that a free falling object or light follows a geodesic in spacetime;
> - explain that spacetime represents the three dimensions of space and one dimension of spacetime;
> - interpret spacetime diagrams for stationary objects, those moving at a constant speed and accelerating objects;
> - state that the greater the gradient of a worldline, the smaller the velocity;
> - explain that curved lines on spacetime diagrams correspond to non-inertial (accelerating) frames of reference i.e. accelerations are represented by worldlines of changing gradient;
> - explain that general relativity allows understanding that mass curves spacetime and that gravity is causes by the curvature of spacetime;
> - state that the event horizon is the boundary of a black hole and that no matter or radiation can escape from within the event horizon.
> - state that the density of a black hole is so great that the escape velocity at the event horizon is equal to the speed of light;
> - explain that to a distant observer, time appears to be frozen at the event horizon of a black hole;

> **Summary continued**
> - explain what is meant by the Schwarzschild radius of a black hole;
> - solve problems using the equation for the Schwarzschild radius of a black hole $r_{Schwarzschild} = \frac{2GM}{c^2}$;

6.12 End of topic 6 test

End of topic 6 test Go online

The following data should be used when required:

Gravitational acceleration on Earth g	9.8 m s^{-2}
Gravitational acceleration on Moon g	1.6 m s^{-2}

Q1: Einstein's theory of general relativity considers frames of reference that are _____.

a) inertial
b) non-inertial

....

Q2: It deals with objects that are _____ .

a) stationary
b) moving at constant velocity
c) accelerating

....

Q3: Einstein identified that there is no way of distinguishing between the effects on an observer of a uniform gravitational field and of constant acceleration.
This is called his _____ .

....

Q4: In a gravitational field time runs more _____ .

a) slowly
b) quickly

....

Q5: When at the rear of an accelerating object, time passes _____ .

a) slowly
b) quickly

An astronaut on a spacecraft suspends a 5.5 kg mass from a newton balance. The reading is 8.8 N.

Q6: The spacecraft might be stationary on the surface of the Earth.

a) True
b) False

....

Q7: The spacecraft might be stationary on the surface of the Moon.

a) True
b) False

....

Q8: The spacecraft might be accelerating away from the surface of the Earth at 1.6 m s^{-2}.

a) True
b) False

....

Q9: The spacecraft might be accelerating away from the surface of the Moon at 1.6 m s^{-2}.

a) True
b) False

....

Q10: The spacecraft might be accelerating in deep space at 1.6 m s^{-2}.

a) True
b) False

....

Q11: Two astronauts awake from a deep sleep onboard a space capsule. They don't know whether they have landed on the surface of a planet or whether they are accelerating in deep space.
Can they perform an experiment inside the capsule to decide which situation they are in?

a) Yes
b) No

....

Q12: A person in freefall is equivalent to a body at rest in deep space since their downwards acceleration exactly cancels out the effect of being in a gravitational field.

a) True
b) False

Complete the following statements to form an explanation for gravitational lensing.

Q13: An astronomical body of large mass warps _____ so that it is curved.

....

Q14: The shortest path for _____ from a distant object is now curved and not straight.

....

Q15: The _____ bends round the large mass.

....

Q16: A line on a spacetime diagram which maps a particle's spatial location at every instant in time is called a _____ .

....

Q17: Curved lines on spacetime diagrams correspond to non-inertial frames of reference. This means objects which are _____ .

....

Q18: Which of the following correctly describes the objects' motions?

a) X: constant speed, Y: accelerating, Z: stationary
b) X: accelerating, Y: decelerating, Z: constant speed
c) X: constant speed, Y: decelerating, Z: accelerating
d) X: constant speed, Y: decelerating, Z: stationary

....

Q19: Accelerating objects are shown on spacetime diagrams as having world lines of changing _____ .

....

TOPIC 6. GENERAL RELATIVITY AND SPACETIME

Q20: Which two labels represent simultaneous events?

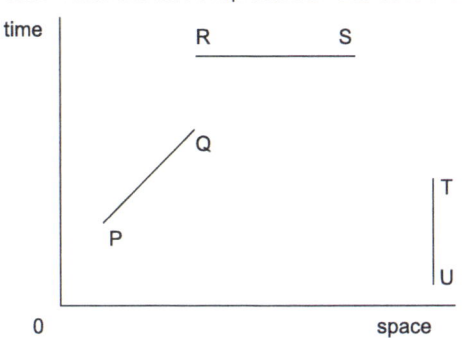

a) T and U
b) R and S
c) P and Q
d) Q and R

....

Q21: Complete the spacetime diagram using the following labels:

The present   **v > c**   **The future**   **v = c**   **The past**

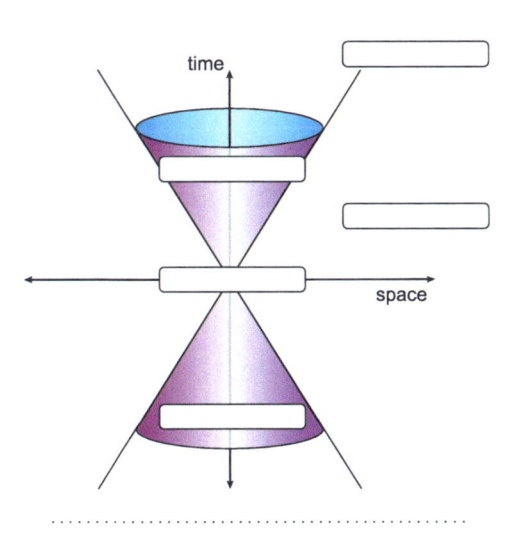

....

Q22: Which of the following statements about black holes are true?

I. The Schwarzschild radius is the distance from the centre of a black hole to the event horizon.
II. The escape velocity for black holes is greater than the speed of light.
III. A black hole results from the extreme curvature of space due to its compact mass.

a) I only
b) III only
c) I and III only
d) II and III only
e) I, II and III

....

Q23: Calculate the Schwarzschild radius for a black hole of mass 5.97×10^{31} kg.

Give your answer to *at least* 2 significant figures.

$r_{Schwarzschild}$ = _____ m

....

Q24: The event horizon is 4.31×10^8 m from a black hole's singularity. Determine the mass of the black hole.

Give your answer to *at least* 2 significant figures.

M = _____ kg

Unit 1 Topic 7

Stellar physics

Contents

7.1 Introduction 163
7.2 Properties of stars 163
    7.2.1 Temperature and colour 163
    7.2.2 Spectral class 164
    7.2.3 The power of a star 165
    7.2.4 Stellar luminosity 166
    7.2.5 Apparent brightness 170
    7.2.6 Summary of properties 171
    7.2.7 Quiz 173
7.3 Star formation 174
7.4 Stellar nucleosynthesis and fusion reactions 175
7.5 Hertzsprung-Russell diagrams 177
7.6 Stellar evolution and life cycles on H-R diagrams 181
7.7 Extended information 185
7.8 Summary 186
7.9 End of topic 7 test 187

Prerequisites

- Irradiance and inverse square law (Higher).
- Nuclear fusion (Higher).
- The relationship between the peak wavelength and the temperature of an object (Higher).

Learning objective

By the end of this topic you should be able to:

- state that:
  - all stellar objects give out a wide range of wavelengths of electromagnetic radiation but that each object gives out more energy at one particular wavelength;
  - the wavelength of this peak wavelength is related to the temperature of the object,

Learning objective continued

  with hotter objects having a shorter peak wavelength than cooler objects;
  - peak wavelengths allow the temperature of stellar objects to be calculated;
  - hotter objects also emit more radiation per unit surface area at all wavelengths than cooler objects;
- carry out calculations using the fact a star's power per unit area = σT^4;
- state that the luminosity of a star:
  - in watts, is a measure of the total power the star emits i.e. the total energy emitted per second;
  - depends on its radius and surface temperature;
- carry out calculations using the equation for luminosity $L = 4\pi r^2 \sigma T^4$;
- state that the apparent brightness of a star:
  - is the amount of energy per second reaching a detector per unit area;
  - depends on its luminosity and its distance from the observer;
- carry out calculations using the equation for apparent brightness $b = \frac{L}{4\pi d^2}$;
- state that the mass of a new star determines its luminosity and surface temperature;
- state that a Hertzsprung-Russell (H-R) diagram is a plot of luminosity versus the surface temperature of stars and that the surface temperature scale is in descending order;
- identify the long diagonal band on a H-R diagram as the main sequence and state that the Sun is on the main sequence;
- state that the main sequence stars are in their long lived stable phase where they are fusing hydrogen into helium in their cores;
- identify the:
  - higher luminosity and lower temperature stars lying to the right of the main sequence as the red giants and red supergiants;
  - lower luminosity and higher temperature stars lying to the left of the main sequence as the white dwarfs;
- predict the colour of a star based upon its position on the Hertzsprung-Russell diagram;
- describe how stars:
  - produce heat energy using the proton-proton chain reaction;
  - are formed in terms of the gravitational effects on cold dense interstellar clouds;
- state that the energy released from nuclear fusion in a star results in an outwards thermal pressure;
- explain:
  - how stars on the main sequence are in gravitational equilibrium;
  - why a star's life cycle is determined by its mass;
- relate the process of stellar evolution for a star to its path on a H-R diagram.

7.1 Introduction

In this topic we are going to examine different types of star, such as **red giants** and **white dwarfs**. We will explore what happens to a star when it gets old and we will find out how astronomers use diagrams to keep track of their progression along the life cycle. We will even find out how it can be argued that we are all made of star material.

7.2 Properties of stars

7.2.1 Temperature and colour

A star is a massive body of gas which emits light. The colour of a star is a good guide as to its approximate surface temperature. For example dull red stars are cool and bluish-white stars are very hot.

You will see in a later topic that an ideal black body is an object that is both a perfect emitter and absorber of electromagnetic radiation. That is, a black body emits and absorbs all frequencies equally and perfectly. Stars are not ideal black bodies, but treating them as such is a good approximation.

You may remember from Higher that the energy which a star emits is spread over a wide range of wavelengths. Astronomers can analyse a star's intensity-wavelength graph and use the peak wavelength to determine the surface temperature with the equation $\lambda_{max} T =$ constant.

The shorter the wavelength of the peak wavelength, the higher the surface temperature of the star.

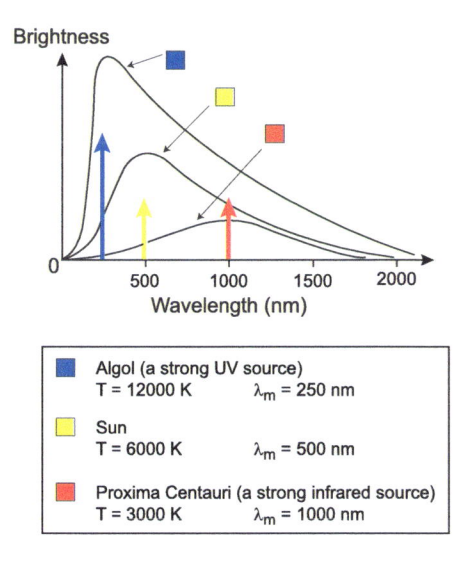

- Algol (a strong UV source)
  T = 12000 K λ_m = 250 nm
- Sun
  T = 6000 K λ_m = 500 nm
- Proxima Centauri (a strong infrared source)
  T = 3000 K λ_m = 1000 nm

7.2.2 Spectral class

The relative strengths of particular absorption lines gives the **spectral class** of a star. O is the hottest and M is the coolest type. So the Sun is a fairly middle of the road G star, but the relationship between surface temperature and spectral class is neither linear nor logarithmic.

Class	Temperature (K)	Colour	Notes	Examples
O	≥ 33,000	blue	Most of the electromagnetic radiation given out is ultraviolet	Several stars in the Orion constellation
B	10,000 - 33,000	blue to blue white	These stars are short lived	Rigel
A	7,500 - 10,000	white	This type of star is fairly common in our part of the galaxy	Sirius (one of the brightest stars in the sky) Deneb
F	6,000 - 7,500	yellowish white	Again these star types are common in our part of the galaxy	Capella
G	5,200 - 6,000	yellow	Our own star, the sun is a class G star	The Sun, Polaris (northern pole star)
K	3,700 - 5,200	orange	May be suitable for sustaining life on solar system in their orbit	Arcturus, Aldebaran
M	≤ 3,700	red	The most common of all types of star	Barnard's star

You can use a mnemonic to remember the order. The standard one is "Oh Be a Fine Girl/Guy, Kiss Me", but some may prefer "Oh Boy, An F Grade Kills Me".

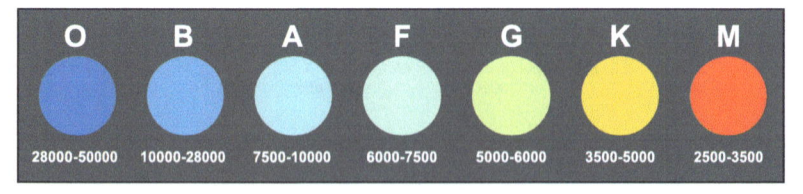

O	B	A	F	G	K	M
28000-50000	10000-28000	7500-10000	6000-7500	5000-6000	3500-5000	2500-3500

The above diagram is clearly not to scale, since the hottest stars are also generally the biggest and brightest. These stars use up their supply of hydrogen very quickly and therefore have shorter lifetimes. In a sense they are a bit like certain movie stars: the brightest largest ones tend to live fast and die young. In fact, all of the stellar properties we shall meet (spectral type, **luminosity**, radius and temperature) depend on a star's mass.

In section 7.6 we will see that once a star evolves beyond its normal hydrogen fusing phase, its size and temperature will change, but the life cycle it will follow is also determined by its mass. So for a star, its mass is really important.

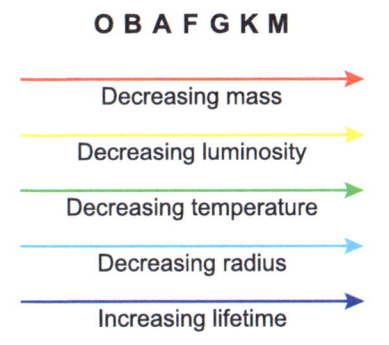

O B A F G K M

→ Decreasing mass
→ Decreasing luminosity
→ Decreasing temperature
→ Decreasing radius
→ Increasing lifetime

7.2.3 The power of a star

The power output P of a radiating object depends on its temperature and surface area. This relationship is known as the Stefan-Boltzmann Law and is written as follows:

$$P = \sigma A T^4$$

(7.1)

where
P is the power output in watts
T is the surface temperature in kelvin
A is the surface area in square metres
σ is the Stefan-Boltzmann constant, which has a value $5.67 \times 10^{-8} W m^{-2} K^{-4}$.

This agrees with everyday observation, since a hot cup of tea cools faster than a lukewarm one. Furthermore, large objects cool faster, since their bigger surface area means they can emit more energy per second than smaller objects.

Using Equation 7.1, a star's power per unit area can be therefore be found from:

$$\text{Power per unit area} = \sigma T^4$$

(7.2)

This relationship means that a small increase in a star's surface temperature results in a very large increase in its power output per unit area.

Examples

1.

Vega has a surface temperature of 9602 K. Calculate the power per unit area for Vega.

power per unit area $= \sigma T^4$
power per unit area $= 5.67 \times 10^{-8} \times 9602^4$
power per unit area $= 4.82 \times 10^8 W m^{-2}$

....

2.

The power per unit area of Rigel is $8.30 \times 10^8 W m^{-2}$. Find the surface temperature of Rigel.

power per unit area $= \sigma T^4$
$8.30 \times 10^8 = 5.67 \times 10^{-8} \times T^4$
$T = 11000 K$

7.2.4 Stellar luminosity

The luminosity L of a star is a measure of the total power the star emits from all wavelengths i.e. the total energy emitted per second. We can therefore see from the Stefan-Boltzmann law (Equation 7.1) that the luminosity can be expressed as:

$$L = \sigma A T^4$$

(7.3)

TOPIC 7. STELLAR PHYSICS

Luminosity is measured in watts. For a spherical star of radius r, the surface area is $4\pi r^2$. So stellar luminosity can be calculated using the equation:

$$L = 4\pi r^2 \sigma T^4$$

(7.4)

Therefore, the luminosity of a star depends on both its radius and surface temperature. A star can be luminous because it is hot or it is large or both. You may find it helpful to think of the luminosity as being a measure of the true brightness of a star. However, it should be appreciated that a star's total luminosity takes account of all the wavelengths of radiation, such as infrared and microwaves, not just the visible range.

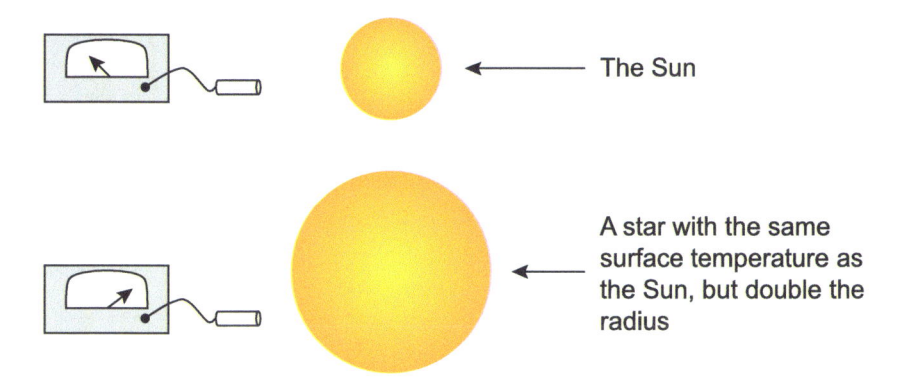

The Sun

A star with the same surface temperature as the Sun, but double the radius

The luminosity is proportional to the surface area of the star.

Consider a star which has a radius double that of the Sun. Its surface area would be four times that of the Sun. If this star and the Sun have the same surface temperature, then this star would have a luminosity four times that of the Sun. Considering Equation 7.3 and Equation 7.4, their luminosities can be compared as follows.

$$\frac{L_{star}}{L_{sun}} = \frac{\text{surface area}_{star}}{\text{surface area}_{sun}} = \frac{2^2}{1^2} = 4$$

© HERIOT-WATT UNIVERSITY

168 UNIT 1. ROTATIONAL MOTION AND ASTROPHYSICS

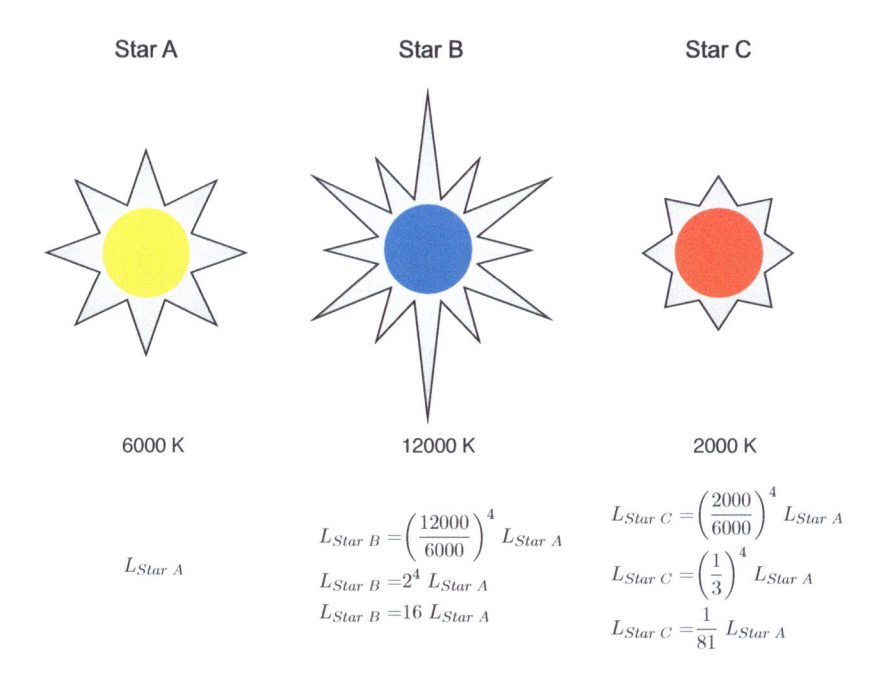

Star A	Star B	Star C
6000 K	12000 K	2000 K
$L_{Star\ A}$	$L_{Star\ B} = \left(\dfrac{12000}{6000}\right)^4 L_{Star\ A}$	$L_{Star\ C} = \left(\dfrac{2000}{6000}\right)^4 L_{Star\ A}$
	$L_{Star\ B} = 2^4\, L_{Star\ A}$	$L_{Star\ C} = \left(\dfrac{1}{3}\right)^4 L_{Star\ A}$
	$L_{Star\ B} = 16\, L_{Star\ A}$	$L_{Star\ C} = \dfrac{1}{81} L_{Star\ A}$

The luminosity is proportional to the fourth power of the surface temperature of the star.

It is worth noting that Equation 7.1 and Equation 7.3 are not provided on the Relationships sheet in the exam, but Equation 7.4 and Equation 7.2 will be included.

Examples

1.

The Sun has a surface temperature of 5780 K. The radius of the Sun is 6.955× 10⁸ m. Calculate the luminosity of the Sun.

$L = 4\pi r^2 \sigma T^4$

$L = 4\pi \left(6.955 \times 10^8\right)^2 \times 5.67 \times 10^{-8} \times 5780^4$

$L = 3.85 \times 10^{26}\, W$

....

© HERIOT-WATT UNIVERSITY

TOPIC 7. STELLAR PHYSICS

2.

Betelgeuse has a surface temperature of 3500 K. The luminosity of Betelgeuse is 7.2×10^{31} W. Calculate the radius of Betelgeuse.

$$L = 4\pi r^2 \sigma T^4$$
$$7.2 \times 10^{31} = 4\pi r^2 \times 5.67 \times 10^{-8} \times 3500^4$$
$$r = 8.2 \times 10^{11} m$$

....

3.

Sirius is the brightest star in the night sky. Sirius has a luminosity of 9.9×10^{27} W. The radius of Sirius is 1.2×10^6 km. Calculate the surface temperature of Sirius.

$$L = 4\pi r^2 \sigma T^4$$
$$9.9 \times 10^{27} = 4\pi (1.2 \times 10^9)^2 \times 5.67 \times 10^{-8} \times T^4$$
$$T = 9900\ K$$

....

4.

Arcturus has a surface temperature of 4300 K and a radius of 1.8×10^{10} m. Another star has the same luminosity but a surface temperature of only 3600K. Determine its radius.

$$r^2 T^4 = r^2 T^4$$
$$(1.8 \times 10^{10})^2 \times (4300)^4 = r^2 \times (3600)^4$$
$$r^2 = 6.59 \times 10^{20}$$
$$r = 2.6 \times 10^{10} m$$

© HERIOT-WATT UNIVERSITY

7.2.5 Apparent brightness

If you look up at the night sky on a clear night, you will notice that the stars appear to have different levels of brightness. Some appear dimmer than others. Now this may be because they genuinely emit less visible light, but it may also be because they are further away.

For example, consider two stars A and B. Star B emits more visible radiation per second than A. However, the two stars may appear equally bright to you an observer on Earth if star B is more distant than star A. What you are judging with your eye is the **apparent brightness** of the stars.

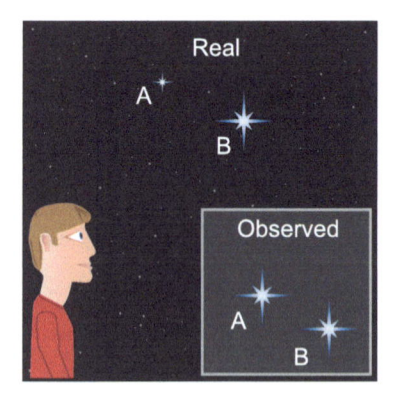

The apparent brightness of a star is a measure of how much electromagnetic radiation actually reaches a detector. Strictly speaking, it takes account of all the wavelengths of the electromagnetic spectrum, not just visible light. The proper astronomical definition for apparent brightness is the amount of energy per second reaching a detector per unit area. It depends upon the star's luminosity and its distance from us. The exact relationship between apparent brightness and distance is called the inverse square law. You may remember it from the Higher course.

Consider the radiation leaving a star and travelling through space. The apparent brightness of the star is the amount of energy per second reaching us per unit area. As you move away from a star, the radiation spreads out to cover the surface of a progressively larger sphere. The area of a sphere is equal to $4\pi r^2$, where r is its radius. Thus, the amount of energy per unit area received by a detector (the star's apparent brightness) will vary inversely as the square of the star's distance from us. Doubling the distance from a star will quarter the apparent brightness. Tripling the distance will make the apparent brightness become one ninth of the previous value etc.

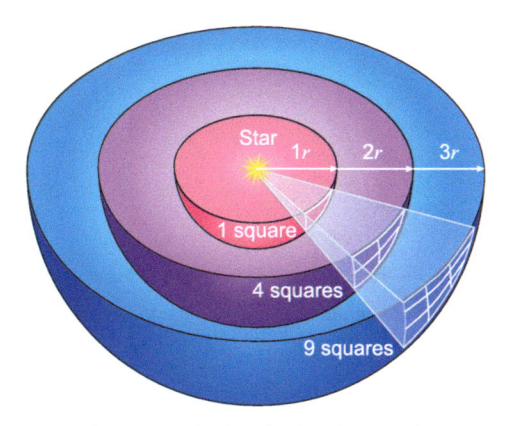

So the apparent brightness b of a star can be found using the equation

$$b = \frac{L}{4\pi d^2}$$

(7.5)

where L is the luminosity and d is the distance between the observer and the star.

Example A star 9.94×10^{16} m from Earth has a diameter of 1.02×10^9 m. It has a luminosity, L, of 1.30×10^{26} W. Calculate the apparent brightness, b, of the star.

$b = \dfrac{L}{4\pi d^2}$

$b = \dfrac{1.30 \times 10^{26}}{4\pi \left(9.94 \times 10^{16}\right)^2}$

$b = 1.047 \times 10^{-25} Wm^{-2}$

7.2.6 Summary of properties

Astronomers use a star's apparent brightness and distance to find its luminosity. They can then use the luminosity and the star's surface temperature to determine the star radius. They find the surface temperature of the star from finding its peak emission wavelength.

Example Pollux is a star in the Gemini constellation with an apparent brightness on Earth of 1.0×10^{-16} Wm^{-2}. Parallax measurements give a distance from the Earth of 3.2×10^{17} km. Pollux has a surface temperature of 4900K.

Find the radius of Pollux.

First we need to find the luminosity, L.

$$b = \frac{L}{4\pi d^2}$$

$$1.0 \times 10^{-16} = \frac{L}{4\pi \times (3.2 \times 10^{20})^2}$$

$$L = 1.0 \times 10^{-16} \times 4\pi \times (3.2 \times 10^{20})^2$$

$$L = 1.29 \times 10^{26} W$$

Now we use the value of L in the formula $L = 4\pi r^2 \sigma T^4$ to find the radius, r, of Pollux.

Note: Remember σ is the Stefan-Boltzmann constant, which has a value $5.67 \times 10^{-8} W m^{-2} K^{-4}$.

$$L = 4\pi r^2 \sigma T^4$$

$$1.29 \times 10^{26} = 4\pi r^2 \times 5.6 \times 10^{-8} \times 4900^4$$

$$r = \sqrt{\frac{1.29 \times 10^{26}}{4\pi \times 5.6 \times 10^{-8} \times 4900^4}}$$

$$r = 5.6 \times 10^8 m$$

Useful units in astronomy

You met the definition of a light year in the National 5 course. One light year is the distance that light travels in one year. Therefore, 1 light year can be found from

$$d = vt$$
$$d = 3 \times 10^8 \times 365.25 \times 24 \times 60 \times 60$$
$$d = 9.47 \times 10^{15} m$$

Another useful unit for distance is the Astronomical Unit (AU). The AU is the average distance between the Earth and the Sun, which is about 150 million kilometres. Astronomical units are usually used to measure distances within our solar system and the value is stated on the AH data sheet as:

$$1 \text{ AU} = 1 \cdot 5 \times 10^{11} m$$

Similarly, the solar radius ($R_\odot$) is a unit of distance used to express the radius of a star in relation to the current radius of the Sun. The data sheet states that 1 solar radius $= 6 \cdot 955 \times 10^8 m$.

The solar mass (M☉) is a unit of mass in astronomy that is used to express the mass of a star in relation to that of the Sun. The data sheet states that 1 solar mass $= 2 \cdot 0 \times 10^{30} kg$.

7.2.7 Quiz

Quiz: Properties of stars Go online

Q1: Using the following data, find the relationship between λ_{max} and T_{max}.

Star	T_{max} (K)	λ_{max} (nm)
Algol	12000	250
Sun	6000	500
Proxima Centauri	3000	1000

Hint: Find the quotient and product of each set of data.

....

Q2: A student making notes creates the incomplete table shown below.

Class	Temperature (K)	Colour
O	>33,000	
	10,000 - 33,000	Blue to blue-white
A		White
	6,000 - 7,500	Yellowish-white
G		Yellow
K	3,700 - 5,200	
		Red

The following table shows some of the physical properties of Capella.

Property of Capella	Value
Surface temperature	4970 K
Radius	8.33×10^{10}
Mass	2.57 solar masses
Distance to Earth	42.9 light years

Q3: Calculate the power per unit area of Capella.

....

Q4: Calculate the luminosity of Capella.

....

Q5: Calculate the apparent brightness of Capella as viewed from Earth.

7.3 Star formation

Stars are born in huge clouds of gas and dust called nebulae. These regions are extremely cold at temperatures of about 10 K to 20 K. Higher temperatures would cause the particles to have too much kinetic energy to stay together.

Image Credit: NASA/STScI Digitized Sky Survey/Noel Carboni

Gravitational forces pull the gas and dust into the centre. This accumulation of mass increases the gravitational attraction and results in even more hydrogen gas being pulled in. This leads to an increase in density. Since the gravitational potential energy of the gas is converted to heat, the temperature and pressure at the core also increase. Hydrogen nuclei are positively charged and therefore repel each other, but eventually the temperature of the core will become sufficiently large for the hydrogen nuclei to have enough kinetic energy to be moving fast enough to overcome their electrostatic repulsion and undergo fusion.

Gravitational attraction pulls material inwards, increasing the temperature and eventually allowing fusion

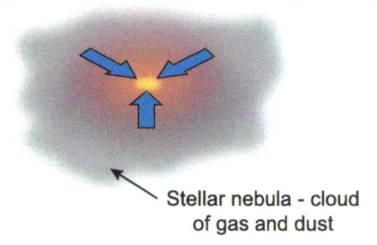

Stellar nebula - cloud of gas and dust

7.4 Stellar nucleosynthesis and fusion reactions

In this section we are going to look at why stars are so hot and bright- the origin of solar energy. You will remember from the Higher course that nuclear fusion involves nuclei joining together to form a larger nucleus and that this results in a release of energy according to $E = mc^2$. The energy released from nuclear fusion in a star is ultimately released as radiation from its surface.

Stellar nucleosynthesis is the name of the process whereby a star carries out nuclear fusion to produce new elements heavier than hydrogen. A star's mass determines what types of nucleosynthesis occur in its core. This is because heavier stars have a stronger gravitational pull in their cores, leading to higher core temperatures. The smallest stars only convert hydrogen into helium by a process called the proton-proton chain. They can't fuse heavier elements, since they are not sufficiently hot to overcome the repulsive forces between larger nuclei.

The proton-proton chain consists of three stages:

1. Two hydrogen nuclei (^1_1H) undergo fusion to produce a deuterium nucleus (^2_1H), a positron and a neutrino. The positron is annihilated by an electron to produce further energy in the form of gamma ray photons.

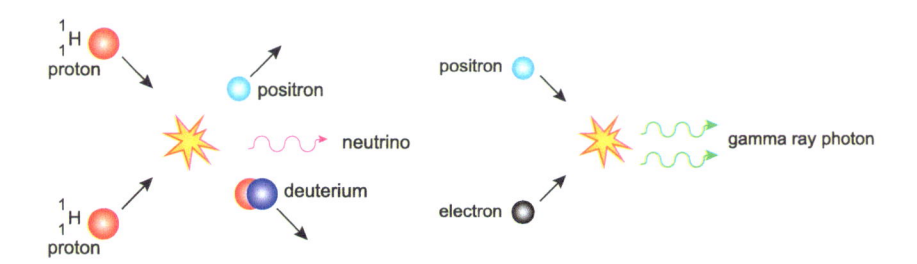

2. The deuterium nucleus (^2_1H) then fuses with a proton to form a helium - 3 nucleus (^3_2He) and a gamma ray photon.

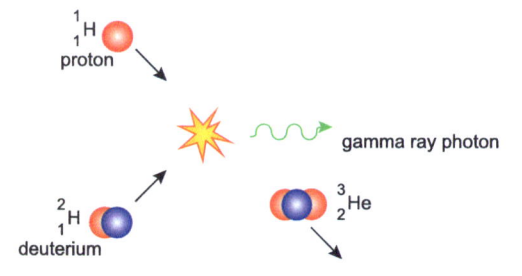

3. Two helium-3 nuclei (^3_2He) fuse to form a helium-4 nucleus (^4_2He) and two free protons, which may cause further proton- proton chains.

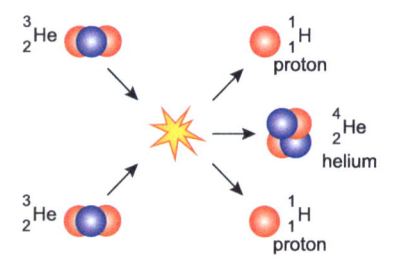

The net effect is to combine 4 protons to form one helium nucleus, with the energy released going into the particles and gamma ray photons produced at each step of the chain.

$$^1_1H + ^1_1H \rightarrow ^2_1H + e^+ + v + \text{energy}$$
$$^1_1H + ^2_1H \rightarrow ^3_2He + \gamma + \text{energy}$$
$$^3_2He + ^3_2He \rightarrow ^4_2He + 2^1_1H + \text{energy}$$

The following is a very brief summary of the proton- proton chain:

1. A proton fuses with another proton to form deuterium.
2. Then deuterium fuses with a proton to form a helium - 3 nucleus.
3. Two helium - 3 nuclei then fuse to form a helium - 4 nucleus.

TOPIC 7. STELLAR PHYSICS

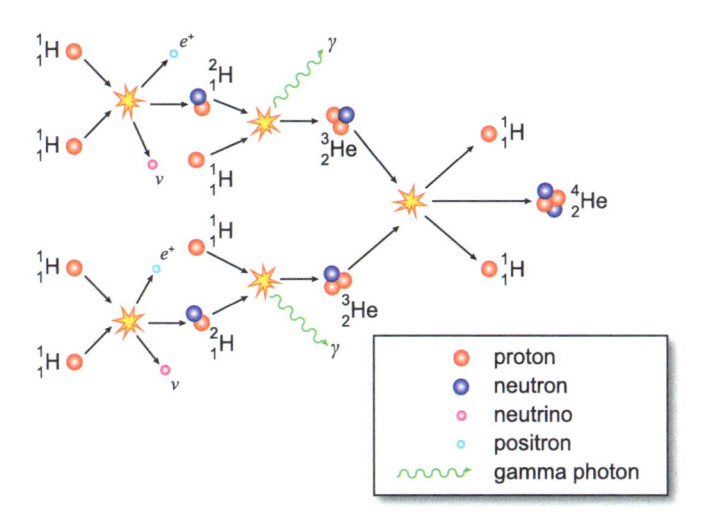

The proton - proton chain reaction Go online

There is an animation online showing the proton-proton chain reaction, which is the dominant fusion reaction in small stars.

7.5 Hertzsprung-Russell diagrams

Astronomers Hertzsprung and Russell developed the technique of plotting luminosity against surface temperature. Note that both the surface temperature and luminosity scales are logarithmic and that the surface temperature is in descending order. For historical reasons, the lowest surface temperature stars are on the right hand side to match the spectral order OBAFGKM.

178 UNIT 1. ROTATIONAL MOTION AND ASTROPHYSICS

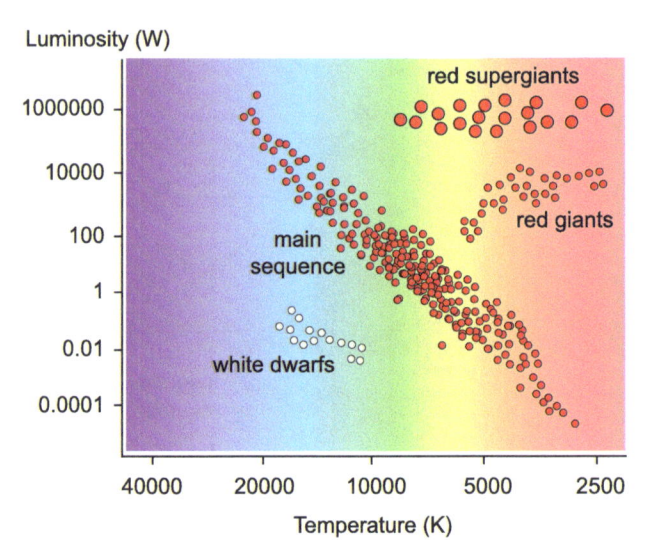

Hot bright stars are shown at the top left of the diagram and cool dim stars are shown at the bottom right.

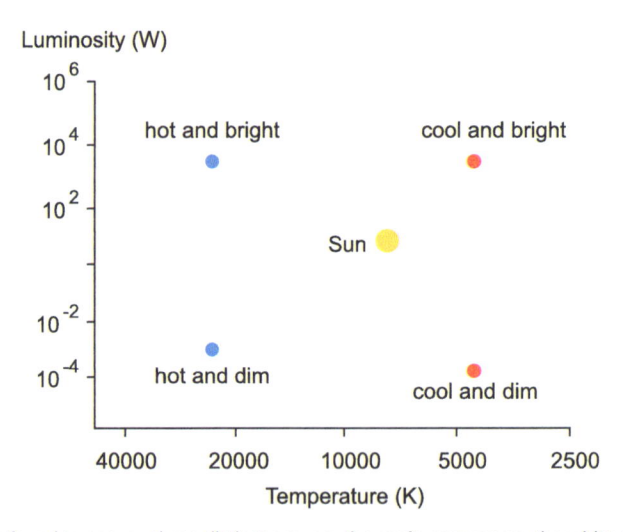

The stars were found to occupy three distinct groups: the **main sequence**, the white dwarfs and the red giants and supergiants.

© HERIOT-WATT UNIVERSITY

TOPIC 7. STELLAR PHYSICS

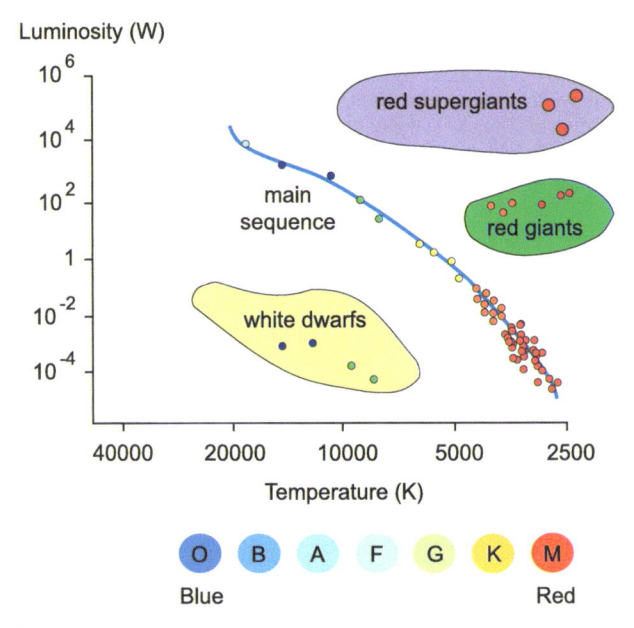

Main sequence

The long diagonal band of stars that runs from the top left to the bottom right is called the main sequence. Most of the stars fall on this curve. Infact about 90% of the stars in our region of the Milky Way belong to the main sequence. Main sequence stars are still in their long-lived stable phase where they are fusing hydrogen into helium in their cores.

The mass of such a new star determines its luminosity and surface temperature. This is because a high mass star has a larger inwards gravitational pull and therefore a higher core temperature. So, the most massive main sequence stars are shown at the upper left. They have a high luminosity and are hot (blue). The least massive ones are shown at the lower right. They are less luminous and are cooler (yellow or red). Since the Sun is neither especially large nor small in mass when compared to the rest of the main sequence, it is neither especially hot, cold, bright or dim. This means that the Sun lies fairly close to the middle of the main sequence.

We will see in section 7.6 that stars begin their life on the main sequence and then evolve to different parts of the H-R diagram. Most of the lifetime of a star is spent on the main sequence.

Red giants and red supergiants

You may recall the luminosity of a star can be found from $L = 4\pi r^2 \sigma T^4$. So stars which have a lower surface temperature and a higher luminosity than the main sequence stars must also have larger radii. These stars are called red giants and **red supergiants**. They are found in the top right corner of the H-R diagram. They are stars which have moved off the main sequence, having started to fuse heavier elements. Red supergiant stars are the brightest of all stars.

© HERIOT-WATT UNIVERSITY

White dwarfs

The equation $L = 4\pi r^2 \sigma T^4$ also leads to the conclusion that stars which have a higher surface temperature and a lower luminosity than the main sequence stars must also have smaller radii. These are the white dwarfs. They lie in the bottom left corner of the H-R diagram and are typically about the size of the Earth. White dwarfs are faint hot stars at the end of their lives. All of their fusion reactions have stopped and they are just slowly cooling and dimming to ultimately form a **black dwarf**.

You could argue that the white dwarfs are a bit like old faded Hollywood actors in that they are not as bright or as large as they once were. They have stopped working and are slowly but surely fading away. However, you may contest that the analogy breaks down with regards their hotness!

It is also worth noting that the word white dwarf is a bit of a misnomer, because some are actually blue. They are just named this way since the vast majority are white.

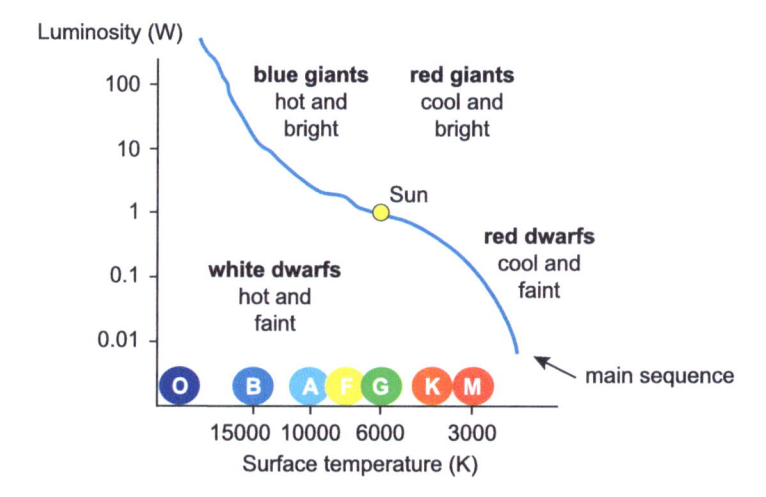

As you can see, the colour of a star can be predicted based upon its position in the Hertzsprung-Russell diagram.

TOPIC 7. STELLAR PHYSICS

Hertzsprung-Russell diagram matching task

Q6: Use the labels below to match the correct location on the Hertzsprung-Russell diagram.

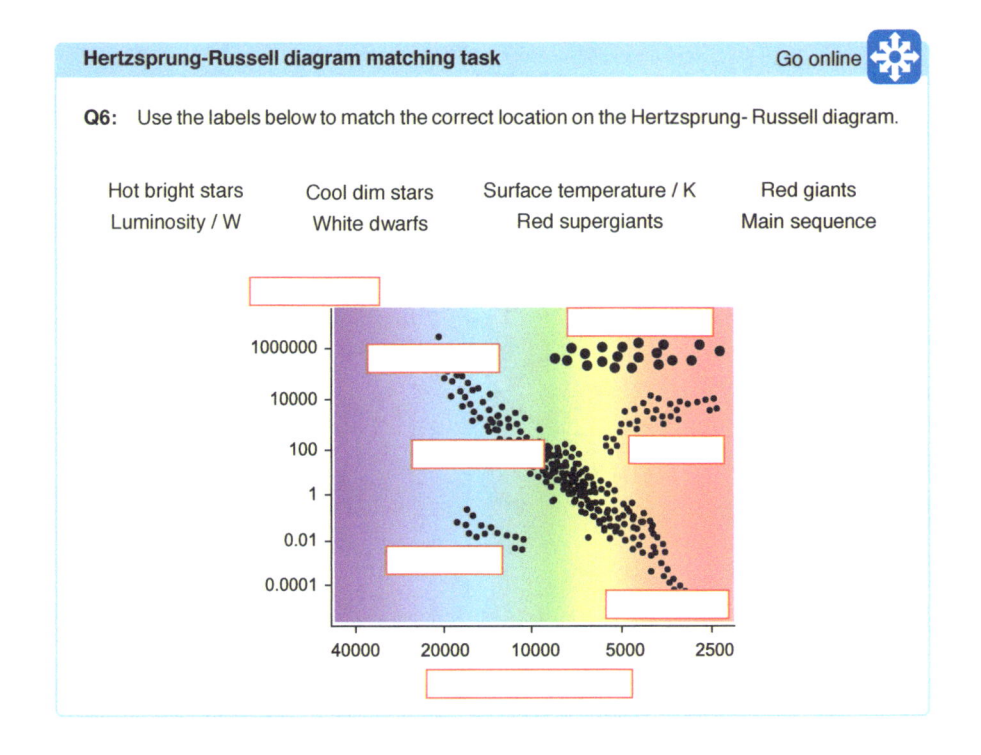

Hot bright stars Cool dim stars Surface temperature / K Red giants
Luminosity / W White dwarfs Red supergiants Main sequence

7.6 Stellar evolution and life cycles on H-R diagrams

The mass of a star ranges in size from roughly 0.08 solar masses to 150 solar masses. If an object has a mass less than 0.08 solar masses, then it would be unable to achieve a high enough core temperature to initiate fusion. These "failed stars" are called brown dwarfs. A star cannot have a mass greater than 150 solar masses: the outwards force due to the thermal pressure from fusion would be so large in comparison to the inwards gravitational pull that the extra mass would be driven off into space.

The H-R diagram is a useful tool for tracing the evolution of stars. All main sequence stars will eventually evolve to become either red giants or red supergiants. Which of these they become, the time they spend on the main sequence and their eventual fate also depend on their mass. Even though stars with a large mass have a lot of fuel, they use it up more quickly and therefore don't spend as long on the main sequence. For instance, a star with a mass comparable to that of the Sun will spend about 10 billion years on the main sequence. In comparison, a star that has a mass about 15 times that of the Sun will only belong to the main sequence for approximately 10 million years.

During the stable stage whilst a star is still on the main sequence, the outwards force due to the thermal pressure produced from hydrogen fusion in the core balances the gravitational pull trying to compress it.

© HERIOT-WATT UNIVERSITY

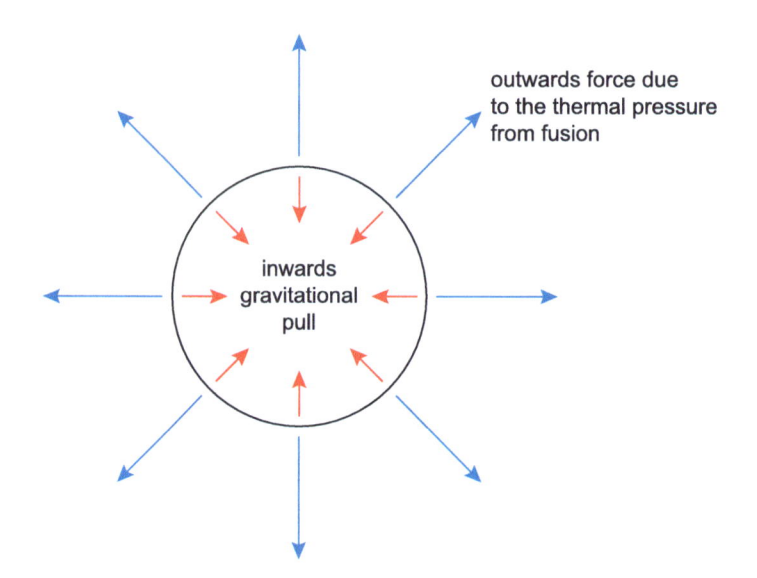

Equilibrium: The outwards force due to thermal pressure from fusion balances the inwards gravitational pull.

Eventually, the hydrogen in the core runs out and nuclear fusion stops. Then, since the outward pressure is removed, the core contracts and heats up. Eventually the temperature is high enough for the core to fuse helium into carbon and oxygen. This releases a huge amount of energy, which pushes the outer layers of the star outwards. As the outer layers expand and cool, the star leaves the main sequence and becomes either a red giant or a red supergiant, depending on its mass.

Death of a low or medium mass star

The core temperature of a low or medium mass star is not sufficiently high to allow the fusion of elements beyond helium. Therefore, once all the helium has been converted into carbon and oxygen, the core once more contracts due to gravitational attraction. The outer layers become more and more unstable as the core contracts. The star pulsates and ejects the outer layers into space as a planetary **nebula**. This leaves behind a dense core-a white dwarf. Remember the white dwarfs lie below the main sequence as they have low luminosities. A white dwarf will then simply cool down and fade away, ultimately forming a black dwarf.

Death of a high mass star

Really massive stars become red supergiants, which can fuse progressively heavier elements all the way up to iron. Once they are depleted, the core collapses in less than a second and a **supernova** explosion occurs. A supernova event can temporarily outshine a whole galaxy. It provides the energy needed to produce all the elements heavier than iron up to uranium. In fact, every human being is made from atoms that were formed in a supernova. So we're all made of star material after all! Shock waves blow off the outer layers, leaving behind either a **neutron star** or a black hole.

TOPIC 7. STELLAR PHYSICS

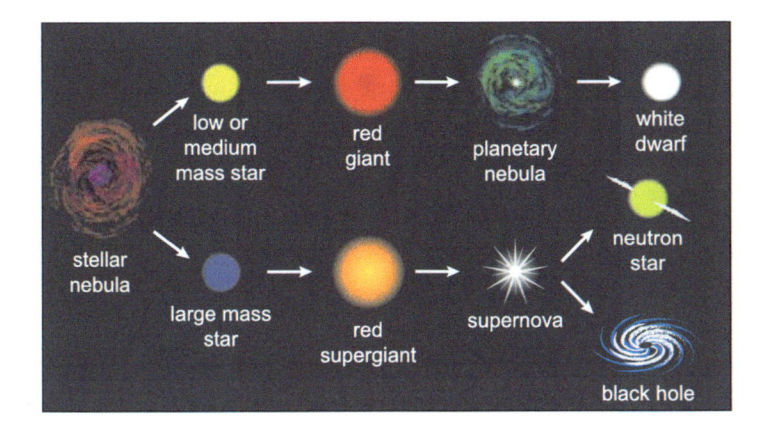

The life cycle of a star

The evolution of a star can be mapped out on a H-R diagram. The one below shows the life cycle of a star of mass equal to that of the Sun. As you can see it starts on the main sequence and then moves diagonally up to the right to become a red giant, before eventually becoming a white dwarf.

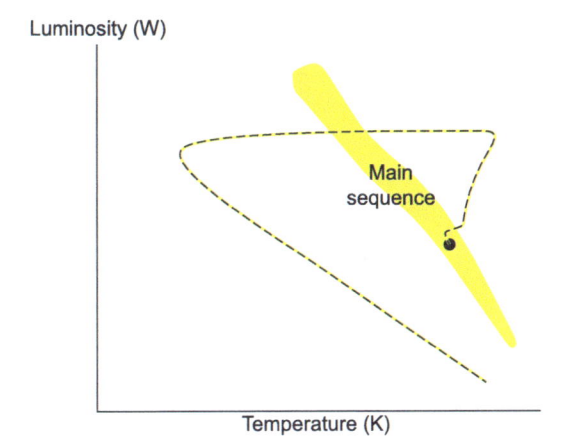

Interactive H-R diagram   Go online

There is an interactive version of H-R diagram available online where you will find out more details about Stellar evolution and life cycles.

© HERIOT-WATT UNIVERSITY

Stellar matching task

Q7: Match each evolution stage (1 - 6) with their descriptions (A - F).

1. **Nebula**
2. **Black dwarf**
3. **White dwarf**
4. **Supernova**
5. **Neutron star**
6. **Black hole**

A) The remains of a supernova. It exerts such a strong gravitational pull that light cannot escape.
B) Faint hot stars that lie below and to the left of the main sequence. They are the remnants of a medium or low mass star.
C) A high density small star left over from a supernova.
D) A large cloud of gas and dust from which a star is formed.
E) The remains of a white dwarf after it has cooled.
F) The explosion of a red supergiant.

The life cycle of a low or medium mass star

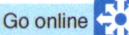

Q8: Place the following statements about the life cycle of a low or medium mass star into the correct sequence 1 - 6.

- The centre of the nebula's temperature increases.
- Hydrogen runs out in the core and a red giant forms.
- The outer layers of the red giant drift off into space.
- Nuclear fusion starts.
- A white dwarf is formed.
- Nebula contracts due to gravitational attraction.

The life cycle of a large mass star   Go online

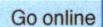

Q9: Place the following statements about the life cycle of a low or medium mass star into the correct sequence 1 - 6.

- Hydrogen runs out in the core and a red supergiant forms.
- Nuclear fusion starts.
- The core collapses and a supernova explosion occurs.
- A dense neutron star or a black hole is formed.
- Nebula contracts due to gravitational attraction.
- The centre of the nebula's temperature increases.

7.7 Extended information

Web links   Go online

These web links should serve as an insight to the wealth of information available online and allow you to explore the subject further.

- http://astro.unl.edu/naap/blackbody/animations/blackbody.swf
- https://www.youtube.com/watch?v=m8GXpk8PZ-o
- http://www.spacetelescope.org/videos/heic1017b/
- http://astronomy.swin.edu.au/cosmos/h/hertzsprung-russell+diagram
- http://www.pa.msu.edu/~baldwin/carol/Lifecycleofstars.swf
- https://www.youtube.com/watch?v=WP5HA7fKDXk
- http://hubblesite.org/gallery/movie_theater/starslife/
- http://hubblesite.org/newscenter/archive/releases/2010/28/video/d/
- https://www.youtube.com/watch?v=PM9CQDIQI0A
- http://www.bbc.co.uk/education/clips/zwqgkqt
- http://www.bbc.co.uk/science/space/universe/collections/great_space_explanations#p006t02b
- http://www.bbc.co.uk/science/space/universe/sights/stars#p009md0x
- http://www.bbc.co.uk/science/space/universe/sights/black_holes/#p00bq8pr

7.8 Summary

Summary

You should now be able to:

- state that:
  - all stellar objects give out a wide range of wavelengths of electromagnetic radiation but that each object gives out more energy at one particular wavelength;
  - the wavelength of this peak wavelength is related to the temperature of the object, with hotter objects having a shorter peak wavelength than cooler objects;
  - peak wavelengths allow the temperature of stellar objects to be calculated;
  - hotter objects also emit more radiation per unit surface area at all wavelengths than cooler objects;
- carry out calculations using the fact a star's power per unit area = σT^4;
- state that the luminosity of a star:
  - in watts, is a measure of the total power the star emits i.e. the total energy emitted per second;
  - depends on its radius and surface temperature;
- carry out calculations using the equation for luminosity $L = 4\pi r^2 \sigma T^4$;
- state that the apparent brightness of a star:
  - is the amount of energy per second reaching a detector per unit area;
  - depends on its luminosity and its distance from the observer;
- carry out calculations using the equation for apparent brightness $b = \frac{L}{4\pi d^2}$;
- state that the mass of a new star determines its luminosity and surface temperature;
- state that a Hertzsprung-Russell (H-R) diagram is a plot of luminosity versus the surface temperature of stars and that the surface temperature scale is in descending order;
- identify the long diagonal band on a H-R diagram as the main sequence and state that the Sun is on the main sequence;
- state that the main sequence stars are in their long lived stable phase where they are fusing hydrogen into helium in their cores;
- identify the:
  - higher luminosity and lower temperature stars lying to the right of the main sequence as the red giants and red supergiants;
  - lower luminosity and higher temperature stars lying to the left of the main sequence as the white dwarfs;
- predict the colour of a star based upon its position on the Hertzsprung-Russell diagram;

TOPIC 7. STELLAR PHYSICS

> **Summary continued**
>
> - describe how stars:
>   - produce heat energy using the proton-proton chain reaction;
>   - are formed in terms of the gravitational effects on cold dense interstellar clouds;
> - state that the energy released from nuclear fusion in a star results in an outwards thermal pressure;
> - explain:
>   - how stars on the main sequence are in gravitational equilibrium;
>   - why a star's life cycle is determined by its mass;
> - relate the process of stellar evolution for a star to its path on a H-R diagram.

7.9 End of topic 7 test

End of topic 7 test Go online

The following data should be used when required:

Gravitational acceleration on Earth g	$9.8\ m\ s^{-2}$

Q10: The luminosity of a star depends upon:

a) its surface temperature, its radius and the distance from the observer.
b) its surface temperature and the distance from the observer.
c) its radius and the distance from us.
d) only its surface temperature.
e) its surface temperature and its radius.

....

Q11: The apparent brightness of a star depends on distance from the star.

a) True
b) False

....

Q12: The apparent brightness of a star depends on surface temperature of the star.

a) True
b) False

....

© HERIOT-WATT UNIVERSITY

Q13: Alpha Cassiopeiae has an apparent brightness of 4.4×10^{-9} W m^{-2} and is at a distance of 2.2×10^{18} m from Earth.

Calculate its luminosity.

L = _____ W

....

Q14: The Sun has an apparent brightness of 1370 W m^{-2} and is 1.5×10^{11} m from Earth. The star Procyon A has an apparent brightness of 1.78×10^{-8} W m^{-2} and is 1.0×10^{17} m from Earth.

How many times bigger is the luminosity of Procyon A relative to that of the Sun?

$L_{Procyon\ A}$ = _____ × L_{Sun}

....

Q15: The Sun has an apparent brightness of 1.4×10^{3} W m^{-2} and is 1.5×10^{11} m from Earth. The radius of the Sun is 6.96×10^{8} m.

Sirius A has an apparent brightness of 1.23×10^{-7} W m^{-2}. The luminosity of Sirius A is 25.4 times that of the Sun.

Determine the distance from Earth to Sirius A.

$r_{Sirius\ A}$ = _____ m

....

Q16: When compared to the other stars, the stars which emit the shortest peak wavelength are:

a) red and cool.
b) red and hot.
c) blue and cool.
d) blue and hot.

....

TOPIC 7. STELLAR PHYSICS

Q17: The long diagonal band on a Hertzsprung-Russell diagram is made up of stars in their long lived stable phase.

This is called the _____ .

....

Q18: As a star ages it progresses _____ the main sequence.

a) up
b) down

....

Q19: All main sequence stars will eventually become a:

a) red giant / red supergiant.
b) black hole.
c) black dwarf.
d) neutron star.
e) white dwarf.

....

Q20: The Sun is a:

a) white dwarf.
b) neutron star.
c) supernova.
d) red dwarf.
e) main sequence star.

....

Q21: When a star is on the main sequence the outwards force due to the thermal pressure from _____ the inwards gravitational pull.

a) fission is greater than
b) fission balances
c) fusion is greater than
d) fusion balances
e) fusion is less than
f) fission is less than

....

Q22: Which type of star has a high surface temperature but a low luminosity?

....

Q23: White dwarf stars have a higher surface temperature and are more dense than a star on the main sequence.

a) True
b) False

Q24: White dwarf stars carry out nuclear fission.

a) True
b) False

Q25: White dwarf stars carry out nuclear fusion.

a) True
b) False

Q26: As a star moves from the main sequence to become a red giant:

a) its surface temperature decreases and luminosity decreases.
b) its surface temperature decreases and luminosity increases.
c) its surface temperature increases and luminosity decreases.
d) its surface temperature increases and luminosity increases.

Q27: Complete the chart using the following words:

Main sequence   **Supernova**   **Black dwarf**
Black hole   **Planetary nebula**   **White dwarf**

Unit 1 Topic 8

End of section 1 test

End of section 1 test

Quantity	Symbol	Value
Gravitational acceleration on Earth	g	9.8 m s^{-2}
Radius of the Earth	r_E	6.4×10^6 m
Mass of the Earth	M_E	6.0×10^{24} kg
Mass of the Moon	M_M	7.3×10^{22} kg
Radius of the Moon	r_M	1.7×10^6 m
Mean radius of Moon orbit		3.84×10^8 m
Solar radius	R_S	6.955×10^8 m
Mass of the Sun	M_S	2.0×10^{30} kg
1 AU		1.5×10^{11} m
Stefan - Boltzmann constant	σ	5.67×10^{-8} W m^{-2} K^{-4}
Universal constant of gravitation	G	6.67×10^{-11} m^3 kg^{-1} s^{-2}
Planck's constant	h	6.63×10^{-34} J s
Speed of light in a vacuum	c	3.00×10^8 m s^{-1}

Q1: A moving particle has a varying force applied to it. Its acceleration in m s^{-2} is given by the expression a = 4.0 t where t is the time in seconds.
Its velocity is 3.0 m s^{-1} at 0 seconds.
Calculate its velocity at 5.0 seconds.
v = _____ m s^{-1}

....

Q2: A mass of 0.420 kg is being rotated by a string in a vertical circle of radius 1.20 m at a constant angular velocity ω rad s^{-1}. The tension in the string when the mass is at the bottom of the circle is 12.0 N.

1. Calculate ω.
   _____ rad s^{-1}
2. Calculate the tension when the mass is at the top of the circle.
   _____ N
3. Calculate the tension when the string is horizontal.
   _____ N

....

Q3: The moon has mass 7.3×10^{22} kg and radius 1.7×10^6 m.
Calculate the escape velocity of the Moon.

_____ m s^{-1}

....

Q4: Calculate the Schwarzschild radius for a black hole of mass 2.51×10^{31} kg.

$r_{Schwarzschild}$ = _____ m

....

Q5: The surface temperature of the star Wolf 489 is 5030 K. Its luminosity is 3.7×10^{22} W. What is the radius of Wolf 489?

r = _____ m

....

Q6: A star has radius 1.08×10^8 km and its luminosity is 1.27×10^{31} W. What is the surface temperature of the star?

T = _____ K

....

Q7: After hydrogen fusion ends in a star's core, its position on the Hertzsprung-Russell diagram moves towards the:

a) upper right.
b) upper left.
c) lower left.
d) lower right.

....

Q8: Which star is most likely to become a supernova?

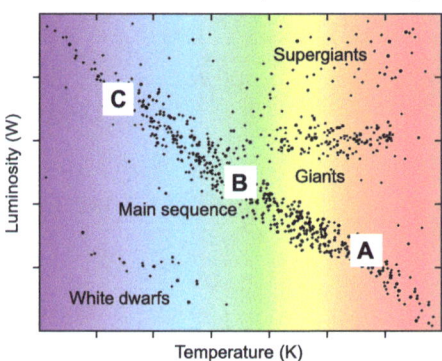

a) A
b) B
c) C

....

Q9: Stars with greater mass have _____ life cycles.

a) longer
b) shorter

....

Q10: On a Hertzsprung-Russell diagram, the main sequence stars that lie on the top left of the diagonal line have a:

a) smaller mass and have higher temperatures.
b) larger mass and have higher temperatures.
c) smaller mass and have lower temperatures.
d) larger mass and have lower temperatures.

Quanta and Waves

1	**Introduction to quantum theory**	**199**
	1.1 Introduction	201
	1.2 Atomic models	201
	1.3 Quantum mechanics	211
	1.4 Heisenberg uncertainty principle	211
	1.5 Extended information	213
	1.6 Summary	214
	1.7 End of topic 1 test	215
2	**Wave particle duality**	**217**
	2.1 Introduction	218
	2.2 Wave-particle duality of waves	218
	2.3 Wave-particle duality of particles	224
	2.4 Quantum tunnelling	230
	2.5 Extended information	231
	2.6 Summary	231
	2.7 End of topic 2 test	232
3	**Magnetic fields and particles from space**	**235**
	3.1 Introduction	236
	3.2 Force on a moving charge in a magnetic field	237
	3.3 The path of a charged particle in a magnetic field	239
	3.4 Helical motion in a magnetic field	243
	3.5 The Solar Wind	248
	3.6 Extended information	252
	3.7 Summary	253
	3.8 End of topic 3 test	253
4	**Simple harmonic motion**	**255**
	4.1 Introduction	256
	4.2 Defining SHM	256
	4.3 Equations of motion in SHM	259
	4.4 Energy in SHM	264
	4.5 Applications and examples	267

4.6	Damping	272
4.7	Extended information	274
4.8	Summary	274
4.9	End of topic 4 test	275

5 Waves ... 277

5.1	Introduction	278
5.2	Definitions	278
5.3	Travelling waves	282
5.4	Superposition of waves and phase difference	286
5.5	Stationary waves	294
5.6	Beats	298
5.7	Extended information	301
5.8	Summary	301
5.9	End of topic 5 test	302

6 Interference ... 305

6.1	Introduction	306
6.2	Coherence and optical path difference	306
6.3	Reflection of waves	310
6.4	Extended information	313
6.5	Summary	313
6.6	End of topic 6 test	314

7 Division of amplitude ... 315

7.1	Introduction	316
7.2	Thin film interference	316
7.3	Wedge fringes	322
7.4	Extended information	327
7.5	Summary	327
7.6	End of topic 7 test	328

8 Division of wavefront ... 331

8.1	Introduction	332
8.2	Interference by division of wavefront	332
8.3	Young's slits experiment	333
8.4	Extended information	338
8.5	Summary	338
8.6	End of topic 8 test	339

© HERIOT-WATT UNIVERSITY

9 Polarisation 341
9.1 Introduction 342
9.2 Polarised and unpolarised waves 342
9.3 Polaroid 343
9.4 Polarisation by reflection 346
9.5 Applications of polarisation 348
9.6 Extended information 353
9.7 Summary 353
9.8 End of topic 9 test 354

10 End of section 2 test 355

Unit 2 Topic 1

Introduction to quantum theory

Contents

1.1 Introduction ... 201
1.2 Atomic models ... 201
    1.2.1 The Bohr model of the hydrogen atom ... 202
    1.2.2 Atomic spectra ... 206
    1.2.3 Black body radiation and the UV catastrophe ... 208
    1.2.4 The photoelectric effect ... 209
1.3 Quantum mechanics ... 211
    1.3.1 Limitations of the Bohr hydrogen model ... 211
    1.3.2 Quantum mechanics ... 211
1.4 Heisenberg uncertainty principle ... 211
1.5 Extended information ... 213
1.6 Summary ... 214
1.7 End of topic 1 test ... 215

Prerequisites

- Structure of atoms.
- Irradiance, photons and formation of line and emission spectra.
- The EM spectrum and properties of waves.
- Angular momentum.

Learning objective

By the end of this topic you should be able to:

- state that the angular momentum of an electron about a nucleus is quantised in units of $h/2\pi$;
- state the equation $mvr = nh/2\pi$ and perform calculations using this equation;
- qualitatively describe the Bohr model of the atom;
- understand what is meant by black body radiation and the Ultra-Violet Catastrophe;
- understand that classical physics cannot explain Black body radiation, the photoelectric effect, and absorption and emission spectra - that quantum theory is required;
- state that quantum mechanics is used to provide a wider-ranging model of the atom than the Bohr model, and state that quantum mechanics can be used to determine probabilities;
- understand the Heisenberg Uncertainty Principle and be able to use the relationships
$\Delta x \Delta p \geqslant \dfrac{h}{4\pi}$
$\Delta E \Delta t \geqslant \dfrac{h}{4\pi}$.

1.1 Introduction

Towards the end of the 19th century, physical phenomena were described in terms of "classical" theory, as either particles or waves. However, some new discoveries (such as the photoelectric effect) could not be explained using classical theory. As we have seen, such phenomena required a theory that included a particle-like description of light. Analysis of Black Body Radiation curves showed us trends that could not be predicted by classical theory. We will see in this topic that a quantum approach was also used to answer questions about the structure of the atom. The hydrogen atom has the simplest structure, and we will look at a model of the hydrogen atom based on quantum theory and wave-particle duality. We will see that the emission spectrum of hydrogen is due to the electron moving between its allowed orbits, which can be determined by treating the electron as a wave.

Whilst the model of the atom dealt with here gives good agreement with experiments performed on the hydrogen atom, it is found to be unsuitable for larger atoms and molecules. The theory of quantum mechanics, which is used to describe such atoms, is introduced and we finish the topic by looking at Heisenberg's Uncertainty Principle.

1.2 Atomic models

The widely accepted view at the beginning of the 20th century was that an atom consisted of a large positively-charged mass with negatively-charged electrons embedded in it at random positions. This "plum pudding" model (suggested by J J Thomson) was consistent with experimental data available at the time. But the Rutherford scattering experiments first performed early in the 20th century were not consistent with that model, and required a major rethink. Rutherford interpreted the results of his experiments as evidence that the atom consisted of a relatively massive positively charged nucleus with electrons of far lower mass orbiting around it.

The study of atomic spectra also gave interesting results. The emission spectra of elements such as hydrogen, shown in Figure 1.1, consist of a series of discrete lines, rather than a continuous spectrum. In terms of photons, there was no explanation as to why a particular element would only emit photons with certain energies.

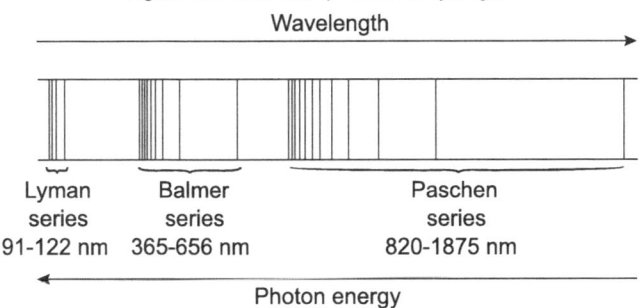

Figure 1.1: Emission spectrum of hydrogen

Lyman series
91-122 nm

Balmer series
365-656 nm

Paschen series
820-1875 nm

The next stage in the development of atomic theory was to try to build up a picture in which these experimental results could be explained.

1.2.1 The Bohr model of the hydrogen atom

In 1913 the Danish physicist Neils Bohr proposed an alternative model for the hydrogen atom. In the **Bohr model**, the electron orbits the nucleus (consisting of one proton) in a circular path, as shown in Figure 1.2. (This figure is not drawn to scale).

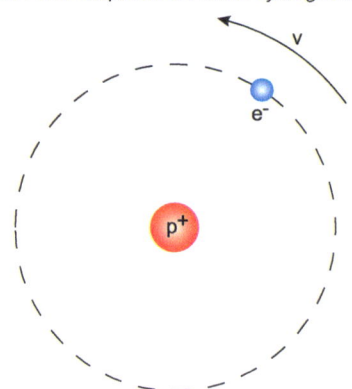

Figure 1.2: Simple model of the hydrogen atom

Bohr assumed that the electron could only have certain allowed values of energy, the total energy being made up of kinetic energy (due to its circular motion) and potential energy (due to the electrical field in which it was located). Each value of energy corresponded to a unique electron orbit, so the orbit radius could only take certain values. Using classical and quantum theory, Bohr calculated the allowed values of energy and radius, matching the gaps between electron energies to the photon energies observed in the line spectrum of hydrogen.

Since the electron is moving with speed v in a circle of radius r, its centripetal acceleration is equal to v^2/r. This presents a problem, as classical physics theory states that an accelerating charge emits electromagnetic radiation. A 'classical' electron would therefore be losing energy, and would spiral into the nucleus, rather than continue to move in a circular path. Bohr suggested that certain orbits in which the electron had an allowed value of angular momentum were stable. The angular momentum L of any particle of mass m moving with speed v in a circle of radius r is $L = mvr$. According to Bohr, so long as the angular momentum of the electron is a multiple of $h/2\pi$, the orbit is stable (h is Planck's constant). Thus the Bohr model proposed the concept of **quantisation of angular momentum** of the electron in a hydrogen atom. That is to say, the electron must obey the condition

$$\text{Angular momentum} = \frac{nh}{2\pi}$$

(1.1)

where n is an integer.

TOPIC 1. INTRODUCTION TO QUANTUM THEORY

This quantisation of angular momentum fitted in with the predicted energy levels, but left a crucial question unanswered. Why were these particular orbits allowed? In other words, what made this value of angular momentum so special, and why did having angular momentum of $nh/2\pi$ make the orbit stable? The answer to this came when de Broglie's ideas on wave-particle duality were published a decade later. Treating the electron as a stationary wave, let us suppose the smallest allowed circumference of the electron's orbit corresponds to one wavelength λ. The next allowable orbit corresponds to 2λ, and so on. The orbit will be stable, then, if the circumference is equal to $n\lambda$. A stationary wave does not transmit energy, so if the electron is acting as a stationary wave then all its energy is confined within the atom, and the problem of a classical electron radiating energy does not occur.

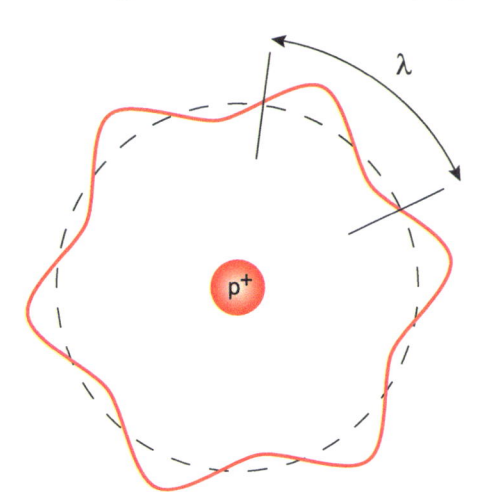

Figure 1.3: Standing wave orbit of the electron in a hydrogen atom

Figure 1.3 shows an electron in the $n = 6$ orbit. There are six de Broglie wavelengths in the electron orbit. *(De Broglie's work will be covered in more detail in a later topic.)*

The circumference of a circle of radius r is $2\pi r$, so the condition for a stable orbit is

$$n\lambda = 2\pi r$$

(1.2)

Rearranging the de Broglie equation (explained fully in Topic 2) $p = h/\lambda$

$$\lambda = \frac{h}{p} = \frac{h}{mv}$$

Now, we can substitute this expression for λ into Equation 1.2.

$$n\lambda = 2\pi r$$
$$\therefore \frac{nh}{mv} = 2\pi r$$
$$\therefore \frac{nh}{2\pi} = mvr$$
$$\therefore mvr = \frac{nh}{2\pi}$$

(1.3)

Thus the angular momentum mvr is a multiple of $h/2\pi$, as predicted by Bohr (Equation 1.1). Treating the electron as a stationary wave gives us the quantisation of angular momentum predicted by Bohr. The unit of allowed angular momentum is sometimes given the symbol $\hbar$ ('h bar'), where $\hbar = h/2\pi$.

It should be pointed out that this result gives a good model for the hydrogen atom, but does not give us the complete picture. For instance, the electron is actually moving in three dimensions, whereas the Bohr atom only considers two dimensions. We will see later in this Topic that a full 3-dimensional wave function is used nowadays to describe an electron orbiting in an atom.

TOPIC 1. INTRODUCTION TO QUANTUM THEORY

Example

In the $n = 2$ orbit of the hydrogen atom, the electron can be considered as a particle travelling with speed 1.09×10^6 m s^{-1}. Calculate:

1. the angular momentum of the electron;
2. the de Broglie wavelength of the electron;
3. the radius of the electron's orbit.

1. Since we have $n = 2$

$$L = \frac{nh}{2\pi}$$
$$\therefore L = \frac{2 \times 6.63 \times 10^{-34}}{2\pi}$$
$$\therefore L = 2.11 \times 10^{-34} \text{ kg m}^2 \text{ s}^{-1}$$

2. The de Broglie wavelength is

$$\lambda = \frac{h}{p} = \frac{h}{mv}$$
$$\therefore \lambda = \frac{6.63 \times 10^{-34}}{9.11 \times 10^{-31} \times 1.09 \times 10^6}$$
$$\therefore \lambda = 6.68 \times 10^{-10} \text{ m}$$

3. The angular momentum is quantised, so we can use Equation 1.3 to find r

$$mvr = \frac{nh}{2\pi}$$
$$\therefore r = \frac{2 \times 6.63 \times 10^{-34}}{2\pi mv}$$
$$\therefore r = \frac{1.326 \times 10^{-33}}{2\pi \times 9.11 \times 10^{-31} \times 1.09 \times 10^6}$$
$$\therefore r = 2.13 \times 10^{-10} \text{ m}$$

The Bohr model of the hydrogen atom   Go online

At this stage there is an online activity which allows you to view the electron as a particle or wave in the Bohr atom.

© HERIOT-WATT UNIVERSITY

1.2.2 Atomic spectra

The Bohr model tells us the allowed values of angular momentum for the electron in a hydrogen atom in terms of the quantum number n. For each value of n, the electron has a specific value of angular momentum L and total electron energy E. The Bohr model of the hydrogen atom allows E to be calculated for any value of n. When the electron moves between two orbits, it either absorbs or emits energy (in the form of a photon) in order to conserve energy, as shown in Figure 1.4.

Figure 1.4: (a) Emission and (b) absorption of a photon

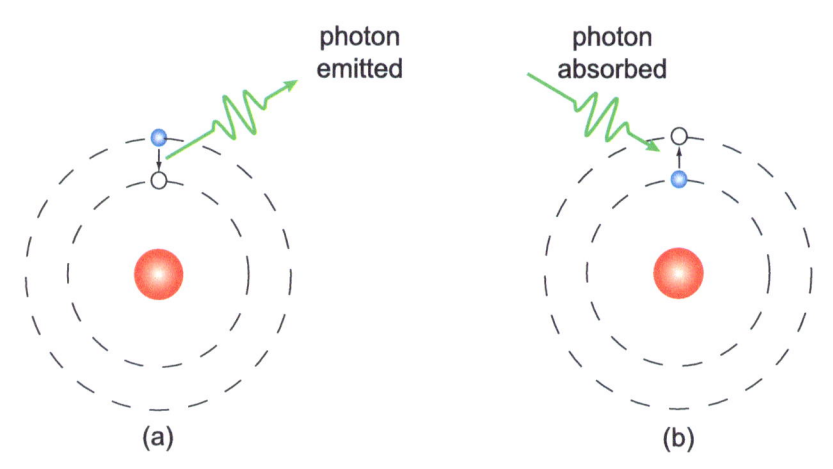

At Higher, in the 'Spectra' topic of the 'Particles and waves' unit, we discussed the Bohr Model and line spectra. There we used energy level diagrams to calculate the difference between energy levels to study the line spectrum produced by atomic hydrogen. The orbit closest to the nucleus is the ground state we discussed at Higher.

An electron loses energy when it moves from a higher n orbit to a lower n orbit. The energy is emitted in the form of a photon (Figure 1.4(a)). Only photons of the exact energies between each the orbits are emitted. The photons are of the specific frequencies that match these energies. This will create a line spectrum unique to hydrogen in this case. Other elements have electrons in different orbits to hydrogen or each other and so their spectrum acts like a fingerprint.

Similarly, absorption of a photon raises the electron to a higher n orbit (Figure 1.4(b)). Only photons with exactly the correct energy/ frequency will be absorbed. This creates an absorption spectrum.

Crucially it is only photons with the correct energy that can be emitted or absorbed. The photon energy must be exactly equal to the energy difference between two allowed orbits. Therefore, the emission spectrum of hydrogen consists of a series of lines rather than a continuous spectrum.

Hydrogen line spectrum

Go online

At this stage there is an online activity which explores how some of the lines in the visible part of the hydrogen spectrum are produced.

Q1: If the electron makes a transition to a higher orbital e.g. $n = 2$ to $n = 3$ is a photon absorbed or emitted?

Explain your answer.

....

Q2: Why does a transition from the $n = 1$ orbit produce Ultraviolet not visible light?

Quiz: Atomic models

Go online

Useful data:

Planck's constant h	6.63×10^{-34} J s
Mass of an electron m_e	9.11×10^{-31} kg

Q3: In Bohr's model of the hydrogen atom,

a) photons orbit the atom.
b) the electron's angular momentum is quantised.
c) the electron constantly emits electromagnetic radiation.
d) the electron constantly absorbs electromagnetic radiation.
e) the angular momentum is constantly changing.

....

Q4: What is the angular momentum of an electron orbiting a hydrogen atom, if the quantum number n of the electron is 4?

a) 1.67×10^{-32} kg m^2 s^{-1}
b) 8.33×10^{-33} kg m^2 s^{-1}
c) 8.44×10^{-34} kg m^2 s^{-1}
d) 4.22×10^{-34} kg m^2 s^{-1}
e) 2.64×10^{-35} kg m^2 s^{-1}

....

Q5: If an electron is orbiting a hydrogen atom with a de Broglie wavelength of λ and quantum number n, the radius r of the orbit is given by the equation

a) $r = 2\pi n \lambda$
b) $r = n\lambda/h$
c) $r = 2\pi/n\lambda$
d) $r = h/n\lambda$
e) $r = n\lambda/2\pi$

....

Q6: The emission spectrum of hydrogen consists of a series of lines. This is because

a) photons can only be emitted with specific energies.
b) photons are re-absorbed by other hydrogen atoms.
c) the photons have zero momentum.
d) hydrogen atoms only have one electron.
e) once a photon is emitted it is immediately re-absorbed.

1.2.3 Black body radiation and the UV catastrophe

When objects are heated they can be thought of as **Black Bodies**, which radiate large amounts of energy in the form of infra-red radiation. Black Bodies have a cavity and a small hole to emit the infra-red radiation in a similar way to a point source of light. One example of a black body is a furnace. A small hole on the outside allows heat to enter and reflect off the inside before being absorbed by the walls of the furnace which then emit heat of a certain wavelength. The radiation reflects off the inside of the furnace and finally escapes and contains all frequencies of EM radiation with varying intensities.

Figure 1.5: A Furnace behaving like a black body

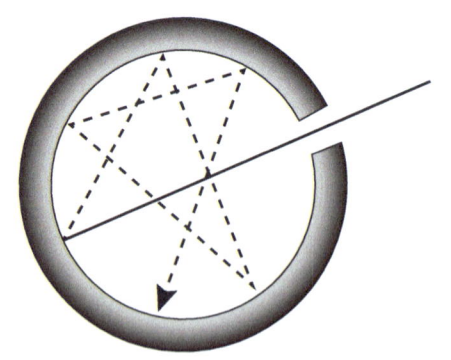

Stars are also approximate black body radiators. Stars absorb most incident light as well as other wavelengths of electromagnetic radiation and stars are able to emit all wavelengths of electromagnetic radiation too. Depending on the amount of heat the object will glow different colours ranging from cooler reds and oranges, through yellow and white up to a blueish white colour at very

TOPIC 1. INTRODUCTION TO QUANTUM THEORY

high temperatures.

It was discovered that there is a relationship between the amount of black-body radiation produced (more precisely called Specific Intensity when emitted by an object and Irradiance when incident on a surface) and the temperature of the black body. This graph shows the irradiance plotted against wavelength and it can be seen that at higher temperatures, the peak of the curve moves towards lower wavelengths or higher frequencies making the light appear closer to the blue end of the spectrum. Lord Rayleigh tried to use classical mechanics to devise an equation to solve this problem but this caused problems as at higher frequencies this predicted infinite energy emitted - this was called the Ultraviolet Catastrophe and wasn't solved until 1900 when Planck produced a relationship which agreed with the black body curves found from experiments. Even Planck took many further years to believe the clear evidence for photons he'd discovered, Einstein's work on Photons (which he called "wave packets") in 1905 won him the Nobel prize and set the pattern for this exploration of the quantum world. Einstein's equations will be covered in more detail in Topic 2.

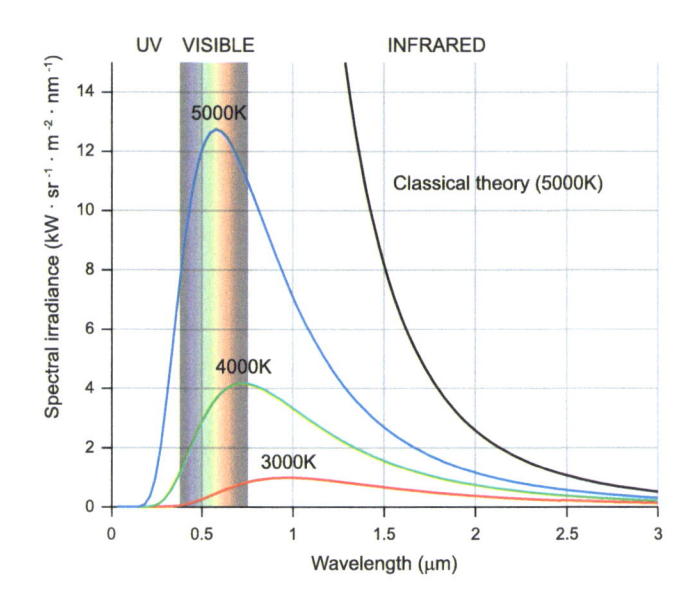

Black body radiation   Go online

There is an online activity allowing you to explore different predictions.

1.2.4 The photoelectric effect

When light hits a material, particularly more reactive metals, electrons can be emitted. This process is called the **photoelectric effect**. Electrons emitted in this process are called *photoelectrons*.

In classical physics, the photoelectric effect is caused by the transfer of energy from the light to an electron. Changing the irradiance from the light source would change the kinetic energy of the electrons emitted from the metal. According to classical physics a dim light would show a time lag

© HERIOT-WATT UNIVERSITY

from the initial incidence of the light and the emission of an electron. Experimental results do not match the two predictions made by classical theory.

In experiments electrons are only dislodged by photons whose energy reach or exceed a threshold value. If the photons have insufficient energy, no electrons are emitted no matter of the light intensity or the length of exposure.

We will discover more on the Photoelectric Effect in the second topic of this unit.

TOPIC 1. INTRODUCTION TO QUANTUM THEORY

1.3 Quantum mechanics

In this section we will cover:

- Limitations of the Bohr hydrogen model
- Quantum mechanics

1.3.1 Limitations of the Bohr hydrogen model

The Bohr model gives an accurate prediction of the spectrum of atomic hydrogen. It also works well for hydrogen-like ions which have a single electron orbiting a nucleus, such as the helium (He^+) and lithium (Li^{2+}) ions. In these cases the larger central charge means that the electrons have different energies associated with the $n = 1,2,3...$ levels, but the underlying principle is the same - the electron has discrete energy levels, and the spectra of these ions are series of lines.

The model soon becomes inadequate when more electrons are added to the system. The motion of one electron is affected not only by the static electric field due to the nucleus, but also by the moving electric field due to the other electrons. The picture becomes even more complicated when we start dealing with molecules such as carbon dioxide. A more sophisticated theory is required to describe atoms and molecules with more than one electron.

1.3.2 Quantum mechanics

Nowadays **quantum mechanics** is used to describe atoms and electrons. In this theory the motion of an electron is described by a **wave function** Ψ. The idea of using a wave function was proposed by the German physicist Erwin Schrodinger, and Ψ is often referred to as the Schrodinger wave function. The motion is described in terms of probabilities, and the wave function is used to determine the probability of finding an electron at a particular location in three dimensional space. The electron cannot be thought of as a point object at a specific position; instead we can calculate the probability of finding the electron within a certain region, within a certain time period. Another German physicist, Werner Heisenberg, was also amongst the first to propose this concept as we will see later in the topic when looking at his Uncertainty Principle.

How does this relate to the Bohr hydrogen atom? The radii corresponding to different values of n predicted by Bohr are the positions of maximum probability given by quantum mechanics, so for the hydrogen atom the two theories are compatible. The allowed energy and angular momentum values calculated using the Bohr model are still correct. As we have mentioned already though, more complicated atoms and molecules require quantum mechanics.

1.4 Heisenberg uncertainty principle

The uncertainty principle was first developed in 1926 by Heisenberg who realised it was impossible to observe sub-atomic particles like electrons with a standard optical microscope no matter how powerful, because an electron has a diameter less than the wavelength of visible light. He came up with a thought experiment using a gamma ray microscope instead. Gamma rays with their higher frequency are much more energetic than visible light, so would change the speed and direction of the electron in an unpredictable way. Heisenberg then realised that in fact a standard optical microscope behaves similarly. To measure the position and velocity of a particle, a light can be shone on it, and then the reflection detected. On a macroscopic scale this method works fine, but on sub-atomic scales the photons of light that hit the sub-atomic particle will cause it to move significantly. So, although the position may have been measured accurately, the velocity or momentum of the particle is changed, and by finding out the position we lose any information we have about the particle's velocity. This can be said as "The very act of observation affects the observed."

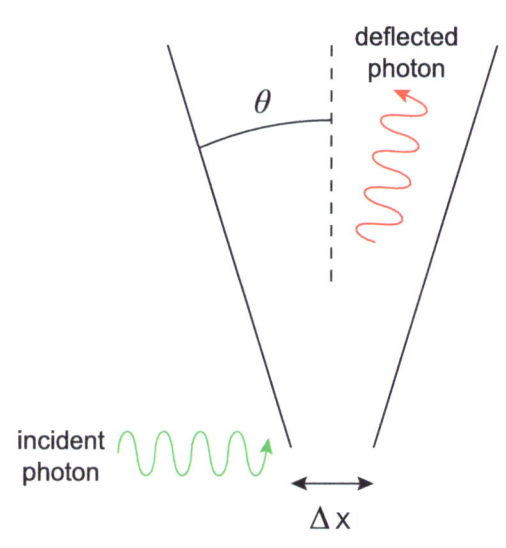

Heisenberg took this further and said that the values of certain pairs of variables cannot both be known exactly, so that the more precisely one variable is known, the less precisely the other can be known. An example of this is if the speed or momentum of a particle is known, then its location must be uncertain and vice versa. Similarly if we know the amount of energy a particle has, then we will not be able to determine how long it will have that energy for. In equation form this can be written as:

$$\Delta x \Delta p \geqslant \frac{h}{4\pi}$$
$$\Delta E \Delta t \geqslant \frac{h}{4\pi}$$

The minimum uncertainty in a particle's position **x** multiplied by the minimum uncertainty in a particle's momentum **p** has a minimum value about equal to the Planck constant divided by 4π. A similar relationship exists with Energy and time. Both can be shown to have the same units of Js.

This meant that particles could no longer be said to have separate, well-defined positions and velocities or energies and times, but they now exist in a "quantum state" which is a combination of their position and velocity or energy and time. You cannot know all the properties of the system at the same time, and any you do not know about in detail can be expressed in terms of probabilities.

A key question that needed answering about atoms was why do electrons orbiting the nucleus not lose energy and spiral inwards if they have a mass? Rutherford's model which had been accepted for the structure of the atom did not address this issue.

TOPIC 1. INTRODUCTION TO QUANTUM THEORY

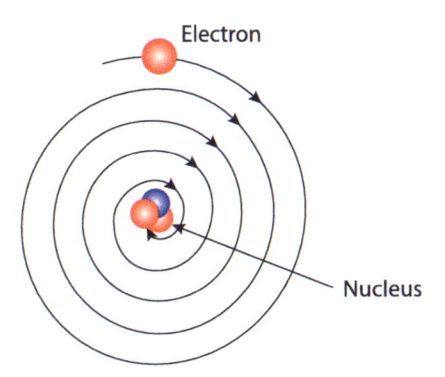

The answer is it is the uncertainty principle that prevents the electrons from getting too close to the nucleus. If they got too close we would precisely know their position and hence we would be very uncertain about their velocity. This uncertainty allows it to continue orbiting at the high speeds needed to stay in orbit. The uncertainty principle also helps explain why alpha particles can escape from the atom despite the strong force holding the nucleus together. The alpha particles use "quantum tunnelling" to escape (explained in more detail in Topic 2). Classical Physics says that they should not be able to escape.

Another example of this effect is nuclear fusion reactions taking place in the Sun. The Sun's temperatures are in actual fact not hot enough to give the protons the energy and speed they require to beat the massive repulsive electromagnetic force of the hydrogen atoms in the sun that they are trying to collide with. Thanks to quantum tunnelling and Heisenberg's uncertainty principle, the protons can apparently pass through the barrier despite the temperature and energy being approximately a thousand times too low for fusion to happen.

1.5 Extended information

Web links Go online

These web links should serve as an insight to the wealth of information available online and allow you to explore the subject further.

- https://www.youtube.com/watch?v=8JF6lvPBAzk
- http://en.wikipedia.org/wiki/Introduction_to_quantum_mechanics
- http://www.bbc.co.uk/science/space/universe/questions_and_ideas/quantum_mechanics
- http://www.theguardian.com/science/2013/nov/10/what-is-heisenbergs-uncertainty-principle
- http://hyperphysics.phy-astr.gsu.edu/hbase/uncer.html

© HERIOT-WATT UNIVERSITY

1.6 Summary

Summary

You should now be able to:

- state that the angular momentum of an electron about a nucleus is quantised in units of $h/2\pi$;
- state the equation $mvr = nh/2\pi$ and perform calculations using this equation;
- qualitatively describe the Bohr model of the atom;
- understand what is meant by black body radiation and the Ultra-Violet Catastrophe;
- understand that classical physics cannot explain Black body radiation, the photoelectric effect, and absorption and emission spectra - that quantum theory is required;
- state that quantum mechanics is used to provide a wider-ranging model of the atom than the Bohr model, and state that quantum mechanics can be used to determine probabilities;
- understand the Heisenberg Uncertainty Principle and be able to use the relationships $\Delta x \Delta p \geqslant \dfrac{h}{4\pi}$ $\Delta E \Delta t \geqslant \dfrac{h}{4\pi}$.

1.7 End of topic 1 test

Q7: Consider an electron orbiting in a Bohr hydrogen atom.

Calculate the angular momentum of the electron, if the quantum number n of the electron is 4.

_____ kg m^2 s^{-1}

....

Q8: An electron in the $n = 1$ orbit of the hydrogen atom moves in a circle of radius 5.29 × 10^{-11} m around the nucleus.

Calculate the de Broglie wavelength of the electron.

_____ m

....

Q9: An electron in a hydrogen-like atom in the $n = 3$ orbital has a de Broglie wavelength of 2.05 × 10^{-10} m.

Calculate the orbit radius of the electron.

_____ m

An electron orbiting in the $n = 4$ level of a hydrogen-like atom drops to the $n = 2$ level, emitting a photon. The energy difference between the two levels is 3.19 × 10^{-19} J.

Q10: What is the energy of the emitted photon?

_____ J

....

Q11: Calculate the wavelength of the emitted photon.

_____ m

....

Q12: Which of the following statements are correct?

- A) Perfect black bodies absorb and emit all wavelengths of radiation.
- B) The ultraviolet catastrophe predicted energies tending to zero at higher frequencies of radiation emitted.
- C) Black body radiation experiments showed higher frequencies of radiation emitted at higher temperatures.
- D) Black Body radiation curves were more evidence that protons existed.
- E) The Sun can be approximated to a black body.

....

Q13: A bullet is travelling at a speed of 210 m s^{-1} and has a mass of 0.100 kg.

If the percentage uncertainty in the momentum of the bullet is 0.01%, what is the minimum uncertainty in the bullet's position if is measured at the exact same time as the speed is measured?

_____ m

Unit 2 Topic 2

Wave particle duality

Contents

- 2.1 Introduction 218
- 2.2 Wave-particle duality of waves 218
    - 2.2.1 The photoelectric effect 218
    - 2.2.2 Compton scattering 221
- 2.3 Wave-particle duality of particles 224
    - 2.3.1 Diffraction 226
    - 2.3.2 The electron microscope 228
- 2.4 Quantum tunnelling 230
- 2.5 Extended information 231
- 2.6 Summary 231
- 2.7 End of topic 2 test 232

Prerequisites

- Wave particle duality, photons and a basic understanding of waves.
- Introduction to Quantum Theory.
- The Photoelectric Effect and Einstein's Energy Equation.

Learning objective

By the end of this topic you should be able to:

- understand some examples of waves behaving as particles - the photo-electric effect and Compton Scattering;
- understand some examples of particles behaving as waves - electron diffraction and quantum tunnelling;
- be able to calculate the de Broglie wavelength using the equation $\lambda = h/p$ and to investigate the formation of standing waves around atoms and in wire loops.

2.1 Introduction

Early in the 20th century, physicists realised that there were certain phenomena which could not be explained using the physical laws known at the time. Principally, phenomena involving light (thought to be composed of electromagnetic waves) could not be explained using this wave model. This topic begins with a look at some of the experiments that required a new theory - quantum theory - to explain their results. This theory explains that in some circumstances a beam of light can be considered to be made up of a stream of particles called photons. Light is said to exhibit "**wave-particle duality**",as both the wave model and the particle model can be used to describe the behaviour of light.

In the second part of the topic, this idea of wave-particle duality is explored further when it is shown that particles can sometimes exhibit behaviour associated with waves. Concentrating mainly on electrons, it will be shown that a stream of electrons can act in the same way as a beam of light passing through a narrow aperture. We finish the topic by looking at an interesting effect called quantum tunnelling which has many uses.

2.2 Wave-particle duality of waves

2.2.1 The photoelectric effect

The photoelectric effect was first observed in experiments carried out around 100 years ago. An early experiment carried out by Hallwachs showed that a negatively-charged insulated metal plate lost its charge when exposed to ultraviolet radiation. Other experiments, using equipment such as that shown in Figure 2.1, showed that electrons can be emitted from the surface of a metal plate when the plate is illuminated with light. This phenomenon is called the **photoelectric effect**.

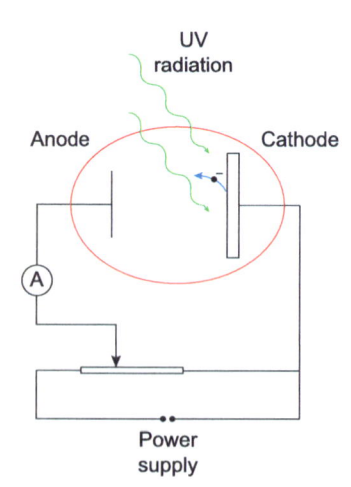

Figure 2.1: Apparatus for photoelectric effect experiments

The equipment shown in Figure 2.1 can be used to observe the photoelectric effect. The cathode and anode are enclosed in a vacuum, with a quartz window to allow the cathode to be illuminated using an ultraviolet lamp (quartz is used because ordinary glass does not transmit ultraviolet light). A sensitive ammeter records the current in the circuit (the photocurrent). The potentiometer can be used to provide a "stopping potential" to reduce the photocurrent to zero. Under these conditions, the stopping potential provides a measure of the kinetic energy of the electrons emitted from the cathode (the **photoelectrons**), since when the photocurrent drops to zero, even the most energetic of the photoelectrons does not have sufficient energy to reach the anode.

In classical wave theory, the irradiance (power per unit area) of a beam of light is proportional to the square of the amplitude of the light waves. This means that the brighter a beam of light, the more energy is falling per unit area in any period of time, as you would expect. Yet experiments showed that the speed or kinetic energy of the emitted electrons did not depend on the irradiance of the beam. In fact the energy of the photoelectrons depended on the frequency (or wavelength) of the light. Below a certain frequency, no electrons were emitted at all, whatever the irradiance of the beam.

These observations cannot be explained using the wave theory of light. In 1905 Einstein proposed a quantum theory of light, in which a beam of light is considered to be a stream of particles ('quanta') of light, called **photons**, each with energy E, where

$$E = hf$$

(2.1)

In Equation 2.1, f is the frequency of the beam of light, and h is a constant, called Planck's constant after the German physicist Max Planck. The value of h is
6.63×10^{-34} J s. The equivalent expression in terms of the speed c and wavelength λ of the light waves is

$$E = \frac{hc}{\lambda}$$

Example

Calculate the energy of a photon in the beam of light emitted by a helium-neon laser, with wavelength 633 nm.

The frequency of the photon is given by the equation

$$f = \frac{c}{\lambda}$$
$$\therefore f = \frac{3.00 \times 10^8}{6.33 \times 10^{-7}}$$
$$\therefore f = 4.74 \times 10^{14} \text{ Hz}$$

The photon energy E is therefore

$$E = hf = 6.63 \times 10^{-34} \times 4.74 \times 10^{14}$$
$$\therefore E = 3.14 \times 10^{-19} \text{ J}$$

The quantum theory explains why the kinetic energy of the photoelectrons depends on the frequency, not the irradiance of the incident radiation. It also explains why there is a 'cut-off frequency', below which no electrons are emitted.

When a photon is absorbed by the cathode, its energy is used in exciting an electron. A photoelectron is emitted when that energy is sufficient for an electron to escape from an atom. The cut-off frequency corresponds to the lowest amount of energy required for an electron to overcome the attractive electrical force and escape from the atom. The conservation of energy relationship for the photoelectric effect is given by Einstein's photoelectric equation

$$hf = hf_0 + \tfrac{1}{2}m_e v_e^2$$

(2.2)

In Equation 2.2, hf is the energy of the incident photon and $\tfrac{1}{2}m_e v_e^2$ is the kinetic energy of the photoelectron. hf_0 is called the **work function** of the material, and is the minimum photon energy required to produce a photoelectron.

TOPIC 2. WAVE PARTICLE DUALITY

Example

Monochromatic ultraviolet radiation of frequency 1.06×10^{15} Hz is focused onto a magnesium cathode in a photoelectric effect experiment. The kinetic energy of the photoelectrons produced is 1.13×10^{-19} J. What is the work function of magnesium?

The photon energy is

$$E = hf = 6.63 \times 10^{-34} \times 1.06 \times 10^{15}$$
$$\therefore E = 7.03 \times 10^{-19} \text{ J}$$

Rearranging Einstein's equation, the work function is

$$hf_0 = hf - \tfrac{1}{2}m_e v_e^2$$
$$\therefore hf_0 = (7.03 \times 10^{-19}) - (1.13 \times 10^{-19})$$
$$\therefore hf_0 = 5.90 \times 10^{-19} \text{ J}$$

2.2.2 Compton scattering

Experiments carried out in 1923 by the American physicist Arthur Compton provided further evidence to support the photon theory. He studied the scattering of a beam of X-rays fired at a thin sheet of graphite and found that some of the emergent beam was scattered into a wide arc. Furthermore the scattered X-rays had a slightly longer wavelength than the incident beam.

Compton explained his observations in terms of a collision between a photon and an electron in the graphite sheet, as shown in Figure 2.2.

Figure 2.2: Compton scattering from the collision of a photon with an electron

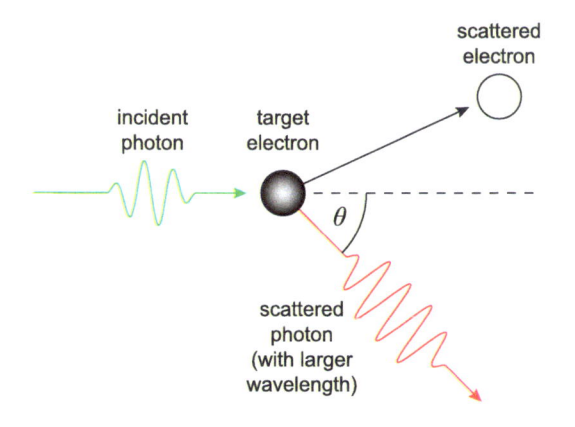

The incident X-rays have frequency f, so an individual photon has energy $E = hf$.

It was found that the scattered photons had frequency f', where $f' < f$.

We can compare the Compton scattering experiment to a moving snooker ball colliding with a stationary snooker ball. The X-ray is acting like a particle colliding with another particle, rather than a wave. Assuming the electron is initially at rest, the equation of conservation of energy is

$$hf = hf' + E_{k\ of\ recoil\ electron}$$

The conservation of energy relationship shows that $f' < f$, since the recoil energy of the electron $E_{k\ recoil}$ clearly has to be greater than zero as it is moving away from the collision. The wavelength of the scattered photon is $\lambda' = c/f'$, so the scattered wavelength is larger than the incident wavelength.

Compton performed experiments that confirmed this wavelength shift, which cannot be explained using a wave model of light. **Compton scattering** is therefore another example of the particle nature of light.

Compton also used the conservation of linear momentum in his analysis. Momentum is a property associated with particles, rather than waves, so Compton scattering is further proof of the dual nature of light. Using Einstein's energy-mass relationship from special relativity, the energy of a particle is related to its mass by the equation

$$E = mc^2$$

We can also express the energy of a photon in terms of its frequency f or wavelength λ

$$E = hf = \frac{hc}{\lambda}$$

Combining these two equations

$$mc^2 = \frac{hc}{\lambda}$$
$$\therefore mc = \frac{h}{\lambda}$$
$$\therefore p = \frac{h}{\lambda}$$

(2.3)

The momentum p of a photon is given by Equation 2.3.

TOPIC 2. WAVE PARTICLE DUALITY

Equation 2.3 is a particularly important equation, as it demonstrates the wave-particle duality of light. The equation relates one property associated with waves, the wavelength λ, to a property associated with particles, namely the momentum p.

Quiz: Wave-particle duality of waves   Go online

Useful data:

Planck's constant h	6.63×10^{-34} J s
Speed of light c	3.00×10^8 m s^{-1}

Q1: An argon laser produces monochromatic light with wavelength 512 nm. What is the energy of a single photon emitted by an argon laser?

a) 3.88×10^{-19} J
b) 1.29×10^{-27} J
c) 1.02×10^{-31} J
d) 3.39×10^{-40} J
e) 1.13×10^{-48} J

....

Q2: The work function of aluminium is 6.53×10^{-19} J. What is the minimum frequency of light which can be shone on a piece of aluminium to produce photoelectrons?

a) 1.01×10^{-15} Hz
b) 9.85×10^{14} Hz
c) 2.95×10^{23} Hz
d) Cannot say without knowing the wavelength of the light.
e) Cannot say without knowing the kinetic energy of the electrons.

....

Q3: In a Compton scattering experiment,

a) the momentum of the photons is unchanged after scattering.
b) the kinetic energy of the photons is unchanged after scattering.
c) the wavelength of the scattered photons decreases.
d) the wavelength of the scattered photons increases.
e) the wavelength of the scattered photons is unchanged.

....

224 UNIT 2. QUANTA AND WAVES

Q4: A source of light contains blue (λ = 440 nm) and red (λ = 650 nm) light. Which *one* of the following statements is true?

a) The blue photons have greater photon energy and greater photon momentum.
b) The blue photons have greater photon energy, but the red photons have greater photon momentum.
c) The red photons have greater photon energy, but the blue photons have greater photon momentum.
d) The red photons have greater photon energy and greater photon momentum.
e) All photons have the same photon energy.

....

Q5: Calculate the momentum of a single microwave photon emitted with wavelength 25.0 mm.

a) 7.96×10^{-24} kg m s^{-1}
b) 7.96×10^{-27} kg m s^{-1}
c) 2.65×10^{-32} kg m s^{-1}
d) 2.65×10^{-35} kg m s^{-1}
e) 5.53×10^{-44} kg m s^{-1}

2.3 Wave-particle duality of particles

We have seen that light can be described in terms of a beam of particles (photons) rather than as transverse waves, under certain circumstances. The equation $\lambda = h/p$ (Equation 2.3) relates a property of waves (the wavelength λ) to a property of particles (the momentum p).

In 1924, Louis de Broglie (pronounced 'de Broy') proposed that this equation could also be applied to particles. That is to say, he suggested that under certain circumstances particles would behave as if they were waves, with a wavelength given by the above equation. Within three years, experiments with electron beams had proved that this was indeed the case. The wavelength given by $\lambda = h/p$ for a particle is now known as the **de Broglie wavelength**. As the following example shows, in most everyday cases the de Broglie wavelength is extremely small.

Example

Find the de Broglie wavelength of:

1. an electron (m_e = 9.11 $\times$ 10^{-31} kg) travelling at 4.00 $\times$ 10^5 m s^{-1};
2. a golf ball (m_b = 0.120 kg) travelling at 20 m s^{-1}.

1. The momentum p_e of the electron is

$$p_e = m_e v_e = 9.11 \times 10^{-31} \times 4.00 \times 10^5 = 3.64 \times 10^{-25} \text{ kg m s}^{-1}$$

© HERIOT-WATT UNIVERSITY

The de Broglie wavelength of the electron is therefore

$$\lambda_e = \frac{h}{p_e} = \frac{6.63 \times 10^{-34}}{3.64 \times 10^{-25}} = 1.82 \times 10^{-9}\,\text{m}$$

2. The momentum p_b of the golf ball is

$$p_b = m_b v_b = 0.120 \times 20 = 2.40\,\text{kg m s}^{-1}$$

The de Broglie wavelength of the golf ball is therefore

$$\lambda_b = \frac{h}{p_b} = \frac{6.63 \times 10^{-34}}{2.40} = 2.76 \times 10^{-34}\,\text{m}$$

The de Broglie wavelength we have calculated for the golf ball is extremely small, so small that we do not observe any wave-like behaviour for such an object.

Whilst Equation 2.3 can be used to calculate the de Broglie wavelength, it should be noted that if a particle is travelling at greater than about 0.1 c, the momentum must be determined using a relativistic calculation. Such a calculation is beyond the scope of this course.

De Broglie also suggested that electrons have orbits similar to standing waves and the electron waves form into integer number of wavelengths around the circumference as below:

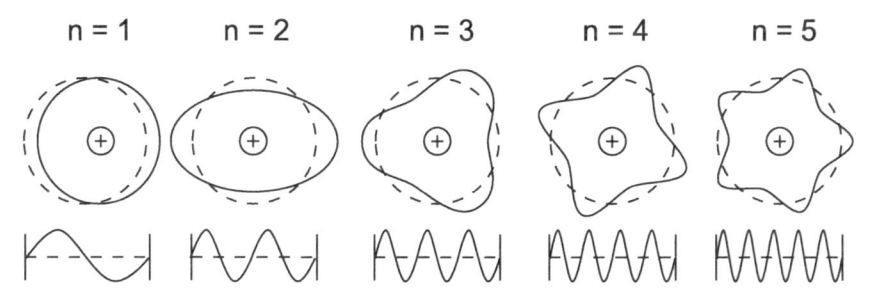

n = 1   n = 2   n = 3   n = 4   n = 5

This can be demonstrated nicely by using a metal wire loop or even a metal "Chladni" plate on top of a vibrating object. A sequence of standing waves (see topic 5 for an explanation of standing waves) will be seen around the loop similar to electron standing waves around an atom.

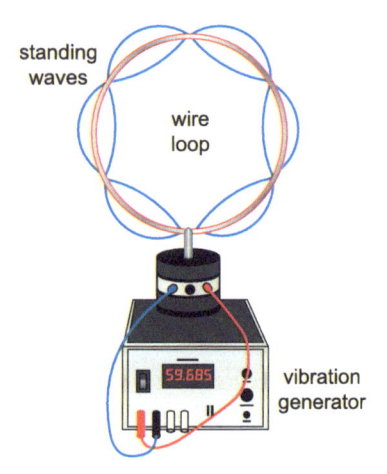

2.3.1 Diffraction

Since we can now calculate the wavelength associated with a moving particle, under what circumstances can we expect a particle to exhibit wave-like behaviour?

There are several aspects of the behaviour of light which can only be explained using a wave model. Interference, for example, occurs when two waves overlap in time and space, and the irradiance of the resultant wave depends on whether they interfere constructively or destructively.

Figure 2.3: (a) Constructive and (b) destructive interference of two sine waves

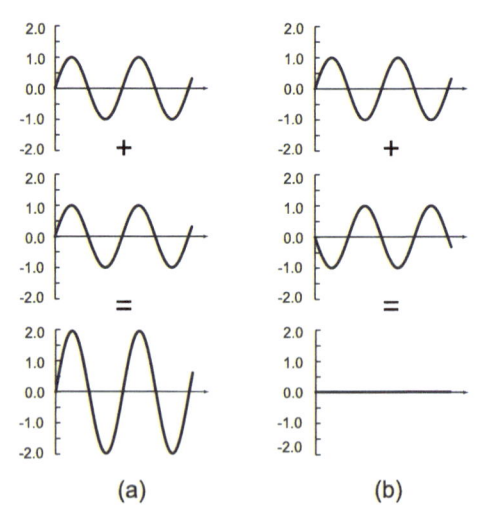

(a)   (b)

Constructive interference (Figure 2.3(a)) of two identical light waves would result in very bright illumination at the place where they overlap. Destructive interference (Figure 2.3(b)) results in

TOPIC 2. WAVE PARTICLE DUALITY

darkness - the two waves "cancel each other out". Such an effect cannot usually be seen with particles - if two people kick footballs at you, you will feel the effects of both of them hitting you - they won't cancel each other out!

Another property of waves, which particles do not usually exhibit is **diffraction**. If a wave passes through an aperture, the size of which is about the same as the wavelength of the light, the light "spreads out" as it emerges from the aperture. This effect is shown in Figure 2.4, where the wavefronts of two sets of waves approaching two different apertures are shown.

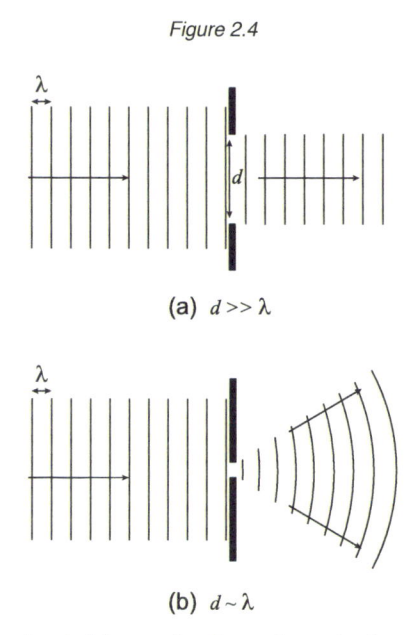

Figure 2.4

(a) $d \gg \lambda$

(b) $d \sim \lambda$

In Figure 2.4(a), the diameter d of the aperture is much greater than the wavelength λ, and the waves incident on the aperture pass through almost unaffected. In Figure 2.4(b), the waves have a wavelength that is of the same order of magnitude as the aperture diameter, so the waves are diffracted and spread out as they emerge from the aperture.

An analogy for particles would be a train travelling through a tunnel. The width of the tunnel is only slightly larger than the width of the train, so one might expect to observe diffraction. Of course this doesn't happen - the train passes through the tunnel and continues travelling along the track. We don't observe diffraction because the de Broglie wavelength associated with the train is many orders of magnitude smaller than the tunnel width, as we have already seen with the example of the golf ball.

Our earlier example showed that for electrons, the de Broglie wavelength is much larger, although still a very small number. But this wavelength ($\sim 10^{-9}$ to 10^{-10} m) is about the same as the distance between atoms in a crystalline solid. The first proof of de Broglie's hypothesis came in 1927 when the first **electron diffraction** experiments were performed independently by Davisson and Germer in the United States and G. P. Thomson in Aberdeen.

© HERIOT-WATT UNIVERSITY

228 UNIT 2. QUANTA AND WAVES

Figure 2.5: Electron diffraction (a) schematic diagram of the experiment; (b) diffraction pattern recorded on the photographic plate

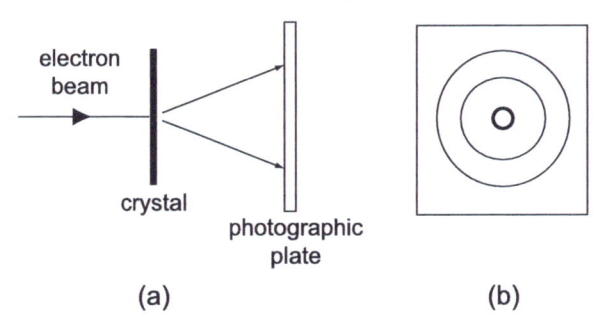

The spacing between atoms in a solid is typically around 10^{-10} m, so an electron travelling with a suitable momentum between two atoms in a crystal could be diffracted by the atoms. In a crystalline solid the atoms are arranged in a regular pattern or array, and act like a diffraction grating. Strong diffraction occurs in specific directions, which are determined by the atomic spacing within the crystal and the wavelength of the incident beam. Figure 2.5 shows the experimental arrangement and a typical diffraction pattern obtained. If the wavelength of the beam is known, the diffraction pattern can be used to determine the atomic spacing and crystal structure.

2.3.2 The electron microscope

In an optical microscope a sample is illuminated with white light and viewed through a series of lenses. The resolving power, or resolution, of the microscope is limited by the wavelength of the light. A microscope allows us to view a magnified image of very small objects, or view the fine structure of a sample. The resolving power tells us the smallest particles that can be distinguished when viewing through a microscope. The resolution is diffraction-limited, because light will be spread out when passing between two objects placed about one wavelength apart, making the two objects indistinguishable.

An electron microscope has a much smaller wavelength than visible light. Typically, the de Broglie wavelength of electrons in an electron microscope is around 10^{-10} m whilst the wavelength of visible light is around 10^{-7} m. Using an electron beam instead of a light source to 'illuminate' the sample means smaller features of the sample can be distinguished.

Quiz: Wave-particle duality of particles Go online

Useful data:

Planck's constant h	6.63×10^{-34} J s
Speed of light c	3.00×10^{8} m s^{-1}

© HERIOT-WATT UNIVERSITY

TOPIC 2. WAVE PARTICLE DUALITY

Q6: de Broglie's theory suggests that
a) particles exhibit wave-like properties.
b) particles have momentum and kinetic energy.
c) only waves can be diffracted.
d) photons always have zero momentum.
e) photons and electrons have the same mass.

....

Q7: What is the de Broglie wavelength of an electron (mass $m_e = 9.11 \times 10^{-31}$ kg) travelling with velocity 6.40×10^6 m s^{-1} ?
a) 1.75×10^{-15} m
b) 2.43×10^{-12} m
c) 2.47×10^{-12} m
d) 1.14×10^{-10} m
e) 1.56×10^{-7} m

....

Q8: A particle has a speed of 3.0×10^6 m s^{-1}. Its de Broglie wavelength is 1.3×10^{-13} m. What is the mass of the particle?
a) 1.5×10^{-38} kg
b) 7.8×10^{-32} kg
c) 1.7×10^{-27} kg
d) 5.1×10^{-23} kg
e) 5.1×10^{-21} kg

....

Q9: The demonstrations of electron diffraction by Davisson & Germer and G P Thomson showed that
a) electrons have momentum.
b) particles can exhibit wave-like properties.
c) light cannot be diffracted.
d) electrons have zero kinetic energy.
e) electrons have zero rest mass.

....

Q10: An electron microscope can provide higher resolution than an optical microscope because
a) electrons cannot undergo diffraction.
b) the electrons have a shorter wavelength than visible light.
c) electrons are particles whereas light is waves.
d) photons cannot be focused properly.
e) photons have a smaller momentum than electrons.

2.4 Quantum tunnelling

One very strange aspect of quantum physics is Quantum Tunnelling. This occurs due to the uncertainty of the position of objects such as electrons. There is a finite probability that an object trapped behind a barrier without enough energy to overcome the barrier may at times appear on the other side of the barrier. This process actually has no effect on the barrier itself - it is not broken down by this action. One example is electrons approaching the nucleus of atoms - there is some probability, no matter how small, that the electron will appear on the other side of the electromagnetic field which repels it.

This occurrence of appearing the other side of the nuclei is known as quantum tunnelling, and is easiest to visualise by thinking of the electron as a wave rather than a particle - a particle clearly cannot pass through the barrier but a wave may slightly overlap the barrier and even though most of the wave is on one side the small part of the wave that does cross the barrier leads to a small chance of the particle that generated the wave appearing on the far side of the barrier. This is shown in the following diagram and the video links below:

Quantum tunnelling through a barrier

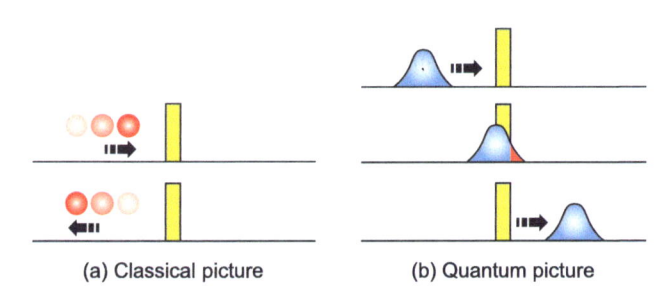

(a) Classical picture   (b) Quantum picture

Note the classical physics particle is repelled by the electromagnetic force, but the quantum physics wave has a small probability of tunnelling through the barrier.

This technique can be used to make quantum tunnelling composites (QTCs) which are materials which can be used as highly sensitive pressure sensors (due to the ability of quantum tunnelling to cross thin barriers and in electronic devices such as modern Tunnel FET or TFET transistors replacing the older MOSFET transistors. Possible uses of the pressure sensors would include measuring the forces of impacts e.g. on fencing helmet or boxing glove or to accurately read a person's blood pressure.

2.5 Extended information

Web links Go online

These web links should serve as an insight to the wealth of information available online and allow you to explore the subject further.

- http://hyperphysics.phy-astr.gsu.edu/hbase/mod1.html
- http://www.livescience.com/24509-light-wave-particle-duality-experiment.html
- http://hyperphysics.phy-astr.gsu.edu/hbase/debrog.html
- https://courses.lumenlearning.com/physics/chapter/29-6-the-wave-nature-of-matter/
- https://www.youtube.com/watch?v=Q1YqgPAtzho
- http://www.engineering.com/Videos/VideoPlayer/tabid/4627/VideoId/2797/What-Is-Quantum-Tunneling.aspx
- https://www.youtube.com/watch?v=6LKjJT7gh9s
- http://upload.wikimedia.org/wikipedia/commons/5/50/EffetTunnel.gif

2.6 Summary

Summary

You should now be able to:

- understand some examples of waves behaving as particles - the photo-electric effect and Compton Scattering;
- understand some examples of particles behaving as waves - electron diffraction and quantum tunnelling;
- be able to calculate the de Broglie wavelength using the equation $\lambda = h/p$ and to investigate the formation of standing waves around atoms and in wire loops.

2.7 End of topic 2 test

Q11: Photons emitted by a monochromatic light source have energy 3.75×10^{-19} J.
Calculate the frequency of the light.
$f = $ _____ Hz

Q12: Photons in a beam of monochromatic infrared radiation each have individual photon energy 5.35×10^{-20} J.
What is the wavelength of the infrared beam?
$\lambda = $ _____ m

Q13: Light with a maximum frequency 8.35×10^{14} Hz is shone on a piece of metal. The work function of the metal is 3.65×10^{-19} J.
Calculate the maximum kinetic energy of the electrons produced due to the photoelectric effect.
$E_k = $ _____ J

Q14: A laser emits monochromatic light with wavelength 425 nm.
Calculate the momentum of a single photon emitted by the laser.
$p = $ _____ kg m s^{-1}

Q15: A filament lamp emits light with a wavelength range of 300 to 725 nm.
Calculate the largest value of photon momentum of the photons emitted by the lamp.
$p = $ _____ kg m s^{-1}

A proton and an electron are both travelling at the same velocity.

Q16: Which particle has the greater momentum?

a) The proton
b) The electron
c) Neither - both have the same momentum

Q17: Which particle has the greater de Broglie wavelength?

a) The proton
b) The electron
c) Neither - both have the same de Broglie wavelength

TOPIC 2. WAVE PARTICLE DUALITY

In a television tube, electrons are accelerated by a high voltage and strike the screen with an average velocity of 6.35×10^6 m s^{-1}.

Q18: On average, what is the momentum of an electron just before it strikes the screen?

p = _____ kg m s^{-1}

....

Q19: Calculate the average de Broglie wavelength of the electrons.

λ_e = _____ m

....

Q20:

Complete the paragraph using some of the following words.

uncertainty	wave	electron
fission	neutron	proton
fusion	barrier	transistors

Quantum tunnelling is when an incident _____ is thought of as a _____ and part of the wave crosses a _____ . Due to the _____ of the wave's position, this allows the electron to pass through the barrier. This is how nuclear _____ is possible in the sun and how modern _____ have become so efficient and tiny.

Unit 2 Topic 3

Magnetic fields and particles from space

Contents

3.1 Introduction 236
3.2 Force on a moving charge in a magnetic field 237
3.3 The path of a charged particle in a magnetic field 239
3.4 Helical motion in a magnetic field 243
        3.4.1 Charged particles in the Earth's magnetic field 246
3.5 The Solar Wind 248
3.6 Extended information 252
3.7 Summary 253
3.8 End of topic 3 test 253

Prerequisites

- Circular Motion.

Learning objective

By the end of this topic you should be able to:

- explain what is meant by magnetic field and magnetic force;
- explain the behaviour of charged particles in a magnetic field;
- have an understanding of cosmic rays and the solar wind and how they interact with the Earth's atmosphere and magnetic field.

3.1 Introduction

We are all familiar with permanent magnets and the fact that they exert forces on each other, as well as on certain types of metal and metallic ores. The first descriptions of magnetic effects were made in terms of **magnetic poles**. Every magnet has two poles. Unlike or opposite magnetic poles exert forces of attraction on each other, while like or similar poles repel each other. In addition, both poles of a magnet exert forces of attraction on unmagnetised iron. One pole is called the north pole or N-pole (actually short for "north-seeking" pole). It points approximately towards the north geographic pole of the Earth. The other end is called the south pole (S-pole). This alignment happens because the Earth is itself a magnet, with a south magnetic pole near to the north geographic pole.

It is tempting to compare north and south magnetic poles with positive and negative charges. While there are similarities, the major difference is that it is possible for isolated positive and negative charges to exist but there is no evidence to suggest that an isolated magnetic pole (a monopole) can exist.

In this topic, we will start by defining what is meant by a magnetic field and magnetic induction. We will then be looking at what happens when charged particles enter magnetic fields at various angles and will finally look at cosmic rays and the solar wind and how they interact with the Earth's atmosphere and magnetic field.

It is usual to describe the interaction between two magnets by applying a field description to them. We can consider that one magnet sets up a magnetic field and that the other magnet is in this magnetic field. A magnetic field is usually visualised by drawing magnetic field lines. Figure 3.1 shows the field pattern around a bar magnet.

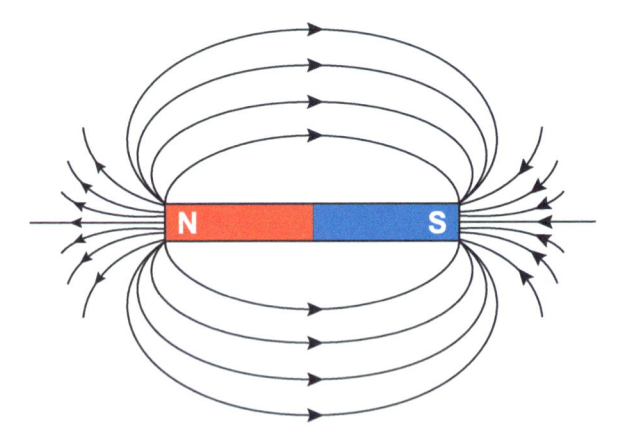

Figure 3.1: The field pattern around a bar magnet

The following points should be noted about magnetic field lines:

- they show the direction in which a compass needle would point at any position in the field;
- they never cross over each other because the direction of the magnetic field is unique at all points - crossing field lines would mean that the magnetic field pointed in more than one direction at the same place - obviously impossible;

- they are three-dimensional, although this is not often seen on a page or screen - because the magnetic field they represent is three-dimensional;
- they indicate the magnitude of the magnetic field at any point - the closer the lines are together, the stronger is the field.
- like electric field lines, magnetic field lines are used to visualise the magnitude and the direction of the field. Also like electric field lines, they do not exist in reality.

An atom consists of a nucleus surrounded by moving electrons. Since the electrons are charged and moving, they create a magnetic field in the space around them. Some atoms have magnetic fields associated with them and behave like magnets. Iron, nickel and cobalt belong to a class of materials that are **ferromagnetic**. In these materials, the magnetic fields of atoms line up in regions called **magnetic domains**. If the magnetic domains in a piece of ferromagnetic material are arranged so that most of their magnetic fields point the same way, then the material is said to be a magnet and it will have a detectable magnetic field.

The Earth's magnetic field is thought to be caused by currents in the molten core of the Earth. A simplified view of the Earth's magnetic field is that it is similar to the field of a bar magnet. This means that the field lines are not truly horizontal at most places on the surface of the Earth, the angle to the horizontal being known as the magnetic inclination. The Earth's magnetic field at the poles is vertical. A compass needle is simply a freely-suspended magnet, so it will point in the direction of the Earth's magnetic field at any point. The fact that the magnetic and geographic poles do not exactly coincide causes a compass needle reading to deviate from geographic north by a small amount that depends on the position on the Earth. This difference is known as the magnetic declination or the magnetic variation.

3.2 Force on a moving charge in a magnetic field

$$F = BQ\,v\sin\theta$$

(3.1)

Where F is the magnetic force experienced by a charged particle in Newtons, B is the magnetic field strength measured in Tesla, Q is the size of the charged particle in Coulombs and θ is the angle between B and v. The component of v that is perpendicular to the magnetic field is $v\sin\theta$.

At Higher, we learned how to work out the direction of force experienced by a negative charge using the "Right Hand Rule".

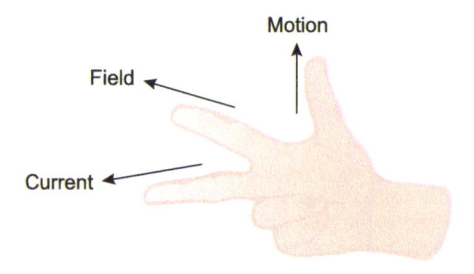

The Right Hand Rule for Forces on a negative charge in a magnetic field. A positive charge would experience a force in the opposite direction.

> **Top tip**
>
> A good way to remember this is that:
>
> - The **F**irst finger represents the magnetic **F**ield.
> - The se**C**ond finger represents the electron **C**urrent.
> - The **Th**umb represents **Th**rust or Motion (i.e. Force).
>
> An even quicker way is to remember the American secret service "**FBI**" for the thumb (**F**), first finger (**B**) and second finger (**I**). Remember it is always the right hand used for electron current.

The relationship given in Equation 3.1 shows the two conditions that must be met for a charge to experience a force in a magnetic field.
These are:

1. The charge must be moving.
2. The velocity of the moving charge must have a component perpendicular to the direction of the magnetic field.

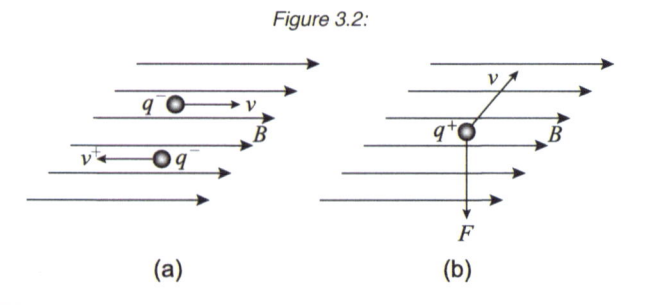

Figure 3.2:

(a)   (b)

TOPIC 3. MAGNETIC FIELDS AND PARTICLES FROM SPACE

If the charge is moving parallel or antiparallel to the field, as shown in Figure 3.2(a), then no magnetic force acts on the moving charge and its velocity is unchanged. If the charge is moving with a velocity that is perpendicular to the direction of the magnetic field, as shown in Figure 3.2(b), then the magnetic force experienced by the charge is a maximum. In this last case, $\sin\theta = \sin 90°= 1$ and so Equation 3.1 becomes

$$F = qvB$$

(3.2)

Example

An electron enters a uniform magnetic field of magnetic induction 1.2 T with a velocity of 3.0 × 10^6 m s^{-1} at an angle of 30° to the direction of the magnetic field.

a) Calculate the magnitude of the magnetic force that acts on the electron.

b) If a proton instead of an electron enters the same magnetic field with the same velocity, what difference, if any, would there be to the force experienced by the proton?

a) Using Equation 3.1 we have

$$F = Bqv\sin\theta$$
$$\therefore F = 1.2 \times 1.6 \times 10^{-19} \times 3.0 \times 10^6 \times \sin 30°$$
$$\therefore F = 2.9 \times 10^{-13} \text{ N}$$

b) Since the charge on the proton has the same magnitude as the charge on the electron, the force on the proton also has the same magnitude.
The sense of the direction of the force is different, since a proton has a positive charge while an electron has a negative charge.

3.3 The path of a charged particle in a magnetic field

We have seen that if a charged particle enters a magnetic field in a direction perpendicular to the magnetic field, then the field exerts a force on the particle in a direction that is perpendicular to both the field and the velocity of the particle. Because the magnetic force is always perpendicular to the velocity, v, of the particle, this force cannot change the *magnitude* of the velocity, only its *direction*. Since the magnetic force never has a component in the direction of v, then it follows that this force does no work on the moving charge. In other words, the *speed* of a charged particle moving in a magnetic field does not change although its *velocity* does because of the change in direction.

240　UNIT 2. QUANTA AND WAVES

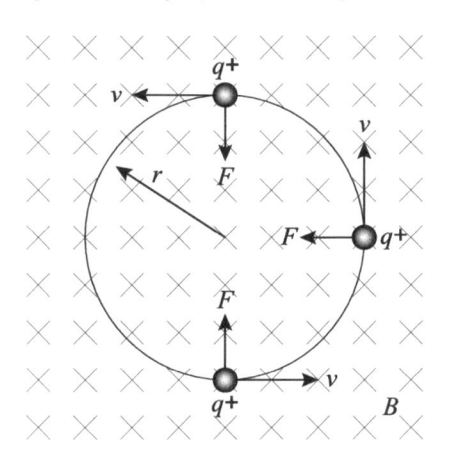

Figure 3.3: Charged particle in a magnetic field

Consider Figure 3.3 showing a uniform magnetic field directed into the viewing plane. (Think of the field as looking at the end of an arrow travelling away from you.) A positively charged particle that, at the bottom of the figure, has a velocity from left to right will experience a force directed to the top of the figure, as shown. Although this force has no effect on the speed of the particle, it does change its direction. Since the constant-magnitude magnetic force always acts at right angles to the velocity, it can be seen that the particle follows a circular path, with the force always directed towards the centre of the circle.

We have met this type of motion before, where the force is always perpendicular to the direction of motion. It is similar to the force exerted by a rope on an object that is swung in a circle above the head. It is also the same as the gravitational force that keeps a planet or a satellite in orbit around its parent body.

The centripetal or radial acceleration of the particle is given by v^2/r where r is the radius of the circular path of the particle.

Using Newton's second law, we have

$$F = ma$$
$$\therefore qvB = m \times \frac{v^2}{r}$$
$$\therefore qB = \frac{mv}{r}$$

(3.3)

where m is the mass of the particle.

© HERIOT-WATT UNIVERSITY

TOPIC 3. MAGNETIC FIELDS AND PARTICLES FROM SPACE

We can rearrange Equation 3.3 to find the radius of the circular path.

$$qB = \frac{mv}{r}$$
$$\therefore r = \frac{mv}{qB}$$

(3.4)

Example

An electron enters a uniform magnetic field of magnetic induction 1.2 T with a velocity of 3.0 × 10^6 m s^{-1} perpendicular to the direction of the magnetic field.

a) Calculate the radius of the circular path followed by the electron.

b) If a proton instead of an electron enters the same magnetic field with the same velocity, what difference, if any, would there be to the path followed by the proton?

a) Using Equation 3.4, we have for the electron

$$r_e = \frac{m_e v_e}{q_e B}$$
$$\therefore r_e = \frac{9.11 \times 10^{-31} \times 3 \times 10^6}{1.6 \times 10^{-19} \times 1.2}$$
$$\therefore r_e = 1.4 \times 10^{-5} \text{ m}$$

b) For the proton

$$r_p = \frac{m_p v_p}{q_p B}$$
$$\therefore r_p = \frac{1.67 \times 10^{-27} \times 3 \times 10^6}{1.6 \times 10^{-19} \times 1.2}$$
$$\therefore r_p = 0.026 \text{ m}$$

It can be seen that the radius of the circular path of the proton is much greater than that of the electron. This is because of the difference in masses of the two particles.

What is not shown by the calculation is that the particles will rotate in opposite directions because of the difference in the sign of their charges.

It is worth pointing out that Equation 3.4 only holds for charges moving with non-relativistic velocities. A value of v of about 10% of c is usually taken as the limit for the validity of this expression.

© HERIOT-WATT UNIVERSITY

Charges entering a magnetic field

Go online

At this stage there is an online activity which shows the paths taken by charges that enter a uniform magnetic field perpendicular to the direction of the field.

This activity shows the paths taken by charges that enter a uniform magnetic field perpendicular to the direction of the field.

Quiz: The force on a moving charge

Go online

Useful data:

Fundamental charge e	1.6×10^{-19} C
Mass of an electron m_e	9.11×10^{-31} kg
Mass of proton	1.67×10^{-27} kg

Q1: A proton and an electron both enter the same magnetic field with the same velocity. Which statement is correct?

a) Both particles experience the same magnetic force.
b) The particles experience the same magnitude of force but in opposite directions.
c) The proton experiences a larger force than the electron, in the same direction.
d) The proton experiences a smaller force than the electron, in the same direction.
e) The proton experiences a larger force than the electron, in the opposite direction.

....

Q2: An alpha particle that has a charge of $+2e$ enters a uniform magnetic field of magnitude 1.5 T, with a velocity of 5×10^5 m s^{-1}, perpendicular to the field.
What is the magnitude of the force on the particle?

a) 2.5×10^{-21} N
b) 1.3×10^{-15} N
c) 2.4×10^{-13} N
d) 1.2×10^{-13} N
e) 1.2×10^{-7} N

....

Q3: Which of these bodies can move through a magnetic field without experiencing any net magnetic force?

i. a negatively-charged electron
ii. a positively-charged proton
iii. an uncharged billiard ball

TOPIC 3. MAGNETIC FIELDS AND PARTICLES FROM SPACE

a) (i) only
b) (ii) only
c) (iii) only
d) none of them
e) all of them

....

Q4: A charged particle moves in a magnetic field only.

Which statement is **always** true of the motion of this particle?

The particle moves with

a) constant speed.
b) constant velocity.
c) zero acceleration.
d) increasing acceleration.
e) decreasing acceleration.

....

Q5: A beam of electrons is bent into a circle of radius 3.00 cm by a magnetic field of magnitude 0.60 mT.

What is the velocity of the electrons?

a) 3.16×10^6 m s^{-1}
b) 1.98×10^5 m s^{-1}
c) 1.72×10^3 m s^{-1}
d) 1.08×10^2 m s^{-1}
e) 2.88×10^{-24} m s^{-1}

3.4 Helical motion in a magnetic field

Consider again the situation we looked at in previous Topic, where a charged particle q moves with a velocity v at an angle θ to a magnetic field B. We showed that the particle experiences a force F of magnitude $F = B\,q\,v\sin\theta$. This force is perpendicular to the plane containing B and v and so causes circular motion.

In that analysis, we ignored the component of v that is parallel to the magnetic field B. This component is $v\cos\theta$ as shown in Figure 3.4. In this figure, B and v are in the x-z plane, with B in the x-direction in this plane.

244 UNIT 2. QUANTA AND WAVES

Figure 3.4: Charged particle moving in a magnetic field

The component of the particle's velocity in the z-direction, $v \sin \theta$, perpendicular to the magnetic field that causes the circular motion, keeps the same magnitude but constantly changes direction. The component of the velocity in the x-direction, $v \cos \theta$, parallel to the magnetic field, remains constant in magnitude and direction because there is no force parallel to the field. The resulting motion of the charged particle is made up of two parts - circular motion with constant angular velocity at right angles to the magnetic field, and linear motion with constant linear velocity parallel to the magnetic field. This causes the particle to follow a helical path, with the axis of the **helix** along the direction of the magnetic field. This is shown in Figure 3.5.

Figure 3.5: Helical path of a charged particle in a magnetic field

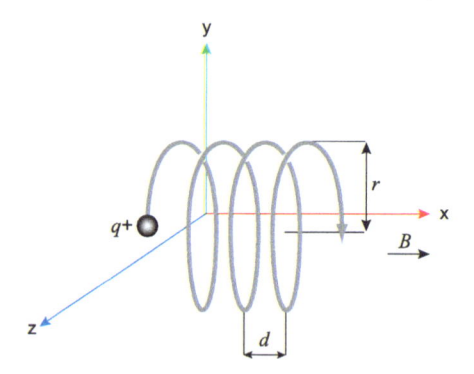

The radius of the circular path of the particle is given by

$$r = \frac{mv \sin \theta}{qB}$$

(3.5)

© HERIOT-WATT UNIVERSITY

TOPIC 3. MAGNETIC FIELDS AND PARTICLES FROM SPACE

This is the equation derived in topic 6 using the component of the velocity perpendicular to the field, $v \sin \theta$.

The **pitch**, d, of the helix, or the distance travelled in the direction of the magnetic field per revolution, can be found as follows. (Remember that periodic time T is the time taken to complete one revolution and it can be found from $\omega = 2\pi/T$)

The speed at which the particle is moving in the x-direction is $v \cos \theta$, so

$$d = v \cos \theta \times T$$
$$\therefore d = v \cos \theta \times \frac{2\pi}{\omega}$$
$$\therefore d = v \cos \theta \times \frac{2\pi r}{v \sin \theta}$$
$$\therefore d = \frac{2\pi r}{\tan \theta}$$

(3.6)

Example

A proton moving with a constant velocity of 3.0 x 10^6 m s^{-1} enters a uniform magnetic field of magnitude 1.5 T. The path of the proton makes an angle of 30° with the magnetic field as shown in Figure 3.6.

Figure 3.6: Path of a proton

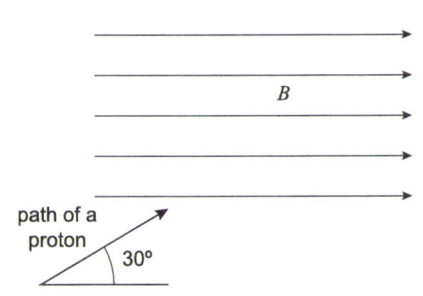

a) Describe and explain the shape of the path of the proton in the magnetic field.

b) Calculate the radius of the path of the proton.

c) Calculate the distance travelled by the proton in the direction of the magnetic field during one revolution.

a) The path is a helical shape.

This is because the proton has both circular and linear motion in the magnetic field. The circular motion is due to the component of its velocity $v \sin 30°$ perpendicular to the field. The linear motion exists because of the component $v \cos 30°$ parallel to the field.

b) To find the radius

$$r = \frac{mv \sin \theta}{qB}$$
$$\therefore r = \frac{1.67 \times 10^{-27} \times 3.0 \times 10^6 \times \sin 30}{1.6 \times 10^{-19} \times 1.5}$$
$$\therefore r = 0.010 \text{ m}$$

c) To find the distance travelled per revolution

$$\text{distance travelled (pitch)} = \frac{2\pi r}{\tan \theta}$$
$$= \frac{2 \times \pi \times 0.010}{\tan 30}$$
$$= 0.11 \text{ m}$$

The path of a charged particle   Go online

At this stage there is an online activity which provides extra practice.

3.4.1 Charged particles in the Earth's magnetic field

We have already seen that there is a non-uniform magnetic field around the Earth, caused by its molten metallic core. This magnetic field is stronger near to the poles than at the equator. The Sun emits charged particles, both protons and electrons. Some of these charged particles enter the Earth's magnetic field near to the poles and in doing so, they spiral inwards. The magnetic field of the Earth traps these charged particles in regions known as the **Van Allen radiation belts** (Figure 3.7). Protons with relatively high masses are trapped in the inner radiation belts, while the electrons, with lower mass and greater speed, are trapped in the outer radiation belts.

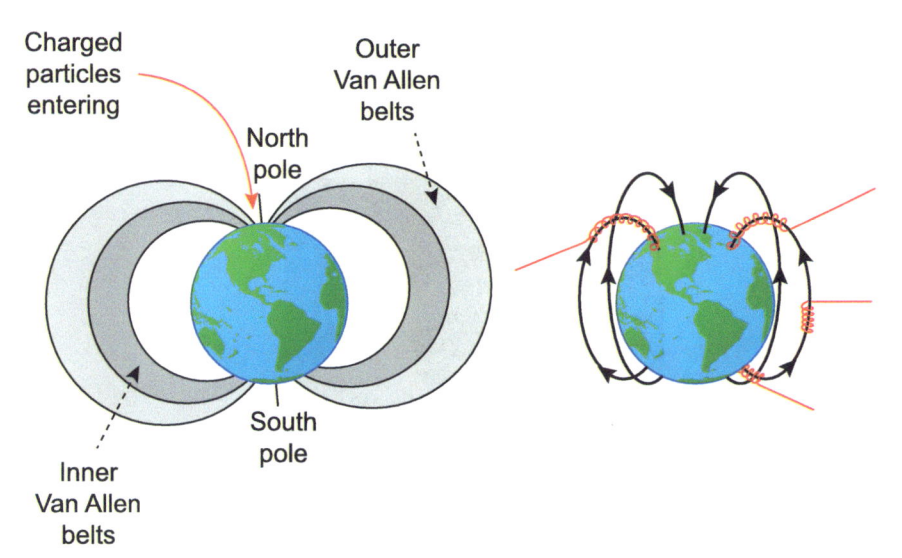

TOPIC 3. MAGNETIC FIELDS AND PARTICLES FROM SPACE 247

Figure 3.7: The Van Allen radiation belts

When they enter the Earth's atmosphere the charged particles can collide with the gases in the atmosphere. In doing so, some of their energy is emitted in the form of light. The light emitted from excited oxygen atoms is green in colour and that emitted from excited nitrogen atoms is pinkish-red. The resulting dramatic displays of coloured lights often seen dancing in the northern and southern night skies are called aurora - aurora borealis or the 'Northern Lights' and aurora australis or the 'Southern Lights'.

3.5 The Solar Wind

The solar wind is made up from a stream of plasma released from the upper layers of the Sun's atmosphere. Plasma is considered to be a fourth state of matter, where the electrons have detached from the atoms leaving a cloud of protons, neutrons and electrons and allowing the plasma to move as a whole. This upper layer of the Sun's atmosphere is correctly called the Corona which can only be seen with the naked eye during a solar eclipse.

The solar wind consists of mostly electrons and protons with energies usually between 1.5 and 10 keV. (An eV is an electrical unit of energy which is covered in more detail in a later section). These charged particles can only escape the immense pull of the Sun's gravity because of their high energy caused by the high temperature of the corona and the sun's powerful magnetic field.

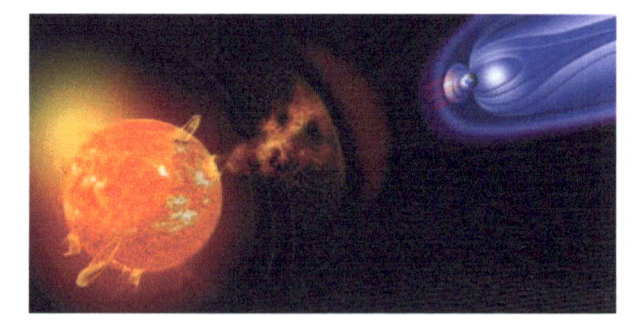

The solar wind flows from the sun and can travel great distances at speeds as high as 900 km s^{-1}. The region the solar wind occupies is known as the **heliosphere**.

The solar wind causes the tails of comets to trail behind the bodies of comets as they go through the solar system.

TOPIC 3. MAGNETIC FIELDS AND PARTICLES FROM SPACE

Comet Hale-Bopp. Photo Credit: A. Dimai and D. Ghirardo, (Col Druscie Obs.), AAC

A comet starts out as a lump of rock and ice contained in a region called an Oort Cloud but when the gravitational field of a nearby star such as the Sun attracts the comet, the heat of the star melts some of the ice contained in the comet which creates a gaseous tail that extends away from the source of the heat due to the solar wind. As a comet gets closer to the Sun it increases in speed due to the ever increasing gravitational field and as more ice evaporates the comet's tail will grow in length.

The upper atmosphere of the sun is incredibly unstable and often intense variations in brightness are seen, called a **solar flare**. A solar flare occurs when the magnetic energy that has built up in the solar atmosphere is suddenly released throwing radioactive material into space and releasing radiation across the full range of frequencies contained in the electromagnetic spectrum. The Sun's surface has huge magnetic loops called **prominences**.

These can cause geomagnetic storms which are induced by coronal mass ejections (CMEs) which have a big impact on equipment and astronauts in space causing effects such as corrosion of equipment, overloading of observation cameras and also exposing astronauts to dangerous levels of radiation. However, at the surface of the Earth we are well protected from the effects of solar flares and other solar activity by the Earth's magnetic field and atmosphere. See Figure 3.8.

© HERIOT-WATT UNIVERSITY

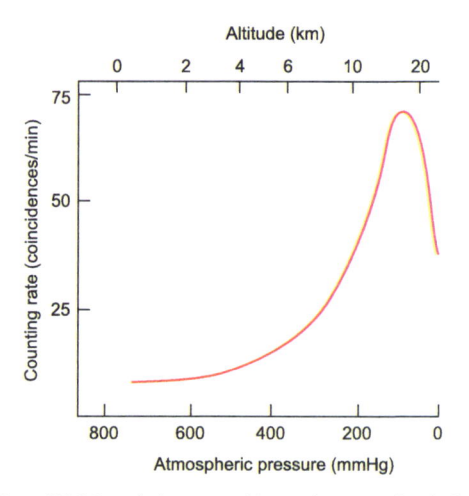

Figure 3.8: Variation of cosmic rays with altitude

In 2007, observations from NASA and Japanese X-ray observatories helped clarify the origin of cosmic rays.

Outer space contains large numbers of cosmic rays and were first discovered in 1912. Cosmic Rays are made up of subatomic particles and ions (such as protons and electrons). These particles travel through space at speed close to the speed of light in all directions. Their energies are thousands of times greater than the energies reached by particles produced in the largest terrestrial particle accelerators. Cosmic rays are constantly colliding with atoms and molecules high in the Earth's atmosphere. The effects are seen at the Earth's surface as cascades of secondary particles.

In the 1960s, scientists recognised that most cosmic rays came from gaseous supernova remnants. These left-over pieces of expired stars expand into the interstellar gas surrounding them. This is a highly energetic process. The expansion creates a shock front that contains magnetic fields. These fields accelerate the charged particles to the enormous energies and produces cosmic rays.

Studies from NASA's Chandra X-ray Observatory and the Japan Aerospace Exploration Agency (JAXA), evidence that the electrons energies in the supernova remnant Cassiopeia A would suggest they are accelerated as quickly as the theory would allow.

Like all planets with magnetic fields the Earth's magnetic field will interact with the solar wind to deflect the charged particles and form an elongated cavity in the solar wind. This cavity is called the **magnetosphere** of the planet. Near to the Earth the particles of the solar wind are travelling about 400 km s^{-1} as they are slowed by the magnetic field to produce a bow shaped shock wave around the earth.

TOPIC 3. MAGNETIC FIELDS AND PARTICLES FROM SPACE

Quiz: Motion of charged particles in a magnetic field — Go online

Useful data:

Fundamental charge e	1.6×10^{-19} C
Mass of an electron m_e	9.11×10^{-31} kg

Q6: A charged particle moving with a velocity v in a magnetic field B experiences a force F.

Which of the quantities v, B and F are *always* at right angles to each other?
(i) F and B
(ii) F and v
(iii) B and v

a) (i) only
b) (ii) only
c) (iii) only
d) (i) and (ii) only
e) (i) and (iii) only

....

Q7: A charged particle enters a magnetic field at an angle of 60° to the field.
Which is the best description of the resulting path of the charged particle through the field?

a) circular, parallel to the magnetic field
b) circular, perpendicular to the magnetic field
c) helical, with the axis in the magnetic field direction
d) helical, with the axis at 60° to the magnetic field
e) straight through the field

© HERIOT-WATT UNIVERSITY

Q8: An electron moving with a constant velocity of 2.0×10^6 m s^{-1} enters a uniform magnetic field of value 1.0 T. The path of the electron makes an angle of 30° with the direction of the magnetic field.

What is the radius of the resulting path of the electron?

a) 8.24×10^{-5} m
b) 1.97×10^{-5} m
c) 1.14×10^{-5} m
d) 6.56×10^{-6} m
e) 5.69×10^{-6} m

Q9: What is the name given to the regions around the Earth where charged particles can become trapped?

a) aurora
b) ionosphere
c) northern lights
d) stratosphere
e) Van Allen belts

3.6 Extended information

Web links — Go online

These web links should serve as an insight to the wealth of information available online and allow you to explore the subject further.

- http://www.schoolphysics.co.uk/age16-19/Atomic%20physics/Electron%20physics/text/Electron_motion_in_electric_and_magnetic_fields/index.html
- https://courses.lumenlearning.com/boundless-physics/chapter/motion-of-a-charged-particle-in-a-magnetic-field/
- https://www.youtube.com/watch?v=GOAEIMx39-w
- https://www.youtube.com/watch?v=ndQEq48IZA4

3.7 Summary

Summary

You should now be able to:

- explain what is meant by magnetic field and magnetic force;
- explain the behaviour of charged particles in a magnetic field;
- have an understanding of cosmic rays and the solar wind and how they interact with the Earth's atmosphere and magnetic field.

3.8 End of topic 3 test

End of topic 3 test Go online

The following data should be used when required:

Fundamental charge e	1.6×10^{-19} C
Mass of an electron m_e	9.11×10^{-31} kg

Q10: A proton enters a magnetic field of magnitude 1.85×10^{-3} T with a velocity of 1.32×10^6 m s^{-1} at an angle of 49.5°.
Calculate the magnitude of the force on the proton.
_____ N

Q11: A beam of electrons travelling at 1.28×10^7 m s^{-1} enters a magnetic field of magnitude 3.62 mT, perpendicular to the field.

1. Calculate the force on an electron while it is moving in the magnetic field.
   _____ N
2. Calculate the radius of the circular path of the electrons in the field.
   _____ mm

Q12: A charged particle has a charge-to-mass ratio of 2.22×10^8 C kg^{-1}.
It moves in a circular path in a magnetic field of magnitude 0.713 T.
Calculate how long it takes the particle to complete one revolution.
_____ seconds

Q13: An ion of mass 6.6 × 10⁻²⁷ kg, moving at a speed of 1 × 10⁷ m s⁻¹, enters a uniform magnetic field of induction 0.67 T at right angles to the field.

The ion moves in a circle of radius 0.31 m within the magnetic field.

What is the charge on the ion, in terms of the charge on an electron?

....

Q14: An electron travelling at velocity 4.9 × 10⁵ m s⁻¹ enters a uniform magnetic field at an angle of 50°, as shown below.

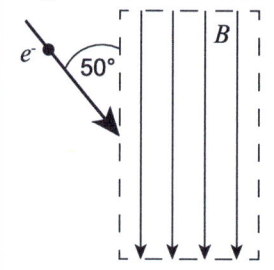

The electron travels in a helical path in the magnetic field, which has induction B = 4.5 × 10⁻³ T.

1. Calculate the radius of the helix.
   _____ m
2. Calculate the pitch of the helix.
   _____ m

....

Q15: Which of the following statements are correct?

A) As a comet approaches a star it slows down due the gravitational field of the star.
B) As a comet approaches a star the comet's tail points away from the star due the solar wind.
C) As a comet approaches a star the comet's tail points towards the star due to the solar wind.
D) As a comet approaches a star the comet enters the heliosphere of the star.
E) As a comet approaches a star the comet's tail increases in size.

....

Q16: What is the name given to the fourth state of matter?

a) bubbles
b) gas
c) liquid
d) plasma
e) solid

Unit 2 Topic 4

Simple harmonic motion

Contents

4.1 Introduction 256
4.2 Defining SHM 256
4.3 Equations of motion in SHM 259
4.4 Energy in SHM 264
4.5 Applications and examples 267
    4.5.1 Mass on a spring - vertical oscillations 267
    4.5.2 Simple pendulum 268
    4.5.3 Loudspeaker cones and eardrums 271
4.6 Damping 272
4.7 Extended information 274
4.8 Summary 274
4.9 End of topic 4 test 275

Prerequisites

- Newton's laws of motion.
- Radian measurement of angles and angular velocity.
- Differentiation of trig functions.

Learning objective

By the end of this topic you should be able to:

- understand what is meant by simple harmonic motion (SHM);
- apply the equations of motion in different SHM systems;
- understand how energy is conserved in an oscillating system;
- explain what is meant by damping of an oscillating system.

4.1 Introduction

In most of the rotational motion and astrophysics topics, we were studying motion where the acceleration was constant, whether it be linear motion or circular motion. In this topic we are looking at a specific situation in which the acceleration is not constant. We will be studying simple harmonic motion (SHM), in which the acceleration of an object depends on its displacement from a fixed point.

The topic begins by defining what is meant by simple harmonic motion, followed by derivations of the equations of motion for SHM. We then consider the kinetic energy and potential energy of an SHM system, as well as examples of different SHM systems. We will return to the subject of energy in the final section of the Topic, when we consider what happens to a 'damped' system.

4.2 Defining SHM

We will begin this topic by looking at the horizontal motion of a mass attached to a spring.

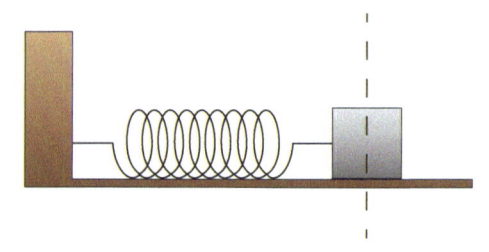

Figure 4.1: Mass attached to a horizontal spring

The mass rests on a frictionless surface, and the spring is assumed to have negligible mass. The spring obeys Hooke's law, so the force F required to produce an extension y of the spring is

$$F = ky$$

(4.1)

k is the spring constant, measured in N m^{-1}. This expression generally holds true for a spring so long as the extension is not large enough to cause a permanent deformation of the spring. If we pull the mass back a distance y and hold it (Figure 4.2), Newton's first law tells us the tension T in the spring (the restoring force) must also have magnitude F. In fact, $T = -F$, since force is a vector quantity, and T and F are in opposite directions. Hence,

$$T = -ky$$

(4.2)

TOPIC 4. SIMPLE HARMONIC MOTION

Figure 4.2: Mass pulled back and held

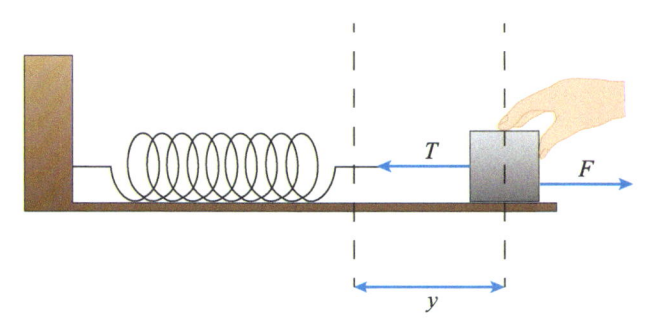

If we now release the mass (Figure 4.3), it is no longer in equilibrium, and we can apply Newton's second law $F = ma$. In this case the force acting on the block is T, so

$$T = ma$$
$$\therefore -ky = m\frac{d^2y}{dt^2}$$
$$\therefore \frac{d^2y}{dt^2} = -\frac{k}{m}y$$

(4.3)

Figure 4.3: Mass accelerated by the tension in the extended spring

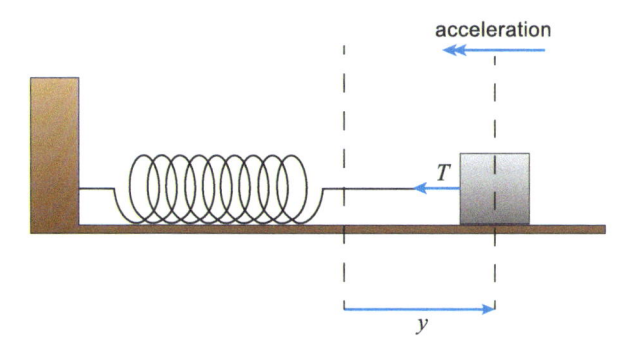

Equation 4.3 tells us that the acceleration of the mass is proportional to its displacement, since m and k are constants. The minus sign means that the acceleration and displacement vectors are in the opposite direction. This equation holds true whatever the extension or compression of the spring, since the spring also obeys Hooke's law when it is compressed (Figure 4.4).

© HERIOT-WATT UNIVERSITY

258 UNIT 2. QUANTA AND WAVES

Figure 4.4: Mass accelerated by the force in the compressed spring

The mass will oscillate around the equilibrium ($T = 0$) position, with an acceleration always proportional to its displacement, and always in the opposite direction to its displacement vector. Such motion is called **simple harmonic motion (SHM)**.

A plot of displacement against time for the mass on a spring shows symmetric oscillations about $y = 0$. The maximum value of the displacement of an object performing SHM is called the **amplitude** of the oscillations.

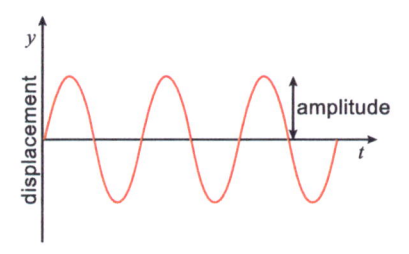

Figure 4.5: Displacement against time for an object undergoing SHM

Equation 4.3 describes a specific example of SHM, that of a mass m attached to spring of spring constant k. In general, for an object performing SHM

$$\frac{d^2y}{dt^2} \propto -y \text{ or } \frac{d^2y}{dt^2} = -\omega^2 y$$

(4.4)

The constant of proportionality is ω^2, where ω is called the angular frequency of the motion. We are used to using ω to represent angular velocity in circular motion, and it is no coincidence that the same symbol is being used again. We will see now how simple harmonic motion and circular motion are related.

TOPIC 4. SIMPLE HARMONIC MOTION

Example

A 1.0 kg mass attached to a horizontal spring (k = 5.0 N m^{-1}) is performing SHM on a horizontal, frictionless surface. What is the acceleration of the mass when its displacement is 4.0 cm?

Using Equation 4.3, the acceleration is:

$$a = -\frac{k}{m}y$$
$$\therefore a = -\frac{5.0}{1.0} \times 0.040$$
$$\therefore a = -0.20 \text{ m s}^{-2}$$

The magnitude of the acceleration is 0.20 m s^{-2}. Normally we do not need to include the minus sign unless we are concerned with the direction of the acceleration.

4.3 Equations of motion in SHM

To try to help us understand SHM, we can compare the motion of an oscillating object with that of an object moving in a circle. Let us consider an object moving anti-clockwise in a circle of radius a at a constant angular velocity ω with the origin of an x-y axis at the centre of the circle. We will be concentrating on the y-component of the motion of this object. Assuming the object is at position y = 0 at time t = 0, then after time t the angular displacement of the object will be $\theta = \omega t$.

Figure 4.6: Object moving in a circle

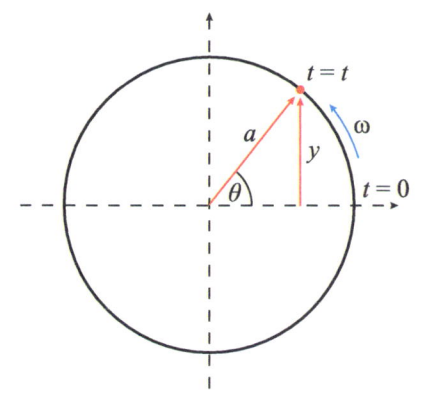

From Figure 4.6, we can see that the y displacement at time t is equal to:

$$a \sin \theta = a \sin \omega t$$

Hence

$$y = a \sin \omega t$$

Differentiating this equation, the velocity in the *y*-direction is:

$$v = \frac{dy}{dt} = \omega a \cos \omega t$$

We can differentiate again to get the acceleration in the *y*-direction.

$$\frac{d^2 y}{dt^2} = \frac{dv}{dt} = -\omega^2 a \sin \omega t$$
$$\therefore \frac{d^2 y}{dt^2} = -\omega^2 y$$

(4.5)

Equation 4.5 gives us the relationship between acceleration and displacement, and is exactly the same as Equation 4.4 for an object undergoing SHM. (Remember it is the projection of the object on the *y*-axis that is undergoing SHM, rather than the object itself.) The angular velocity of the circular motion is therefore equivalent to the angular frequency of the SHM. As with circular motion, we can talk about the period of the SHM as the time taken to complete one oscillation. The relationship between the periodic time T and ω is:

$$T = \frac{2\pi}{\omega}$$

(4.6)

TOPIC 4. SIMPLE HARMONIC MOTION

The **frequency** f of the SHM is the number of complete oscillations performed per second, so

$$f = \frac{1}{T} = \frac{\omega}{2\pi}$$
$$\therefore \omega = 2\pi f$$

(4.7)

Frequency is measured in hertz Hz, equivalent to s^{-1}. Another solution to Equation 4.4 is:

$$y = a \cos \omega t$$

The proof that $y = a \cos \omega t$ represents SHM is given in the following example.

Example

Prove that the equation $y = a \cos \omega t$ describes simple harmonic motion.

For an object to be performing SHM, it must obey the equation $d^2y/dt^2 = -\omega^2 y$. We can differentiate our original equation twice:

$$y = a \cos \omega t$$
$$\therefore \frac{dy}{dt} = -\omega a \sin \omega t$$
$$\therefore \frac{d^2y}{dt^2} = -\omega^2 a \cos \omega t$$
$$\therefore \frac{d^2y}{dt^2} = -\omega^2 y$$

So $y = a \cos \omega t$ does represent SHM. In this case, the displacement is at a maximum when $t = 0$, whereas the displacement is 0 at $t = 0$ when $y = a \sin \omega t$.

There is one more equation we need to derive. We have expressed acceleration as a function of displacement (Equation 4.4 and Equation 4.5). We can also find an expression showing how the velocity varies with displacement. In deriving Equation 4.5 we used the expressions $y = a \sin \omega t$ and $v = \omega a \cos \omega t$. Rearranging both of these equations gives us

$$\sin \omega t = \frac{y}{a} \text{ and } \cos \omega t = \frac{v}{\omega a}$$

Now, using the trig identity $\sin^2\theta + \cos^2\theta = 1$, we have:

$$\sin^2\theta + \cos^2\theta = 1$$
$$\therefore \sin^2(\omega t) + \cos^2(\omega t) = 1$$
$$\therefore \frac{y^2}{a^2} + \frac{v^2}{\omega^2 a^2} = 1$$
$$\therefore y^2\omega^2 + v^2 = \omega^2 a^2$$
$$\therefore v^2 = \omega^2 a^2 - y^2 \omega^2$$
$$\therefore v^2 = \omega^2 (a^2 - y^2)$$
$$\therefore v = \omega\sqrt{(a^2 - y^2)}$$

This equation tells us that for any displacement other than $y = a$, there are two possible values for v, since the square root can give positive or negative values, meaning that the velocity could be in two opposite directions. In fact, to emphasise this point the equation is usually written as,

$$v = \pm\omega\sqrt{a^2 - y^2}$$

(4.8)

Example

An object is performing SHM with amplitude 5.0 cm and frequency 2.0 Hz. What is the maximum value of the velocity of the object?

From Equation 4.8 and the on-screen animation, you should have realised that the velocity has its maximum value when the displacement is zero. So, remembering that $\omega = 2\pi f$.

$$v = \pm\omega\sqrt{a^2 - y^2}$$
$$\therefore v_{max} = 2\pi f \times \sqrt{a^2 - 0}$$
$$\therefore v_{max} = 2\pi f \times a$$
$$\therefore v_{max} = 2\pi \times 2.0 \times 0.050$$
$$\therefore v_{max} = 0.63 \text{ m s}^{-1}$$

Displacement, velocity and acceleration in SHM   Go online

At this stage there is an online activity which allows you to make a thorough investigation of the mass-on-a-spring system.

TOPIC 4. SIMPLE HARMONIC MOTION

Quiz: Defining SHM and equations of motion

Q1: For an object to be performing simple harmonic motion, the net force acting on it must be proportional to its

a) frequency.
b) displacement.
c) velocity.
d) mass.
e) amplitude.

....

Q2: An object is performing SHM with ω = 8.00 rad s^{-1}. What is the frequency of the oscillations?

a) 1.27 Hz
b) 2.55 Hz
c) 25.1 Hz
d) 50.3 Hz
e) 120 Hz

....

Q3: An object is performing SHM oscillations. Which one of these properties is continuously changing as the object oscillates?

a) Period
b) Mass
c) Amplitude
d) Acceleration
e) Frequency

....

Q4: A mass on a spring oscillates with frequency 5.00 Hz and amplitude 0.100 m. What is the maximum acceleration of the mass?

a) 0.063 ms^{-2}
b) 2.50 ms^{-2}
c) 10.0 ms^{-2}
d) 24.7 ms^{-2}
e) 98.7 ms^{-2}

....

Q5: An object is performing SHM with ω = 4.2 rad s^{-1} and amplitude 0.25 m. What is the speed of the object when its displacement is 0.15 m from its equilibrium position?

a) 0.17 ms^{-1}
b) 0.42 ms^{-1}
c) 0.84 ms^{-1}
d) 1.1 ms^{-1}
e) 3.5 ms^{-1}

© HERIOT-WATT UNIVERSITY

4.4 Energy in SHM

Let us return to the first example of SHM we looked at, a mass attached to a spring, oscillating horizontally on a smooth surface. We will now consider the energy in this system. The principle of conservation of energy means that the total energy of the system must remain constant.

As the displacement y of the mass increases, the velocity decreases, and so the kinetic energy of the mass decreases. To keep the total energy of the system constant, this 'lost' energy is stored as potential energy in the spring. The more the spring is stretched or compressed, the greater the potential energy stored in it. We can derive equations to show how the kinetic and potential energies of the system vary with the displacement of the mass.

We already know that the velocity of an object performing SHM is $v = \pm\omega\sqrt{a^2 - y^2}$. We also know that the kinetic energy of a object of mass m is $KE = \frac{1}{2}mv^2$. So

$$v^2 = \omega^2\left(a^2 - y^2\right)$$
$$\therefore KE = \frac{1}{2}m\omega^2\left(a^2 - y^2\right)$$

(4.9)

To obtain the potential energy, consider the work done against the spring's restoring force in moving the mass to a displacement y. We cannot use a simple *work done = force×distance* calculation since the force changes with displacement. Instead we must use a calculus approach, calculating the work done in increasing the displacement by a small amount dy and then performing an integration over the limits of y.

The work done in moving a mass to a displacement y is equal to:

$$\int_{y=0}^{y} F dy$$

This work done is stored as potential energy in the system, so

$$PE = \int_{y=0}^{y} F dy$$

Now, the general equation for SHM is $d^2y/dt^2 = -\omega^2 y$, and Newton's second law of motion tells us that *force = mass × acceleration*, so

$$F = m\frac{d^2y}{dt^2} = -m\omega^2 y$$

TOPIC 4. SIMPLE HARMONIC MOTION

Substituting for F in the potential energy equation

$$PE = \int_{y=0}^{y} F \mathrm{d}y$$
$$\therefore PE = \int_{y=0}^{y} m\omega^2 y \mathrm{d}y$$

Note that the minus sign has been dropped. This is because work is done **on** the spring in expanding it from $y = 0$ to $y = y$. The potential energy stored in the spring is positive, since the PE is at its minimum value (zero) when $y = 0$

$$PE = \int_{y=0}^{y} m\omega^2 y \mathrm{d}y$$
$$\therefore PE = m\omega^2 \int_{y=0}^{y} y \mathrm{d}y$$
$$\therefore PE = \frac{1}{2} m\omega^2 y^2$$

(4.10)

Energy in simple harmonic motion

Consider an SHM system of a 0.25 kg mass oscillating with $\omega = \pi$ rad s⁻¹ and amplitude $a = 1.2$ m. Sketch graphs showing:

1. PE, KE and total energy against displacement.

2. PE, KE and total energy against time.

In both cases sketch the energy over three cycles, assuming the displacement $y = 0$ at $t = 0$.

In summary then, the kinetic energy is at a maximum when the speed of an oscillating object is at a maximum, which is when the displacement y equals zero. Note that at $y = 0$, the force acting on the object is also zero, and hence the potential energy is zero. At the other extreme of the motion, when $y = a$, the velocity is momentarily zero, and so the kinetic energy is zero. The restoring force acting on the object is at a maximum at this point, and so the potential energy of the system is at a maximum. At all other points in the motion of the object, the total energy of the system is partly kinetic and partly potential, with the sum of the two being constant.

© HERIOT-WATT UNIVERSITY

Energy of a mass on a spring   Go online

At this stage there is an online activity which allows you to see the energy-time graph plotted out as the mass oscillates.

Quiz: Energy in SHM   Go online

Q6: A 2.0 kg block is performing SHM with angular velocity 1.6 rad s^{-1}. What is the potential energy of the system when the displacement of the block is 0.12 m?

a) 0.023 J
b) 0.037 J
c) 0.074 J
d) 0.19 J
e) 0.38 J

....

Q7: When the kinetic energy of an SHM system is 25 J, the potential energy of the system is 60 J. What is the potential energy of the system when the kinetic energy is 40 J?

a) 15 J
b) 25 J
c) 37.5 J
d) 45 J
e) 60 J

....

Q8: A mass attached to a spring is oscillating horizontally on a smooth surface. At what point in its motion does the kinetic energy have a maximum value?

a) When its displacement from the rest position is at its maximum.
b) When the mass is midway between its rest position and its maximum displacement.
c) When its displacement from the rest position is zero.
d) The kinetic energy is constant at all points in its motion.
e) Cannot tell without knowing the amplitude and frequency.

....

Q9: A mass-spring system is oscillating with amplitude a. What is the displacement at the point where the kinetic energy of the system is equal to its potential energy?

a) $\pm a/4$
b) $\pm a/2$
c) $\pm a/\sqrt{2}$
d) $\pm a\sqrt{2}$
e) $\pm 2a$

....

TOPIC 4. SIMPLE HARMONIC MOTION

Q10: The total energy of a system oscillating with SHM is 100 J. What is the kinetic energy of the system at the point where the displacement is half the amplitude?

a) 8.66 J
b) 10 J
c) 50 J
d) 75 J
e) 100 J

4.5 Applications and examples

In this section we will cover following applications:

- Mass on a spring - vertical oscillations
- Simple pendulum
- Loudspeaker cones and eardrums.

4.5.1 Mass on a spring - vertical oscillations

We have been using a mass on a spring to illustrate different aspects of SHM. In Section 3.2 we analysed the horizontal mass-spring system. We will look now at the vertical mass-spring system. This is the system we used in the on-screen activities earlier in this Topic. Here we will analyse the forces acting on the mass to show that its motion is indeed SHM.

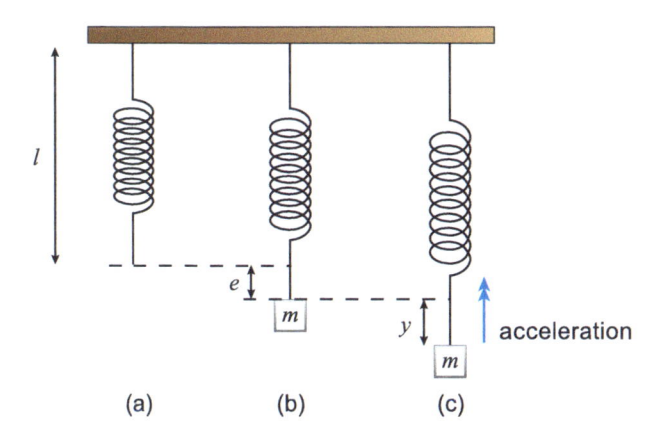

Figure 4.7: Mass hanging from a spring

The spring has a natural length l, spring constant k and negligible mass Figure 4.7(a). When a mass m is attached to the spring Figure 4.7(b), it causes an extension e, and the system is at rest

© HERIOT-WATT UNIVERSITY

in equilibrium. Hence the tension T in the spring ($k \times e$) acting upwards on the mass must be equal in magnitude and opposite in direction to the weight mg. If the mass is now pulled down a distance y and released Figure 4.7(c), the tension in the spring is $k(y+e)$. The resultant force acting upwards is equal to $T - W$. Hence the resultant force is

$$F = k(y+e) - mg$$
$$\therefore F = ky + ke - mg$$

But $ke = mg$, so the resultant force $F = ky$. Using Newton's second law, this resultant force must be equal to (*mass* × *acceleration*), so

$$ky = -m\frac{d^2y}{dt^2}$$
$$\therefore \frac{d^2y}{dt^2} = -\frac{k}{m}y$$

(4.11)

The minus sign appears because the displacement and acceleration vectors are in opposite directions. Equation 4.11 has the form $d^2y/dt^2 = -\omega^2 y$ showing that the motion is SHM, with

$$\omega = \sqrt{\frac{k}{m}}$$

4.5.2 Simple pendulum

Another SHM system is the simple pendulum. A simple pendulum consists of a bob, mass m, attached to a light string of length l. The bob is pulled to one side through a small angle θ and released. The subsequent motion is SHM.

TOPIC 4. SIMPLE HARMONIC MOTION

Figure 4.8: Simple pendulum

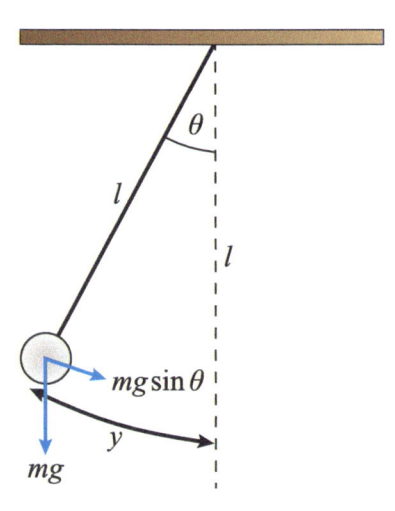

The restoring force on the bob acts perpendicular to the string, so no component of the tension in the string contributes to the restoring force. The restoring force is $mg \sin \theta$, the component of the weight acting perpendicular to the string. Now for small θ, when θ is measured in radians, $\sin \theta \approx \theta$. Also, since we are measuring θ in radians, $\theta = y/l$. Hence the restoring force acting on the bob is $mg \sin \theta = mg \times y/l$.

Now, using Newton's second law

$$mg\frac{y}{l} = -m\frac{d^2y}{dt^2}$$
$$\therefore \frac{d^2y}{dt^2} = -\frac{g}{l}y$$

(4.12)

Again, the minus sign appears because the displacement and acceleration vectors are in the opposite directions. So we have SHM with $\omega = \sqrt{g/l}$. Note that the angular frequency, and hence the period, are both **independent of the mass of the bob**. Note also that we have assumed small θ and hence small displacement y, in deriving Equation 4.12. The simple pendulum only approximates to SHM because of the approximation $\sin \theta \approx \theta$.

© HERIOT-WATT UNIVERSITY

Example

Two simple pendula are set in motion. Pendulum A has string length p and a bob of mass q. Pendulum B has string length $4p$ and the mass of its bob is $5q$. What is the ratio T_A/T_B of the periodic times of the two pendula?

A simple pendulum has $\omega = \sqrt{g/l}$, so the periodic time

$$T = \frac{2\pi}{\omega} = 2\pi\sqrt{\frac{l}{g}}$$

The ratio of periodic times is therefore

$$\frac{T_A}{T_B} = \frac{2\pi\sqrt{\frac{l_A}{g}}}{2\pi\sqrt{\frac{l_A}{g}}}$$

$$\therefore \frac{T_A}{T_B} = \frac{\sqrt{l_A}}{\sqrt{l_B}}$$

$$\therefore \frac{T_A}{T_B} = \frac{\sqrt{p}}{\sqrt{4p}}$$

$$\therefore \frac{T_A}{T_B} = \frac{1}{2}$$

So the ratio of periodic times depends on the square root of the ratio of string lengths. Remember, the mass of the bob does not affect the periodic time of a pendulum.

Simple pendulum   Go online

At this stage there is an online activity which explores the oscillations of a simple pendulum and demonstrates why this is SHM.

Understanding, significance and treatment of uncertainties   Go online

At this stage there is an online activity which will help you understand how to analyse and interpret uncertainties.

4.5.3 Loudspeaker cones and eardrums

A practical situation in which SHM occurs is in the generation and detection of sound waves. Sound waves are longitudinal waves, in which air molecules are made to oscillate back and forth due to high and low pressure regions being created. In a loudspeaker, a cardboard cone is rigidly attached to an electric coil which sits in a magnetic field. A fluctuating electric current in the coil causes the coil and the cone to vibrate back and forth. A pure note causes SHM vibrations of the cone (other notes cause more complicated vibrations). The oscillations of the cone cause the air molecules to oscillate, creating the high and low pressure regions that cause the wave to travel.

The opposite process occurs in the ear, where the oscillations of air molecules cause the tympanic membrane (commonly called the eardrum) to oscillate. These oscillations are converted to an electrical signal in the inner ear, and are transmitted to the brain via the auditory nerve.

Quiz: SHM Systems Go online

Q11: An SHM oscillator consists of a 0.500 kg block suspended by a spring ($k = 2.00$ N m^{-1}), oscillating with amplitude 8.00×10^{-2} m. What is the period T of this system?

a) 0.318 s
b) 1.25 s
c) 2.00 s
d) 3.14 s
e) 12.6 s

....

Q12: What is the frequency f of a simple pendulum consisting of a 0.250 kg mass attached to a 0.300 m string?

a) 0.0278 Hz
b) 0.910 Hz
c) 0.997 Hz
d) 5.20 Hz
e) 35.9 Hz

....

Q13: A 0.19 kg mass is oscillating vertically, attached to a spring. The period of the oscillations is 0.45 s. What is the spring constant of the spring?

a) 2.7 N m^{-1}
b) 5.9 N m^{-1}
c) 17 N m^{-1}
d) 28 N m^{-1}
e) 37 N m^{-1}

....

Q14: A simple pendulum is oscillating with period 0.75 s. If the pendulum bob, mass m, is replaced with a bob of mass $2m$, what is the new period of the pendulum?

a) 0.375 s
b) 0.75 s
c) 1.06 s
d) 1.5 s
e) 3.0 s

....

Q15: A simple pendulum has frequency 5 Hz. How would you increase its frequency to 10 Hz?

a) Decrease its length by a factor of 4.
b) Decrease its length by a factor of 2.
c) Increase its length by a factor of 2.
d) Increase its length by a factor of 4.
e) Increase its length by a factor of 10.

4.6 Damping

In the examples we have looked at so far, we have ignored any external forces acting on the system, so that the oscillator continues to oscillate with the same amplitude and no energy is lost from the system. In reality, of course, this does not happen. For our horizontal mass-spring system, for example, friction between the mass and the horizontal surface would mean that some energy would be 'lost' from the system with every oscillation.

At the maximum displacement, all of the system's energy is potential energy, given by Equation 4.10

$$PE = \tfrac{1}{2} m \omega^2 y^2$$

If the energy of the system is decreasing, the amplitude must also be decreasing to satisfy the above equation. The **damping** of the system describes the rate at which energy is being lost, or the rate at which the amplitude is decreasing.

A system in which the damping effects are small is described as having light damping. A plot of displacement against time for such an oscillator is shown in Figure 4.9

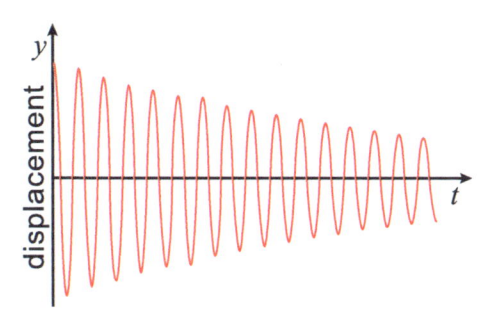

Figure 4.9: Lightly-damped system

In a heavily damped system, the damping is so great that no complete oscillations are seen, and the 'oscillating' object does not travel past the equilibrium point.

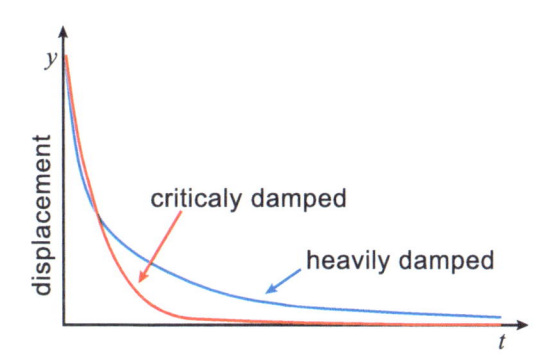

Figure 4.10: Heavily-damped and critically-damped systems

A critically damped system is one with an amount of damping that ensures the oscillator comes to rest in the minimum possible time. Another way of stating this is that critical damping is the minimum amount of damping that completely eliminates the oscillations.

An important application of damping is in the suspension system of a car. The shock absorber attached to the suspension spring consists of a piston in a reservoir of oil which moves when the car goes over a bump in the road. Underdamping (light damping) would mean that the car would continue to wobble up and down as it continued along the street. Overdamping (heavy damping) would make the suspension system ineffective, and produce a bumpy ride. Critical damping provides the smoothest journey, absorbing the bumps without making the car oscillate.

> **How shock absorbers work**
>
> There is a link to the useful animation available online, where you can decide for yourself which type of damping is used in the shock absorber.

4.7 Extended information

> **Web links**   Go online
>
> These web links should serve as an insight to the wealth of information available online and allow you to explore the subject further.
>
> - http://hyperphysics.phy-astr.gsu.edu/hbase/shm.html
> - http://www.animations.physics.unsw.edu.au/mechanics/chapter4_simpleharmonicmotion.html
> - https://www.physics.uoguelph.ca/tutorials/shm/Q.shm.html
> - http://www.nuffieldfoundation.org/practical-physics/simple-harmonic-motion

4.8 Summary

A system is oscillating with simple harmonic motion if its acceleration is proportional to its displacement, and is always directed towards the centre of the motion. We have analysed several SHM systems, and looked at some applications of SHM. Conservation of energy has been discussed, and the effects of damping have been demonstrated.

> **Summary**
>
> You should now be able to:
>
> - define what is meant by simple harmonic motion, and be able to describe some examples of SHM;
> - state the equation $\frac{d^2y}{dt^2} = -\omega^2 y$ and explain what each term in this equation means;
> - show that $y = a\cos\omega t$ and $y = a\sin\omega t$ are solutions of the above equation, and show that $v = \pm\omega\sqrt{a^2 - y^2}$ in both of these cases;
> - derive expressions for the potential and kinetic energies of an SHM system;
> - explain what is meant by damping of an oscillating system.

TOPIC 4. SIMPLE HARMONIC MOTION

4.9 End of topic 4 test

End of topic 4 test — Go online

The following data should be used when required:

| Acceleration due to gravity g | $9.8 \, m \, s^{-2}$ |

The end of topic test is available online. If however you do not have access to the web, you may try the following questions.

Q16: A 1.25 kg mass is attached to a light spring ($k = 40 \, N \, m^{-1}$) and rests on a smooth horizontal surface. The mass is pulled back a distance of 3.2 cm and released.

Calculate the size of the acceleration of the mass at the instant it is released.

_____ $m \, s^{-2}$

....

Q17: A 0.46 kg mass attached to a spring ($k = 26 \, N \, m^{-1}$) is performing SHM on a smooth horizontal surface.

Calculate the periodic time of these oscillations.

_____ s

....

Q18: An SHM system is oscillating with angular frequency $\omega = 3.6 \, rad \, s^{-1}$ and amplitude a = 0.24 m.

1. Calculate the maximum value of the acceleration of the oscillator.

   _____ $m \, s^{-2}$

2. Calculate the maximum value of the velocity of the oscillator.

   _____ $m \, s^{-1}$

....

Q19: An SHM system consists of a 1.04 kg mass oscillating with amplitude 0.257 m.

If the angular frequency of the oscillations is 8.00 rad s^{-1}, calculate the total energy of the system.

_____ J

....

Q20: A body of mass 0.85 kg is performing SHM with amplitude 0.35 m and angular frequency 5.5 rad s^{-1}.

Calculate the displacement of the body when the potential energy of the system is equal to its kinetic energy.

_____ m

....

© HERIOT-WATT UNIVERSITY

Q21: A simple pendulum has a bob of mass 0.55 kg and a string of length 0.47 m.

Calculate the frequency of the SHM oscillations of this pendulum.

_____ Hz

....

Q22: A simple pendulum, consisting of a 0.15 kg bob on a 0.92 m string, is oscillating with amplitude 44 mm.

Calculate the maximum kinetic energy of the pendulum.

_____ J

....

Q23: An SHM system consists of an oscillating 1.5 kg mass. The system is set into motion, with initial amplitude 0.37 m. Damping of the system means that the mass continues to oscillate with an angular frequency of 4.0 rad s^{-1} but 10% of the system's energy is lost in work against friction with every oscillation.

Calculate the amplitude of the second oscillation after the system is set in motion.

_____ m

Unit 2 Topic 5

Waves

Contents

5.1 Introduction	278
5.2 Definitions	278
5.3 Travelling waves	282
5.4 Superposition of waves and phase difference	286
5.4.1 Principle of superposition	286
5.4.2 Fourier series	292
5.5 Stationary waves	294
5.6 Beats	298
5.7 Extended information	301
5.8 Summary	301
5.9 End of topic 5 test	302

Prerequisites

- Radian measurement of angles.
- Simple harmonic motion.

Learning objective

By the end of this topic you should be able to:

- use all the terms commonly employed to describe waves;
- derive an equation describing travelling sine waves, and solve problems using this equation;
- show an understanding of the difference between travelling and stationary waves;
- calculate the harmonics of a number of stationary wave systems;
- apply the principle of superposition;
- be able to explain how beats can be used to tune musical instruments;
- to understand the term phase difference and use the phase angle equation.

5.1 Introduction

A wave is a travelling disturbance that carries energy from one place to another, but with no net displacement of the medium. You should already be familiar with transverse waves, such as light waves, where the oscillations are perpendicular to the direction in which the waves are travelling; and longitudinal waves, such as sound waves, where the oscillations of the medium are parallel to the direction in which the waves are travelling.

We begin this topic with a review of the basic definitions and terminology used to describe waves. We will discuss what terms such as 'frequency' and 'amplitude' mean in the context of light and sound waves. Section 3.3 deals with travelling waves, and we will derive and use mathematical expressions to describe these sorts of waves. Section 3.4 looks at the principle of superposition, which tells us what happens when two or more waves overlap at a point in space, and how they can be used in synthesisers. Section 3.5 looks at Stationary Waves and again we will derive and use mathematical expressions to describe these sorts of waves and look at their application in music. Section 3.6 looks at an interesting phenomenon in music called Beats which are used to tune musical instruments.

5.2 Definitions

The easiest way to explain wave phenomena is to visualise a train of waves travelling along a rope. The plot of displacement (y) against distance (x) is then exactly the same as the rope looks while the waves travel along it. We will be concentrating on sine and cosine waves as the most common sorts of waves, and the simplest to describe mathematically (see Figure 5.1).

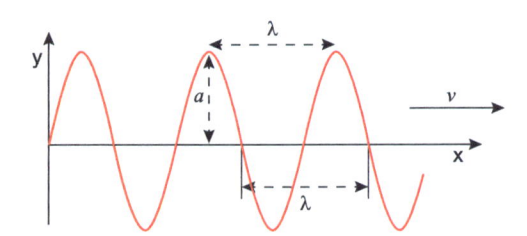

Figure 5.1: Travelling sine wave

As a train of waves passes along the rope, each small portion of the rope is oscillating in the y-direction. There is no movement of each portion of the rope in the x-direction, and when we talk about the **speed** v of the wave, we are talking about the speed at which the disturbance travels in the x-direction.

The **wavelength** λ is the distance between two identical points in the wave cycle, such as the distance between two adjacent crests. The **frequency** f is the number of complete waves passing a point on the x-axis in a given time period. When this time period is one second, f is measured in hertz (Hz), equivalent to s^{-1}. 1 Hz is therefore equivalent to one complete wave per second. The relationship between these three quantities is

TOPIC 5. WAVES

$$v = f\lambda$$

(5.1)

The **periodic time** T (or simply the **period** of the wave) is the time taken to complete one oscillation, in the same way that the periodic time we use to describe circular motion is the time taken to complete one revolution. The period is related to the frequency by the simple equation.

$$T = \frac{1}{f}$$

(5.2)

The **amplitude** a of the wave is the maximum displacement in the y-direction. As the waves pass along the rope, the motion of each portion of the rope follows the simple harmonic motion relationship

$$y = a \sin(2\pi f t)$$

We will use this expression to work out a mathematical relationship to describe wave motion later in this topic.

We normally use the wavelength to describe a light wave, or any member of the electromagnetic spectrum. The visible spectrum extends from around 700 nm (red) to around 400 nm (blue) (1 nm = 10^{-9} m). Longer wavelengths go through the infrared and microwaves to radio waves. Shorter wavelengths lead to ultraviolet, X-rays and gamma radiation (see Figure 5.2). The frequency of visible light is of the order of 10^{14} Hz (or 10^5 GHz, where 1 GHz = 10^9 Hz).

© HERIOT-WATT UNIVERSITY

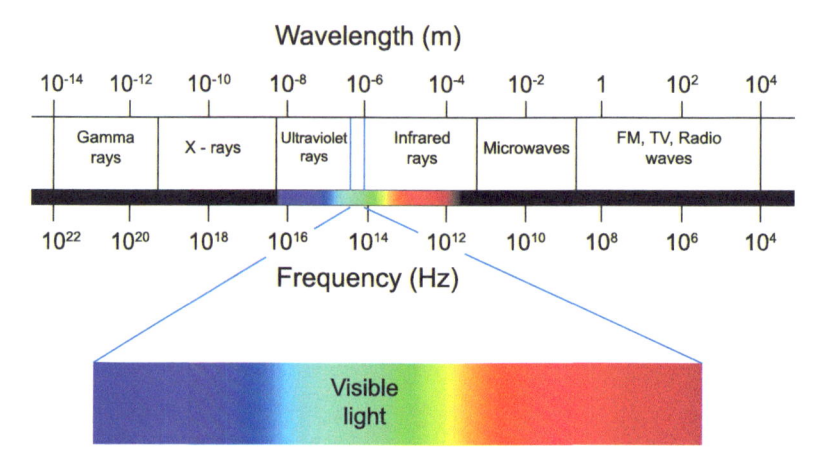

Figure 5.2: The electromagnetic spectrum

Sound waves are usually described by their frequency (or **pitch**). The human ear can detect sounds in the approximate range 20 Hz - 20 000 Hz. Sound waves with frequencies greater than 20 000 Hz are called ultrasonic waves, whilst those with frequencies lower than 20 Hz are infrasonic. The musical note middle C, according to standard concert pitch, has frequency 261 Hz.

The **irradiance** I of a wave tells us the amount of power falling on unit area, and is measured in W m^{-2}. The irradiance is proportional to a^2. In practical terms, this means the brightness of a light wave, or the loudness of a sound wave, depends on the amplitude of the waves, and increases with a^2. This could be seen as the equation $I = \alpha a^2$.

Finally in this Section, let us consider the two waves in Figure 5.3.

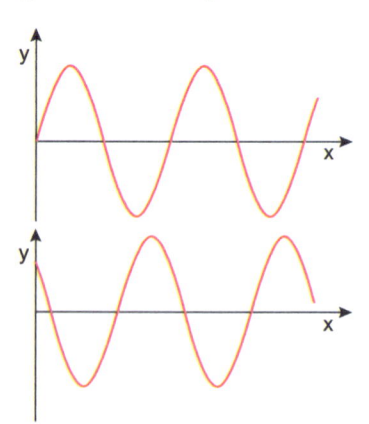

Figure 5.3: Two out-of-phase sine waves

TOPIC 5. WAVES

In terms of amplitude, frequency and wavelength, these waves are identical, yet they are 'out-of-step' with each other. We say they are **out of phase** with each other. A wave is an oscillation of a medium and the **phase** of the wave tells us how far through an oscillation a point in the medium is. Coherent waves **coherent waves** have the same speed and frequency (and similar amplitudes) and so there is a **constant phase relationship between two coherent waves.**

Quiz: Properties of waves   Go online

Useful data:

Speed of light c	3.00×10^8 m s^{-1}

Q1: Which one of the following quantities should be increased to increase the volume (loudness) of a sound wave?

a) frequency
b) wavelength
c) speed
d) phase
e) amplitude

....

Q2: The visible spectrum has the approximate wavelength range 400 - 700 nm. What is the approximate frequency range of the visible spectrum?

a) 2.1×10^{14} - 1.2×10^{15} Hz
b) 4.3×10^{14} - 7.5×10^{14} Hz
c) 1.3×10^{15} - 2.3×10^{15} Hz
d) 4.3×10^{20} - 7.5×10^{20} Hz
e) 1.3×10^{20} - 2.3×10^{20} Hz

....

Q3: Which of these sets of electromagnetic waves are listed in order of **increasing** wavelength?

a) X-rays, infrared, microwaves.
b) radio waves, gamma rays, visible.
c) infrared, visible, ultraviolet.
d) X-rays, infrared, ultraviolet.
e) microwaves, visible, infrared.

....

Q4: What is the frequency of a beam of red light from a helium-neon laser, which has wavelength 633 nm?

a) 190 Hz
b) 2.11×10^5 GHz
c) 4.74×10^5 GHz
d) 2.11×10^8 GHz
e) 4.74×10^8 GHz

....

Q5: A laser produces light waves of average amplitude a m. The irradiance of the beam is 20 W m^{-2}. What is the irradiance if the average amplitude is increased to $3a$ m?

a) 6.7 W m^{-2}
b) 23 W m^{-2}
c) 60 W m^{-2}
d) 180 W m^{-2}
e) 8000 W m^{-2}

5.3 Travelling waves

In this section, we will attempt to find a mathematical expression for a **travelling wave**. The example we shall consider is that of a train of waves being sent along a rope in the x-direction, but the expression we will end up with applies to all transverse travelling waves.

We will start by considering what happens to a small portion of the rope as the waves travel through it. From the definition of a wave we know that although the waves are travelling in the x-direction, there is no net displacement of each portion of the rope in that direction. Instead each portion is performing simple harmonic motion (SHM) in the y-direction, about the $y = 0$ position, and the SHM of each portion is slightly out of step (or phase) with its neighbours.

The displacement of one portion of the rope is given by the SHM equation

$$y = a \sin(2\pi f t)$$

where y is the displacement of a particle at time t, a is the amplitude and f is the frequency of the waves.

The wave disturbance is travelling in the x-direction with speed v. Hence at a distance x from the origin, the disturbance will happen after a time delay of x/v. So the disturbance at a point x after time t is exactly the same as the disturbance at the point $x = 0$ at time $(t - x/v)$.

We can therefore find out exactly what the disturbance is at point x at time t by replacing t by $(t - x/v)$ in the SHM equation

$$y = a \sin 2\pi f \left(t - \frac{x}{v}\right)$$

We can re-arrange this equation, substituting for $v = f\lambda$

$$\begin{aligned} y &= a \sin 2\pi f \left(t - \frac{x}{v}\right) \\ \therefore y &= a \sin 2\pi \left(ft - \frac{fx}{v}\right) \\ \therefore y &= a \sin 2\pi \left(ft - \frac{fx}{f\lambda}\right) \\ \therefore y &= a \sin 2\pi \left(ft - \frac{x}{\lambda}\right) \end{aligned}$$

(5.3)

Note that we are taking the sine of the angle $2\pi \left(ft - {x}/{\lambda}\right)$. This angle is expressed in **radians**. You should also note that this expression assumes that at $t = 0$, the displacement at $x = 0$ is also 0.

For a wave travelling in the $-x$ direction, Equation 5.3 becomes

$$y = a \sin 2\pi \left(ft + \frac{x}{\lambda}\right)$$

(5.4)

We can now calculate the displacement of the rope, or any other medium, at position x and time t if we know the wavelength and frequency of the waves.

> **Travelling waves and the wave equation**　　　　　　　　　　Go online
>
> At this stage there is an online activity which explores how the appearance and speed of a travelling wave depend on the different parameters in the wave equation.
>
> This activity explores how the appearance and speed of a travelling wave depend on the different parameters in the wave equation.

Examples

1. A periodic wave travelling in the x-direction is described by the equation

$$y = 0.2 \sin(4\pi t - 0.1x)$$

What are (a) the amplitude, (b) the frequency, (c) the wavelength, and (d) the speed of the wave? (All quantities are in S.I. units.)

To obtain these quantities, we first need to re-arrange the expression for the wave into a form more similar to the general expression given for a travelling wave. The general expression is

$$y = a \sin 2\pi \left(ft - \frac{x}{\lambda}\right)$$

We are given

$$y = 0.2 \sin(4\pi t - 0.1x)$$

Re-arranging

$$y = 0.2 \sin 2\pi \left(2t - \frac{0.1x}{2\pi}\right)$$

So by comparison, we can see that:

a) the amplitude $a = 0.2$ m
b) the frequency $f = 2$ Hz
c) the wavelength $\lambda = {2\pi}/{0.1} = 63$ m
d) By calculation, the speed of the wave $v = f\lambda = 2 \times 63 = 130$ m s⁻¹

....

2. Consider again the travelling wave in the previous example, described by the equation

$$y = 0.2 \sin(4\pi t - 0.1x)$$

Calculate the displacement of the medium in the y-direction caused by the wave at the point $x = 25$ m when the time $t = 0.30$ s.

To calculate the y-displacement, put the data into the travelling wave equation. Remember to take the sine of the angle measured in *radians*.

$$y = 0.2 \sin(4\pi t - 0.1x)$$
$$\therefore y = 0.2 \sin((4 \times \pi \times 0.30) - (0.1 \times 25))$$
$$\therefore y = 0.2 \sin 1.2699$$
$$\therefore y = 0.2 \times 0.9551$$
$$\therefore y = 0.19 \text{ m}$$

Quiz: Travelling waves

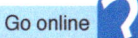

Q6: The equation representing a wave travelling along a rope is

$$y = 0.5 \sin 2\pi \left(0.4t - \frac{x}{12}\right)$$

At time t = 2.50 s, what is the displacement of the rope at the point x = 7.00 m?

a) 0.00 m
b) 0.23 m
c) 0.25 m
d) 0.40 m
e) 0.50 m

....

Q7: Waves are being emitted in the x-direction at frequency 20 Hz, with a wavelength of 1.0 m. If the displacement at x = 0 is 0 when t = 0, which one of the following equations could describe the wave motion?

a) $y = \sin 2\pi (t - 20x)$
b) $y = 2 \sin 2\pi (20t - x)$
c) $y = \sin 2\pi \left(\frac{t}{20} - x\right)$
d) $y = 3 \cos 2\pi (t - 20x)$
e) $y = 20 \sin 2\pi (t - 20x)$

....

Q8: A travelling wave is represented by the equation

$$y = 4 \cos 2\pi (t - 2x)$$

What is the displacement at x = 0 when t = 0?

a) 0 m
b) 1 m
c) 2 m
d) 3 m
e) 4 m

....

Q9: A travelling wave is represented by the equation

$$y = \sin 2\pi (12t - 0.4x)$$

What is the frequency of this wave?

a) 0.4 Hz
b) 0.833 Hz
c) 2.0 Hz
d) 2.5 Hz
e) 12 Hz

....

Q10: A travelling wave is represented by the equation

$$y = 2\sin 2\pi (5t - 4x)$$

What is the speed of this wave?

a) 0.80 m s^{-1}
b) 1.25 m s^{-1}
c) 4.0 m s^{-1}
d) 5.0 m s^{-1}
e) 20 m s^{-1}

5.4 Superposition of waves and phase difference

In this topic we will cover the following:

- The principle of superposition
- The fourier series

5.4.1 Principle of superposition

The **principle of superposition** tells us what happens if two or more waveforms overlap. This might happen when you are listening to stereo speakers, or when two light beams are focused onto a screen. The result at a particular point is simply the sum of all the disturbances at that point.

TOPIC 5. WAVES

Figure 5.4: (a) Constructive interference, (b) destructive interference of two sine waves

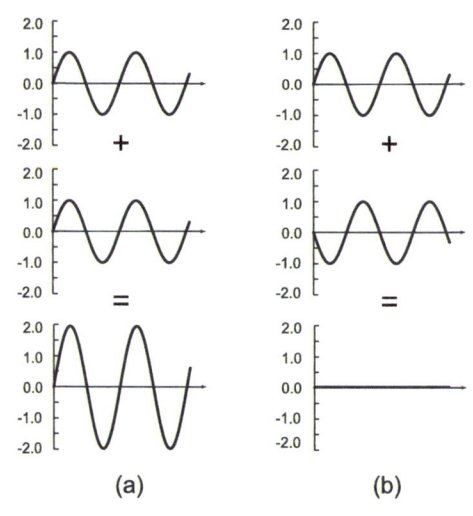

Figure 5.4 shows plots of displacement against **time** at a certain point for two coherent sine waves with the same amplitude. Using the principle of superposition, the lowest graphs show the resultant wave at that point. In both cases, the resultant wave has an amplitude equal to the sum of the amplitudes of the two interfering waves. If the two waves are exactly in phase, as shown in Figure 5.4(a), **constructive interference** occurs, the amplitude of the resultant wave is greater than the amplitude of either individual wave. If they are exactly out-of-phase ('in anti-phase'), the sum of the two disturbances is zero at all times, hence there is no net disturbance. This is called **destructive interference**, as shown in Figure 5.4(b).

The **phase difference** between two waves can be expressed in fractions of a wavelength or as a fraction of a circle, with one whole wavelength being equivalent to a phase difference of 360° or 2π radians. Two waves that are in anti-phase would therefore have a phase difference of $\lambda/2$ or 180° or π radians.

Use the following exercise to investigate the superposition of two coherent waves with phase differences other than 0, π and 2π radians.

The Phase Angle Equation can be used to calculate the phase difference between two points on a single wave or two separate waves.

$$\varphi = 2\pi x/\lambda$$

(5.5)

where φ is the phase difference or phase angle measured in radians and x/λ is the fraction of the wavelength.

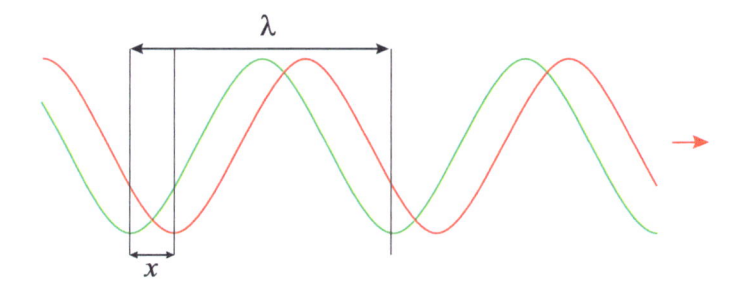

Quiz: Superposition

Q11: Two sine waves are exactly out of phase at a certain point in space, so they undergo destructive interference. If one wave has amplitude 5.0 cm and the other has amplitude 2.0 cm, what is the amplitude of the resultant disturbance?

a) 0 cm
b) 2.0 cm
c) 2.5 cm
d) 3.0 cm
e) 7.0 cm

....

Q12: A listener is standing midway between two loudspeakers, each broadcasting the same signal in phase. Does the listener hear

a) a loud signal, owing to constructive interference?
b) a quiet signal, owing to destructive interference?
c) a quiet signal, owing to constructive interference?
d) a loud signal, owing to destructive interference?
e) no signal at all?

....

Q13: A radio beacon is transmitting a signal (λ = 200 m) to an aeroplane. When the aeroplane is 4.50 km from the beacon what is the separation between the beacon and the aeroplane in numbers of wavelengths?

a) 0.0225 wavelengths
b) 0.044 wavelengths
c) 22.5 wavelengths
d) 44.0 wavelengths
e) 900 wavelengths

....

TOPIC 5. WAVES

Q14: Fourier's theorem tells us that

a) only coherent waves can be added together.
b) any periodic wave is a superposition of harmonic sine and cosine waves.
c) any periodic wave is a superposition of stationary and travelling waves.
d) all sine and cosine waves have the same phase.
e) any periodic wave is made up of a set of sine waves, all with the same amplitude.

....

Q15:

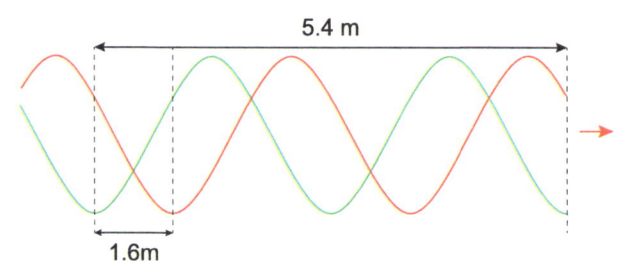

Calculate the phase angle between these two waves.

Superposition of two waves
Go online

Q16: Initially the phase difference is zero. What do you notice about the size of the amplitude of the resulting wave?

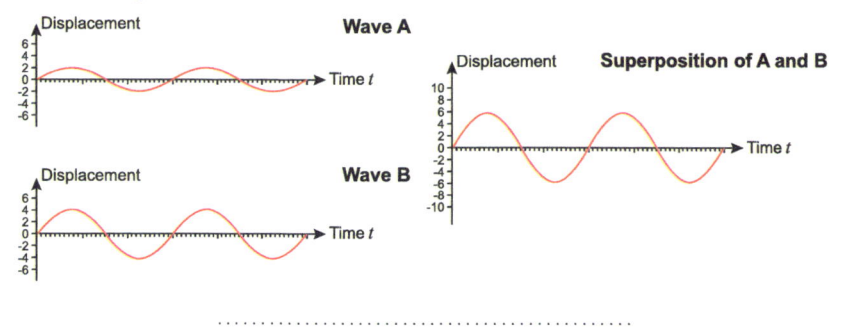

....

© HERIOT-WATT UNIVERSITY

290 UNIT 2. QUANTA AND WAVES

Q17: The amplitudes of A and B are set to 2 units and the phase difference to 1 wavelength. What happens?

[Graphs: Wave A and Wave B shown as identical sine waves with amplitude 2; Superposition of A and B shown as a sine wave with amplitude 4 (scale to ±10).]

....

Q18: The phase difference is set to -0.5 wavelengths and the amplitudes of A and B set to 2 units. What is the amplitude of the resulting wave?

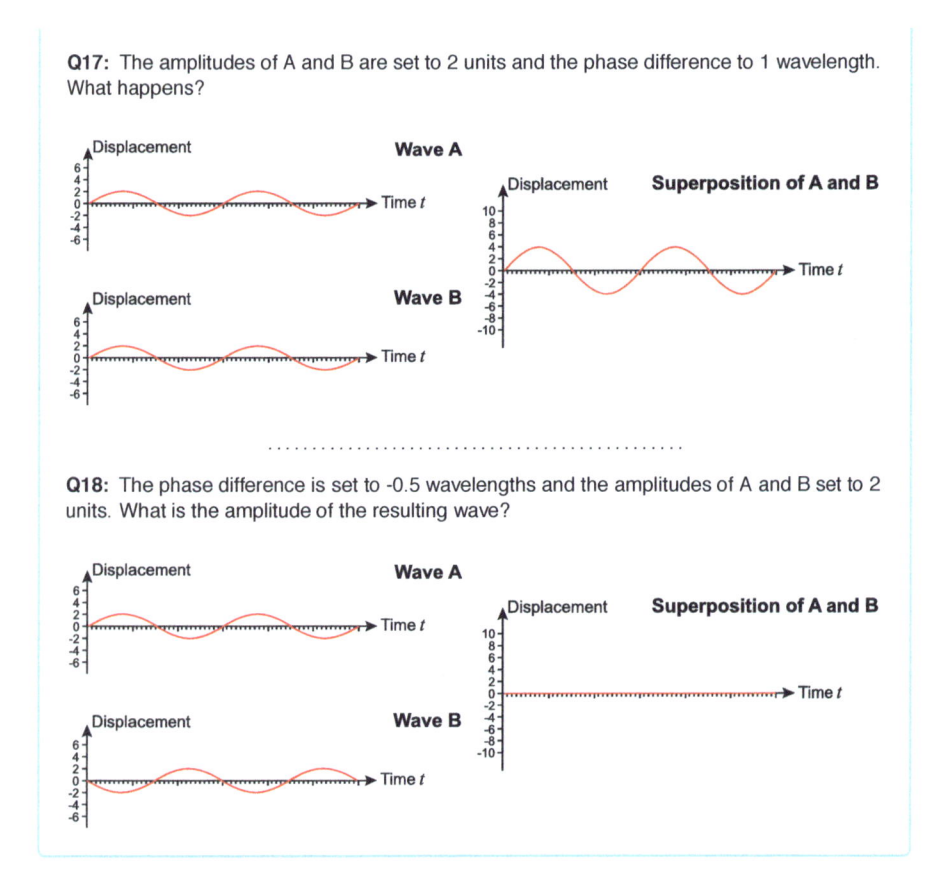

[Graphs: Wave A and Wave B shown as sine waves with amplitude 2; Superposition of A and B shown as a flat line at 0 (scale to ±10).]

As a practical example of interference, let us look at what happens when coherent waves of identical amplitude are emitted in phase by two loudspeakers. It should be clear that at a point equidistant from each speaker, two waves travelling with the same speed will arrive at exactly the same time. Constructive interference will occur, and the amplitude of the resultant wave will be the sum of the amplitudes of the two individual waves.

Figure 5.5 shows the **wavefronts** from two sources S_1 and S_2, producing waves with identical wavelengths. The wavefronts join points of identical phase, such as the crests of a wave, and the distance between adjacent wavefronts from the same source is equal to λ. Where wavefronts from the two sources overlap (shown by the solid black dots), constructive interference occurs.

TOPIC 5. WAVES

Figure 5.5: Interference of waves from two sources

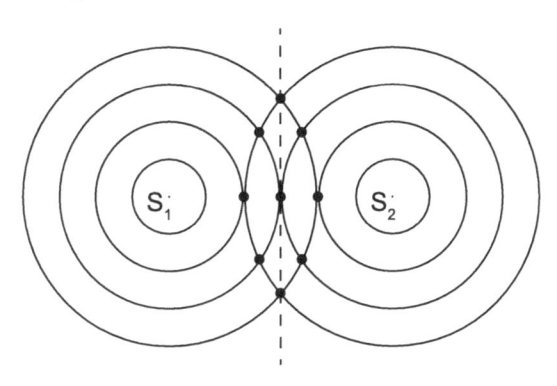

In fact constructive interference will occur at any point where the difference in path length between the waves from each of the two sources is equal to a whole number of wavelengths. At any such point, the arrival of the crest of a wave from the left-hand speaker will coincide with the arrival of a crest from the right-hand speaker, leading to constructive interference.

Put mathematically, the condition for constructive interference of two waves is

$$|l_1 - l_2| = n\lambda$$

(5.6)

where l_1 and l_2 are the distances from source to detector of the two waves, and n is a whole number. (The $||$ around $l_1 - l_2$ means the 'absolute' value, ignoring the minus sign if $l_2 > l_1$.)

If the path difference between the two waves is an odd number of half-wavelengths ($\lambda/2$, $3\lambda/2$, etc.) then destructive interference occurs. The crest of a wave from one speaker will now arrive at the same time as the trough of the wave from the other speaker. If the amplitudes of the two waves are the same, the result of adding them together is zero as in Figure 5.4 (b).

In this case

$$|l_1 - l_2| = \left(n + \frac{1}{2}\right)\lambda$$

(5.7)

© HERIOT-WATT UNIVERSITY

In terms of phase, we can state that constructive interference occurs when the two waves are in phase, and destructive interference occurs when the two waves are in anti-phase.

Check your understanding of constructive and destructive interference using the superposition shown in Figure 5.5. Suppose S_1 and S_2 are emitting coherent sound waves in phase. What would you expect to hear if you walked from S_1 to S_2? (Answer given below the next worked example.)

Example

Two radio transmitters A and B are broadcasting the same signal in phase, at wavelength 750 m. A receiver is at location C, 6.75 km from A and 3.00 km from B. Does the receiver pick up a strong or weak signal?

The distance from A to C is 6750 m, and the wavelength is 750 m, so in wavelengths, the distance A to C is 6750/750 = 9.00 wavelengths.

Similarly, B to C is 3000 m or 3000/750 = 4.00 wavelengths.

So the path difference AC - BC = 5.00 wavelengths, and since this is a whole number of wavelengths, constructive interference will occur and the receiver will pick up a strong signal.

Walking from speaker S_1 to S_2 in Figure 5.5, you would hear the sound rising and falling in loudness, as you moved through regions of constructive and destructive interference.

5.4.2 Fourier series

One very important application of the principle of superposition is in the Fourier analysis of a waveform. Fourier's theorem, first proposed in 1807, states that any periodic wave can be represented by a sum of sine and cosine waves, with frequencies which are multiples (harmonics) of the wave in question. If you had been wondering why so much emphasis has been placed on studying sine and cosine waves, it is because any periodic waveform we might encounter is made up of a superposition of sine and cosine waves.

Put mathematically, any wave can be described by a Fourier series

$$y(t) = \frac{a_0}{2} + \sum_{n=1}^{\infty} a_n \cos(n 2\pi f t) + \sum_{n=1}^{\infty} b_n \sin(n 2\pi f t)$$

(5.8)

Fourier analysis is an important technique in many areas of physics and engineering. For example, the response of an electrical circuit to a non-sinusoidal electrical signal can be determined by breaking the signal down into its Fourier components.

TOPIC 5. WAVES

How do synthesisers work?

The word synthesiser derives from "synthesis." The most common type used by synthesisers is subtractive synthesis. Starting with a wave with various frequencies the synthesiser subtracts components of the wave until the desired tone is achieved. Frequencies can either be removed or minimised or alternatively emphasised and made more prominent. This can make a completely different sound to the original wave. Sounds can be made to sound like a drum, a trumpet, a crash, boom or bang sound effect or even computer generated speech.

A synthesiser has a range of different "oscillators" or sound tone generators which produce waves of different shapes such as square waves, sine waves, triangular waves etc. By combining these waves it can make complicated sounds and add musical effects like sustain, attack, decay and release to the notes.

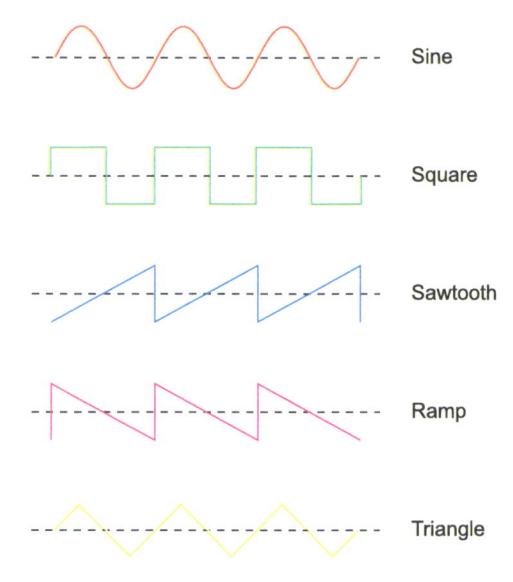

Sine

Square

Sawtooth

Ramp

Triangle

© HERIOT-WATT UNIVERSITY

294   UNIT 2. QUANTA AND WAVES

5.5   Stationary waves

As discussed previously, travelling waves occur when the energy propagation can continue until the end of the medium. Such as a ripple travelling across a pond until it reaches a bank. There are other situations we need to consider, however, that of stationary or standing waves.

Think of a string on a guitar, or violin. The string is stretched taught between two fixed supports. If we pluck the string, it vibrates whilst both ends remain fixed in position (see Figure 5.6).

The wave created reflects from the fix ends and interferes with the reflection from the other end. At the correct frequency for the waves appear still, with constructive interference creating peaks and troughs points called antinodes, and destructive interference creating points with no disturbance called nodes.

This type of wave is called a **stationary wave**. The name arises from the fact that the waves do not travel along the medium. Instead, the points of maximum and zero oscillation are fixed. Many different wave patterns are allowed, with certain properties, as will be discussed shortly.

TOPIC 5. WAVES

At the fixed ends of the medium (the guitar or violin string), no oscillations occur. These points are **nodes**. The points of maximum amplitude oscillations are called **antinodes**. Remember, the nodes are the points where there is 'no-disturbance'.

Figure 5.6: Oscillations of a stretched string held fixed at both ends

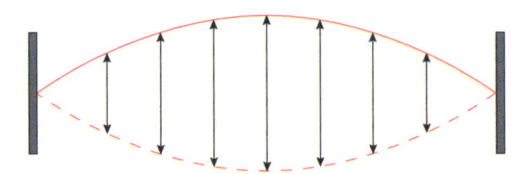

Stationary Waves Go online

Because of the condition of having a node at each end, we can build up a picture of the allowed modes of oscillation, as shown.

Figure 5.7: First four harmonics of a transverse standing wave

The wavelength of the oscillations is twice the distance between adjacent nodes, so the longest possible wavelength in Figure 5.7 is $\lambda_1 = 2L$. This is called the fundamental mode, and oscillates at the fundamental frequency $f_1 = v/\lambda_1 = v/2L$.

Looking at Figure 5.7, we can see that the allowed wavelengths are given by

$$\lambda_1 = 2L, \ \lambda_2 = 2L/2, \ \lambda_3 = 2L/3.....$$

$$\lambda_n = 2L/n = \lambda_1/n$$

(5.9)

The allowed frequencies are therefore given by

$$f_1 = v/2L, \ f_2 = 2v/2L, \ f_3 = 3v/2L.....$$

$$f_n = nv/2L = nf_1$$

(5.10)

These different frequencies are called the **harmonics** of the system, and f_1 is the first harmonic (also called the fundamental mode, as stated earlier). f_2 is called the second harmonic (or the first overtone), f_3 the third harmonic (or second overtone) and so on.

The equation for travelling waves does **not** also describe the motion of stationary waves. It can be proved mathematically that the equation for stationary waves is the superposition of two travelling waves of equal amplitude travelling in opposite directions.

When working out the equation for travelling waves, we stated that every small portion of the medium was performing SHM slightly out of phase with its neighbours, but with the same amplitude. An important difference between stationary and travelling waves is that for stationary waves, each portion of the medium between nodes oscillates **in phase** with its neighbours, but with slightly **different** amplitude.

As with the travelling waves, stationary waves can also be set up for longitudinal as well as transverse waves. Equation 5.9 and Equation 5.10 equally apply for longitudinal and transverse stationary waves.

TOPIC 5. WAVES

Example

An organ pipe has length $L = 2.00$ m and is open at both ends. The fundamental stationary sound wave in the pipe has an antinode at each end, and a node in the centre. Calculate the wavelength and frequency of the fundamental note produced. (*Take the speed of sound in air $v = 340$ m s^{-1}.*)

The distance between two antinodes (like the distance between two nodes) is equal to $\lambda/2$.

$$\therefore \frac{\lambda}{2} = 2.00$$
$$\therefore \lambda = 4.00 \, \text{m}$$

To calculate the frequency of the fundamental, use $n = 1$ in the equation

$$f_n = \frac{nv}{2L}$$
$$\therefore f = \frac{340}{2 \times 2.00}$$
$$\therefore f = \frac{340}{4.00}$$
$$\therefore f = 85.0 \, \text{Hz}$$

We could, of course, have calculated f using $f = v/\lambda$ which gives the same answer.

Longitudinal stationary waves

When someone blows across the top of a bottle, a sound wave is heard. This is because a stationary wave has been set up in the air in the bottle. The oscillating air molecules form stationary longitudinal waves. In this exercise you will work out the wavelength of different longitudinal stationary waves.

a) Consider a pipe of length *L*, with one end open and the other closed. The stationary waves formed by this system have an antinode at the open end and a node at the closed end.
   i. Sketch the fundamental wave in the pipe and calculate its wavelength.
   ii. Sketch the next two harmonics.

© HERIOT-WATT UNIVERSITY

298 UNIT 2. QUANTA AND WAVES

b) If the pipe is open at both ends, the air molecules are free to oscillate at either end, so there will be an antinode at each end.

   i. For a pipe of length L which is open at both ends, sketch the fundamental wave and calculate its wavelength.

   ii. Sketch the next two harmonics.

5.6 Beats

Beats are the regular loud and quiet sounds heard when two sound waves of very similar frequencies interfere with one another. e.g. a note of 350 Hz and a note of 355 Hz would display this effect. The beat frequency is the frequency of the changes between loud and quiet sounds heard by the listener. So if you hear 3 complete cycles of loud and quiet sounds every second, the beat frequency is 3 Hz. Humans can hear beat frequencies of less than or equal to 7 Hz. The difference between the frequencies of two notes that interfere to produce the beats is equal to the **beat frequency**.

For example, if a tuning fork plays middle C which is 262 Hz and a singer makes a continuous sound of 266 Hz then a beat frequency of 4 Hz will be heard.

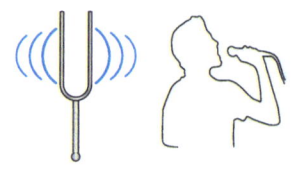

It works because the waves are slightly out of phase and this results in areas of constructive interference creating louder sounds and destructive interference creating quieter sounds. See Figure 5.8 for an visual explanation of how it works.

Figure 5.8: How beats are formed through constructive and destructive interference.

TOPIC 5. WAVES

Constructive interference   Destructive interference

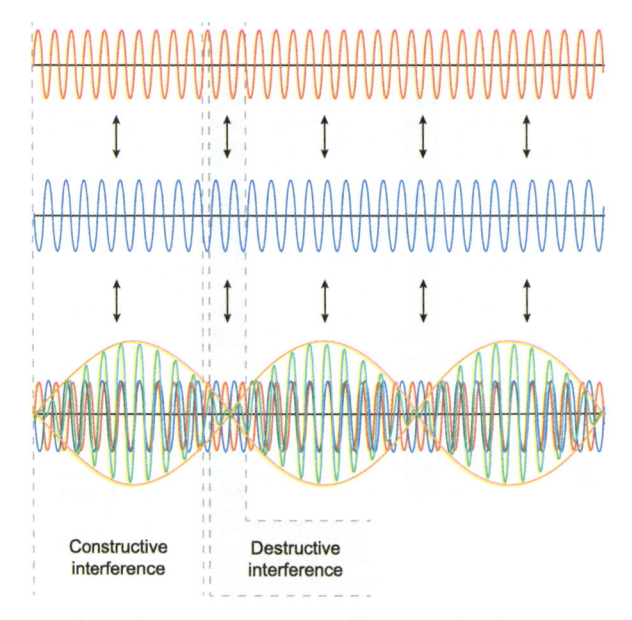

Piano tuners frequently use beats to tune pianos. They open the piano up and pluck the piano strings inside one at a time at the same time as tapping a tuning fork. If no beats can be heard the piano string is in tune as no interference takes place and hence no beat frequency. If beats can be heard the piano tuner tightens or loosens the string until no beats can be heard, if the beat frequency increases, the piano tuner will do the opposite to get nearer to 0 Hz and reduce the beat frequency. The piano tuner needs a full set of tuning forks for all the notes on the piano.

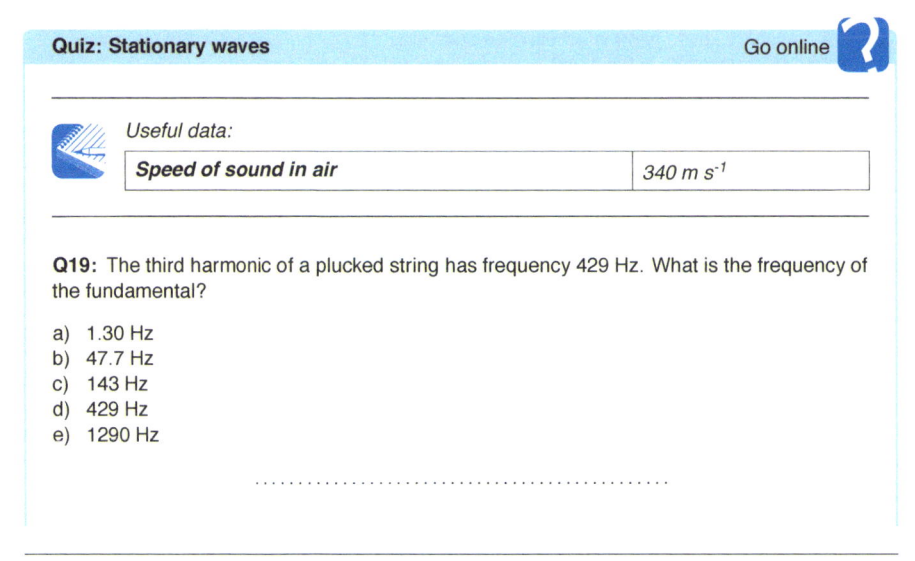

Quiz: Stationary waves Go online

Useful data:

Speed of sound in air	*340 m s^{-1}*

Q19: The third harmonic of a plucked string has frequency 429 Hz. What is the frequency of the fundamental?

a) 1.30 Hz
b) 47.7 Hz
c) 143 Hz
d) 429 Hz
e) 1290 Hz

....

Q20: Which *one* of the following statements is true?

a) Large amplitude oscillations occur at the nodes of a stationary wave.
b) Every point between adjacent nodes of a stationary wave oscillates in phase.
c) The amplitude of the stationary wave oscillations of a plucked string is equal at all points along the string.
d) The distance between adjacent nodes is equal to the wavelength of the stationary wave.
e) Stationary waves only occur for transverse, not longitudinal waves.

．．．

Q21: A string is stretched between two clamps placed 1.50 m apart. What is the wavelength of the fundamental note produced when the string is plucked?

a) 0.67 m
b) 0.75 m
c) 1.50 m
d) 2.25 m
e) 3.00 m

．．．

Q22: The fourth harmonic of a standing wave is produced on a stretched string 2.50 m long. How many antinodes are there?

a) 2.5
b) 3
c) 4
d) 5
e) 10

．．．

Q23: A stationary sound wave is set up in an open pipe 1.25 m long. What is the frequency of the third harmonic note in the pipe?

a) 182 Hz
b) 264 Hz
c) 340 Hz
d) 408 Hz
e) 816 Hz

．．．

Q24: A guitar string with a frequency of 220 Hz is plucked at the same time as a tuning fork with a frequency of 215 Hz.
How many beats will be heard over a period of 5 seconds?

a) 21 beats
b) 25 beats
c) 26 beats
d) 28 beats
e) 29 beats

5.7 Extended information

Web links — Go online

These web links should serve as an insight to the wealth of information available online and allow you to explore the subject further.

- http://www.physicsclassroom.com/class/waves/Lesson-4/Traveling-Waves-vs-Standing-Waves
- http://hyperphysics.phy-astr.gsu.edu/hbase/waves/standw.html
- http://hyperphysics.phy-astr.gsu.edu/hbase/sound/beat.html
- http://www.animations.physics.unsw.edu.au/jw/beats.htm

5.8 Summary

Summary

You should now be able to:

- use all the terms commonly employed to describe waves;
- derive an equation describing travelling sine waves, and solve problems using this equation;
- show an understanding of the difference between travelling and stationary waves;
- calculate the harmonics of a number of stationary wave systems;
- apply the principle of superposition;
- be able to explain how beats can be used to tune musical instruments;
- to understand the term phase difference and use the phase angle equation.

5.9 End of topic 5 test

End of topic 5 test Go online

The following data should be used when required:

Speed of light in a vacuum c	3.00×10^8 m s^{-1}
Speed of sound	340 m s^{-1}
Acceleration due to gravity g	9.8 m s^{-2}

Q25: A laser produces a monochromatic (single wavelength) beam of light with wavelength 542 nm.

Calculate the frequency of the light.

_____ Hz

Q26: A beam of red light (λ = 660 nm) is focussed onto a detector which measures a light irradiance of 1.63×10^{-6} W m^{-2}.

Calculate the measured irradiance when the amplitude of the waves is doubled.

_____ W m^{-2}

Q27: Suppose a knot is tied in a horizontal piece of string. A train of transverse vertical sine waves are sent along the string, with amplitude 8.5 cm and frequency 1.6 Hz.

Calculate the total distance through which the knot moves in 5.0 s.

_____ cm

Q28: A travelling wave is represented by the equation

$$y = 4 \sin \left(2\pi \left(1.6t - \frac{x}{5.5} \right) \right)$$

All the quantities in this equation are in SI units.
What is the value of the speed of the wave?

_____ m s^{-1}

Q29: A transverse wave travelling along a rope is represented by the equation

$$y = 0.55 \sin \left(2\pi \left(0.35t - \frac{x}{2.5} \right) \right)$$

All the quantities in this equation are in SI units.

TOPIC 5. WAVES

Calculate the displacement at the point $x = 0.50$ m when $t = 1.0$ s.

_____ m

Q30: Two coherent sine waves overlap at a point A. The amplitude of one wave is 6.3 cm, and the amplitude of the other wave is 3.2 cm.

Calculate the **minimum** possible amplitude of the resultant disturbance at A.

_____ cm

Q31: Two in-phase speakers A and B are emitting a signal of wavelength 1.04 m. A tape recorder is placed on the straight line between the speakers, 2.69 m from speaker A.

Calculate the **shortest** distance from the tape recorder that speaker B should be placed, to ensure constructive interference where the signal is recorded.

_____ m

Q32: Two loudspeakers are emitting a single frequency sound wave in phase. A listener is seated 4.6 m from one speaker and 2.3 m from the other.

Calculate the **minimum** frequency of the sound waves that would allow constructive interference where the listener is seated.

_____ Hz

Q33: A guitar string is 0.61 m long.

Calculate the wavelength of the 5th harmonic.

_____ m

Q34: A string is stretched between two clamps held 2.25 m apart. The string is made to oscillate at its third harmonic frequency.

Calculate the distance between two adjacent nodes.

_____ m

Q35: A plank of wood is placed over a pit 17 m wide. A girl stands on the middle of the plank and starts jumping up and down, jumping upwards from the plank every 1.3 s. The plank oscillates with a large amplitude in its fundamental mode, the maximum amplitude occuring at the centre of the plank.

Calculate the speed of the transverse waves on the plank.

_____ m s^{-1}

© HERIOT-WATT UNIVERSITY

Q36: A piano is tuned using a 455 Hz tuning fork and a beat frequency of 5 Hz is heard, what was the frequency of the piano string if the piano tuner had to tighten the string to get it into tune?

_____ Hz

Unit 2 Topic 6

Interference

Contents

    6.1 Introduction 306
    6.2 Coherence and optical path difference 306
        6.2.1 Coherence 306
        6.2.2 Optical path difference 306
    6.3 Reflection of waves 310
    6.4 Extended information 313
    6.5 Summary 313
    6.6 End of topic 6 test 314

Prerequisites

- Refractive index.
- Frequency and wavelength.

Learning objective

By the end of this topic you should be able to:

- show an understanding of coherence between light waves;
- explain the difference between path length and optical path length, and calculate the latter;
- understand what happens when waves reflect off media of higher or lower refractive index.

6.1 Introduction

Interference of light waves is responsible for the rainbow colours seen on an oil film on a puddle of water, or in light reflected by a soap bubble. In the next few topics we will be looking at the conditions under which these interference effects can take place.

This topic contains some of the background work necessary to fully understand interference. It begins with the concept of coherence, shows the method of calculating the optical path difference between two light rays and finishes with what happens when waves reflect off media of higher or lower refractive index.

6.2 Coherence and optical path difference

The section begins with the concept of coherence and ends with the method of calculating the optical path difference between two light rays.

6.2.1 Coherence

We briefly discussed coherent waves in the 'Introduction to Waves' topic. Two waves are said to be coherent if they have a constant phase relationship. For two waves travelling in air to have a constant phase relationship, they must have the same frequency and wavelength. At any given point, the phase difference between the two waves will be fixed.

It is easier to produce coherent sound waves or microwaves than it is to produce coherent visible electromagnetic waves. Both sound and microwaves can be generated electronically, with loudspeakers or antennae used to emit the waves. The electronic circuits used to generate these waves can 'frequency lock' and 'phase lock' two signals to ensure they remain coherent. In contrast to this, light waves are produced by transitions in individual atoms, and are usually emitted with random phase.

For us to see interference effects, we require two or more sources of coherent light waves. The best source of coherent radiation is a laser, which emits light at a single wavelength, usually in a collimated (non-diverging) beam. Another way to produce coherent light is to split a wave, for example by reflection from a glass slide. Some of the light will be transmitted, the rest will be reflected, and the two parts must be coherent with each other.

Filament light bulbs and strip lights do not emit coherent radiation. Such sources are called extended sources, or incoherent sources. They emit light of many different wavelengths, and light is emitted from every part of the tube or filament.

6.2.2 Optical path difference

In the Section on Superposition in Topic 5, we solved problems in which waves emitted in phase arrived from two different sources. By measuring the paths travelled by both waves, we could determine whether they arrived in phase (interfering constructively) or out of phase (interfering destructively). We are now going to look at a slightly different situation, in which waves emitted in phase by the same source arrive at a detector via different routes.

Figure 6.1: Two light rays travelling along different paths

TOPIC 6. INTERFERENCE

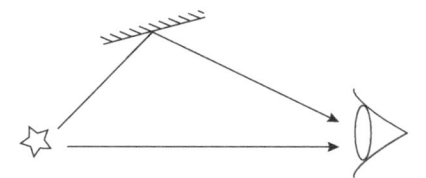

In Figure 6.1 we can see two waves from the same source arriving at the same detector.

The situation is like that we saw in the previous topic. If the **geometrical path difference**, *gpd*, i.e. the difference in how far the waves have travelled, is a whole number of wavelengths (λ, 2λ, 3λ...) then the two waves will arrive in phase. If the geometrical path difference is an odd number of half wavelengths ($\lambda/2$, $3\lambda/2$, $5\lambda/2$...) then the two waves arrive out of phase at the detector and destructive interference takes place.

A further complication can arise if one of the rays passes through a different medium.

Figure 6.2: Two light rays travelling different optical path lengths

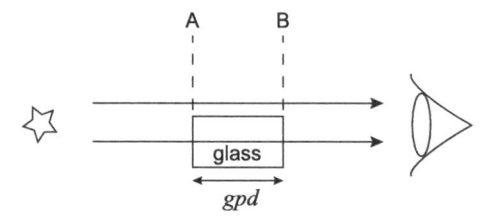

In Figure 6.2, the lower ray passes through a glass block between A and B. An **optical path difference**, *opd*, exists between the two rays, even though they have both travelled the same geometrical distance, because the wavelength of the light changes as it travels through the glass. The refractive index of glass, n_{glass}, is greater than the refractive index of air. So, the waves travel at a slower speed in the glass. The frequency of the waves does not change, therefore the wavelength of the ray in glass, λ_{glass}, must be smaller than the wavelength in air, λ. If the refractive index is 1.00, then λ_{glass} is given by:

$$n_{glass} = \frac{c}{v_{glass}}$$

$$\therefore n_{glass} = \frac{f \times \lambda}{f \times \lambda_{glass}}$$

$$\therefore n_{glass} = \frac{\lambda}{\lambda_{glass}}$$

$$\therefore \lambda_{glass} = \frac{\lambda}{n_{glass}}$$

(6.1)

In Figure 6.2, the upper ray is travelling in air all the way from source to detector. The number of wavelengths contained between A and B will be $\frac{gpd}{\lambda_{glass}}$.

From Figure 6.1, $\lambda_{glass} = \frac{\lambda}{n_{glass}}$ therefore,

$$\frac{gpd}{\lambda_{glass}} = \frac{gpd}{\lambda/n_{glass}} = \frac{n_{glass} \times gpd}{\lambda}$$

So, there will be more wavelengths between A and B in the glass than in the same distance in air.

This is shown in Figure 6.3. For a light ray travelling a distance gpd in a material of refractive index n, the **optical path length** is $n \times gpd$. To determine the phase difference between two waves travelling between the same source and detector, the **optical path difference** must be a whole number of wavelengths for constructive interference, and an odd number of half-wavelengths for destructive interference. In Figure 6.3, the optical path difference is the difference between the optical path lengths of both waves travelling from A to B. For the upper ray, the optical path length of $AB = n_{air} \times gpd = gpd$, since $n_{air} \approx 1.00$. For the lower ray, the optical path length of AB is $n_{glass} \times gpd$.

Thus, the optical path difference is $(n_{glass} - n_{air}) \times gpd$.

Figure 6.3: Optical path difference between air and glass

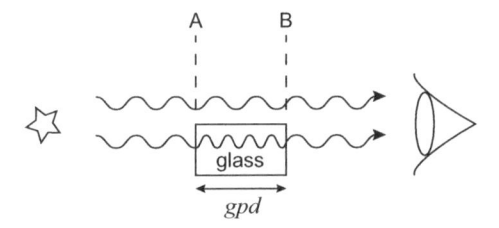

In general, for two rays of light travelling the same distance gpd in media with refractive indices n_1 and n_2, the interference conditions are constructive interference (waves emerge in phase) $(n_1 - n_2)\,gpd = m\lambda$ destructive interference (waves emerge exactly out of phase) $(n_1 - n_2)\,gpd = m + \frac{1}{2}\lambda$ where m is an integer.

Note that we will talk about an "optical path difference" even when studying radiation from outside the visible part of the electromagnetic spectrum.

Example A source emits two beams of microwaves with a wavelength of 6.00×10^{-3} m.

One ray of the waves travels through air to a detector 0.050 m away. Another ray travels the same distance through a quartz plate to the detector.

The quartz has a refractive index of 1.54.

Do the waves interfere constructively or destructively at the detector?

The optical path difference, opd in this case is:

TOPIC 6. INTERFERENCE

$$opd = (n_{quartz} - n_{air}) \times gpd$$
$$\therefore opd = (1.54 - 1) \times 0.050$$
$$\therefore opd = 0.54 \times 0.050$$
$$\therefore opd = 0.027 m$$

To find out how many wavelengths this optical path difference is, divide by the wavelength in air:

$$\frac{0.027}{\lambda} = \frac{0.027}{6.00 \times 10^{-3}} = 4.5 \text{ wavelengths}$$

The .5 means the waves arrive at the detector exactly out of phase, and so interfere destructively.

Quiz: Coherence and optical paths

Q1: Two light waves are coherent if

a) they have the same speed.
b) their amplitudes are identical.
c) the difference in their frequencies is constant.
d) their phase difference is constant.
e) the difference in their wavelengths is constant.

....

Q2: A radio transmitter emits waves of wavelength 500 m, A receiver dish is located 4.5 km from the transmitter. What is the path length, in wavelengths, from the transmitter to the receiver?

a) 0.11 wavelengths
b) 0.5 wavelengths
c) 2 wavelengths
d) 5 wavelengths
e) 9 wavelengths

....

Q3: Light waves of wavelength 450 nm travel 0.120 m through a glass block ($n = 1.50$).
What is the optical path difference travelled?

a) 8.10×10^{-9} m
b) 4.00×10^{-4} m
c) 0.180 m

© HERIOT-WATT UNIVERSITY

d) 0.800 m
e) 4.00×10^5 m

....

Q4: Two light rays travel in air from a source to a detector. Both travel the same distance from source to detector, but one ray travels for 2.50 cm of its journey through a medium of refractive index 1.35.

What is the optical path difference between the two rays?

a) 8.75×10^{-3} m
b) 3.38×10^{-3} m
c) 0.0250 m
d) 14.0 m
e) Depends on the wavelength of the light rays.

....

Q5: In a similar set-up to question 4, two microwaves (λ = 1.50 cm) from the same source arrive with a phase difference of exactly 4.00 wavelengths at the detector. The waves have travelled the same distance, but one has gone through a sheet of clear plastic (n = 1.88) while the other has travelled through air all the way.

What is the thickness of the plastic sheet?

a) 0.0319 m
b) 0.0682 m
c) 0.220 m
d) 4.55 m
e) 14.7 m

6.3 Reflection of waves

The phase of a wave may be changed when it is reflected. For light waves, we must consider the refractive indices of the two media involved. For example, consider a light wave travelling in air (n = 1.00) being reflected by a glass surface (n = 1.50). In this case the wave is travelling in a low refractive index medium, and is being reflected by a higher refractive index medium. Whenever this happens, there is a phase change of 180° (π radians). A wave crest becomes a wave trough on reflection. This is shown in Figure 6.4, where we can see a crest (labelled c) before reflection comes back as a trough (t) after reflection.

Figure 6.4: Phase change upon reflection at a higher refractive index material

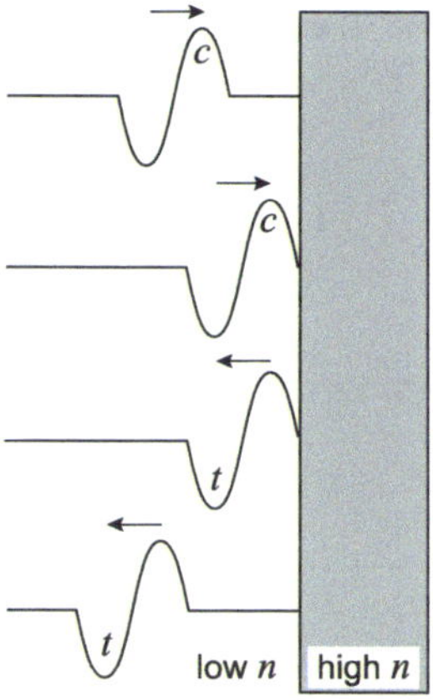

You would see exactly the same effect happening if you sent a wave along a rope that was fixed at one end. As in Figure 6.4, a wave that has a crest leading a trough is reflected back as a wave with a trough leading a crest.

This can be demonstrated effectively by holding a slinky fixed at one end to represent a higher refractive index and sending a transverse wave along the slinky towards the fixed end.

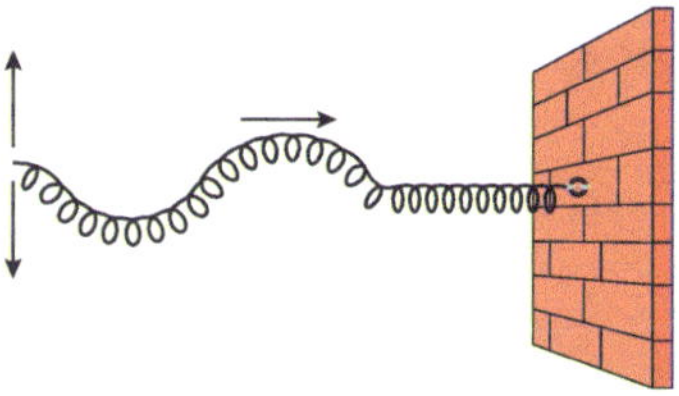

Reflection of a pulse at a fixed end Go online

There is an activity available online demonstrating a pulse moving through the rope fixed to the wall.

There is no phase change when a light wave travelling in a higher index material is reflected at a boundary with a material that has a lower index. If our light wave is travelling in glass, the phase of the wave is not changed when it is reflected at a boundary with air. A wave crest is still a wave crest upon reflection. This reflection is shown in Figure 6.5.

Figure 6.5: No phase change upon reflection at a lower refractive index material

A wave sent along a rope that is unsecured at the end would behave in exactly the same way. A wave which travels as a crest leading a trough remains as a wave with the crest leading the trough upon reflection.

Again, this can be demonstrated using the slinky but this time having the slinky attached to a piece of string to represent the lower medium.

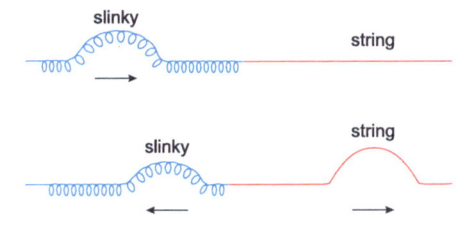

The importance of the phase change in certain reflections is that the optical path of the reflected wave is changed. If the phase of a wave is changed by 180° it is as if the wave has travelled an extra $\lambda/2$ distance compared to a wave whose phase has not been changed by the reflection.

TOPIC 6. INTERFERENCE

6.4 Extended information

Web links — Go online

These web links should serve as an insight to the wealth of information available online and allow you to explore the subject further.

- http://www.physicsclassroom.com/class/waves/Lesson-3/Interference-of-Waves
- http://www.bbc.co.uk/education/guides/z99kkqt/revision
- https://www.youtube.com/watch?v=CAe3lkYNKt8
- http://www.acoustics.salford.ac.uk/feschools/waves/super2.php

6.5 Summary

Summary

You should now be able to:

- state the condition for two light beams to be coherent;
- explain why the conditions for coherence are more difficult to achieve for light than for sound and microwaves;
- define the term "
- optical path difference" and relate it to phase difference.

6.6 End of topic 6 test

End of topic 6 test Go online

The following data should be used when required:

Speed of light in a vacuum c	3.00×10^8 m s^{-1}
Speed of sound	340 m s^{-1}
Acceleration due to gravity g	9.8 m s^{-2}

Q6: Monochromatic light of wavelength 474 nm travels through a glass lens of thickness 21.5 mm and refractive index 1.52.

Calculate the optical path length through the lens.

_____ mm

Q7: Two light rays from a coherent source travel through the same distance to reach a detector. One ray travels 23.7 mm through glass of refractive index 1.52. The other ray travels in air throughout.

Calculate the optical path difference between the rays.

_____ mm

Unit 2 Topic 7

Division of amplitude

Contents

7.1 Introduction 316
7.2 Thin film interference 316
7.3 Wedge fringes 322
7.4 Extended information 327
7.5 Summary 327
7.6 End of topic 7 test 328

Prerequisites

- Waves.
- Interference.

Learning objective

By the end of this topic you should be able to:

- show an understanding of interference by division of amplitude;
- understand how thin film interference works and how it can be used in processes such as creating anti-reflecting coatings;
- understand how wedge fringes work and how they can be used to find the thickness of very small objects.

7.1 Introduction

This topic is all about a process called **interference by division of amplitude**. This is rather a long title for a simple idea. You should already know what we mean by interference from previous Topic 6. Division of amplitude means that each wave is being split, with some of it travelling along one path, and the remainder following a different path. When these two parts are recombined, it is the difference in optical paths taken by the two waves that determines their phase difference, and hence how they interfere when they recombine.

7.2 Thin film interference

One example of interference by division of amplitude with which you are probably familiar is called thin film interference. When oil or petrol is spilt onto a puddle of water, we see a multicoloured film on the surface of the puddle. This is due to the thin film of oil formed on the surface of the puddle. Sunlight is being reflected from the film and the puddle. The film appears multicoloured because of constructive and destructive interference of the sunlight falling on the puddle. In this section we will examine this effect more closely.

Figure 7.1 shows light waves falling on a thin film of oil on the surface of water.

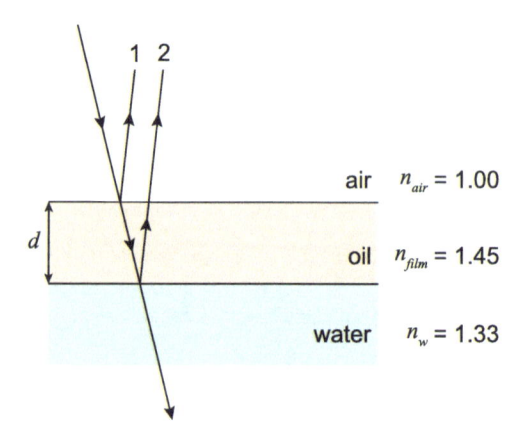

Figure 7.1: Light reflected from a thin film of oil on water

air $n_{air} = 1.00$

oil $n_{film} = 1.45$

water $n_w = 1.33$

TOPIC 7. DIVISION OF AMPLITUDE

A pine cone thrown into an oily puddle

Photo by David Lee / CC BY-SA 2.0

Light is reflected back upwards from both the air-oil boundary (ray 1) and the oil-water boundary underneath it (ray 2). Someone looking at the reflected light will see the superposition of these two rays. Note that this situation is what we have called interference by division of amplitude. For any light wave falling on the oil film, some of the wave is reflected from the surface of the oil film. Some light is reflected by the water surface, and some is transmitted into the water. The two reflected rays travel different paths before being recombined. Under what conditions will the two rays interfere constructively or destructively?

To keep the analysis simple, we will assume the angle of incidence in Figure 7.1 is 0°, so that ray 2 travels a total distance of $2d$ in the film. The optical path difference between rays 1 and 2 is therefore $2n_{film}d$. But there is another source of path difference. Ray 1 has undergone a $\lambda/2$ phase change, since it has been reflected at a higher refractive index medium.

The total optical path difference is therefore equal to

$$2n_{film}d + \frac{\lambda}{2}$$

For constructive interference, this optical path difference must equal a whole number of wavelengths.

$$2n_{film}d + \frac{\lambda}{2} = m\lambda$$

where m = 1, 2, 3...

$$2n_{film}d = m\lambda - \frac{\lambda}{2}$$
$$\therefore 2n_{film}d = \left(m - \frac{1}{2}\right)\lambda$$
$$\therefore d = \frac{\left(m - \frac{1}{2}\right)\lambda}{2n_{film}}$$

(7.1)

Equation 7.1 gives us an expression for the values of film thickness d for which reflected light of wavelength λ will undergo constructive interference, and hence be reflected strongly.

For destructive interference, the optical path difference must equal an odd number of half-wavelengths

$$2n_{film}d + \frac{\lambda}{2} = \left(m + \frac{1}{2}\right)\lambda$$

where m = 1, 2, 3...

$$\therefore 2n_{film}d = \left(m + \frac{1}{2}\right)\lambda - \frac{\lambda}{2}$$
$$\therefore 2n_{film}d = m\lambda$$
$$\therefore d = \frac{m\lambda}{2n_{film}}$$

(7.2)

Equation 7.2 tells us for which values of d reflected light of wavelength λ will undergo destructive interference, and hence be reflected weakly.

Example

Using Figure 7.1, what is the *minimum* thickness of oil film which would result in destructive interference of green light (λ = 525 nm) falling on the film with angle of incidence 0°?

For destructive interference we will use Equation 7.2, with λ = 525 nm = 5.25×10^{-7} m. For the minimum film thickness, we set m equal to 1.

TOPIC 7. DIVISION OF AMPLITUDE

$$d = \frac{m\lambda}{2n_{film}}$$
$$\therefore d = \frac{1 \times 5.25 \times 10^{-7}}{2 \times 1.45}$$
$$\therefore d = \frac{1 \times 5.25 \times 10^{-7}}{2.90}$$
$$\therefore d = 1.81 \times 10^{-7} \text{ m}$$

If we have sunlight falling on an oil film, then we have a range of wavelengths present. Also, an oil film is unlikely to have the same thickness all along its surface. So we will have certain wavelengths interfering constructively where the film has one thickness, whilst the same wavelengths may be interfering destructively at a part of the surface where the film thickness is different. The overall effect is that the film appears to be multi-coloured.

Thin film interference — Go online

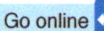

At this stage there is an online activity showing interference as the thickness of a thin film is varied.

Blooming

An important application of thin film interference is anti-reflection coatings used on camera lenses. This process is known as blooming. A thin layer of a transparent material such as magnesium fluoride ($n = 1.38$) is deposited on a glass lens ($n = 1.50$), as shown in Figure 7.2.

Figure 7.2: Anti-reflection coating deposited on a glass lens

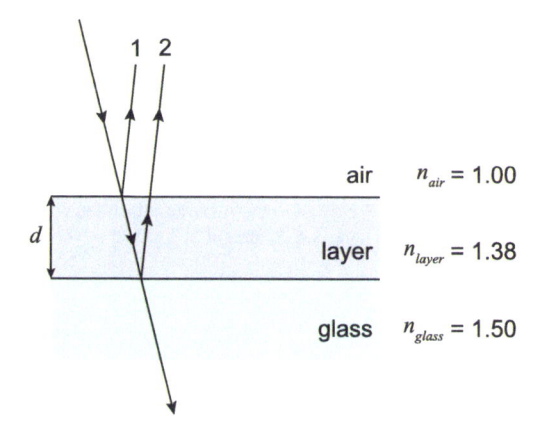

air $\quad n_{air} = 1.00$
layer $\quad n_{layer} = 1.38$
glass $\quad n_{glass} = 1.50$

Once again, if we consider the light falling with an angle of incidence of $0°$ on the lens, then the optical path difference between rays 1 and 2 is $2 \times n_{coating} \times d$. Both rays are reflected by media with a greater value of refractive index, so they **both** undergo the same λ/2 phase change on reflection and there is no extra optical path difference as there was with the oil film on water. So for destructive interference

$$2n_{coating}d = \left(m + \frac{1}{2}\right)\lambda$$
$$\therefore d = \frac{\left(m + \frac{1}{2}\right)\lambda}{2n_{coating}}$$

(7.3)

In this equation, $m = 0, 1, 2...$

The minimum coating thickness that will result in destructive interference is given by putting $m = 0$ in Equation 7.3:

$$d = \frac{\left(0 + \frac{1}{2}\right)\lambda}{2n_{coating}}$$
$$\therefore d = \frac{\lambda}{4n_{coating}}$$

(7.4)

So a coated **(bloomed) lens** can be made non-reflecting for a specific wavelength of light, and Equation 7.4 gives the minimum coating thickness for that wavelength.

Example

For the coated lens shown in Figure 7.2, what is the minimum thickness of magnesium fluoride that can be used to make the lens non-reflecting at $\lambda = 520$ nm?

Use Equation 7.4

$$d = \frac{\lambda}{4n_{coating}}$$
$$\therefore d = \frac{5.20 \times 10^{-7}}{4 \times 1.38}$$
$$\therefore d = 9.42 \times 10^{-8} \text{ m}$$

Since most lenses are made to operate in sunlight, the thickness of the coating is designed to produce destructive interference in the centre of the visible spectrum. The previous example produced an anti-reflection coating in the green part of the spectrum, which is typical of commercial coatings. The extremes of the visible spectrum - red and violet - do not undergo destructive interference upon reflection, so a coated lens often looks reddish-purple under everyday lighting.

Quiz: Thin film interference   Go online

Useful data:

Refractive index of air	1.00
Refractive index of water	1.33

Q1: Two coherent light rays emitted from the same source will interfere constructively if
a) they undergo a phase change on reflection.
b) they travel in materials with different refractive indices.
c) their optical path difference is an integer number of wavelengths.
d) their optical path difference is an odd number of half-wavelengths.
e) they have different wavelengths.

....

Q2: What is the minimum thickness of oil film ($n = 1.48$) on water that will produce destructive interference of a beam of light of wavelength 620 nm?

a) 1.05×10^{-7} m
b) 2.09×10^{-7} m
c) 2.33×10^{-7} m
d) 3.14×10^{-7} m
e) 2.07×10^{-6} m

....

Q3: A soap film, with air on either side, is illuminated by electromagnetic radiation normal to its surface. The film is 2.00×10^{-7} m thick, and has refractive index 1.40. Which wavelengths will be intensified in the reflected beam?

a) 200 nm and 100 nm
b) 560 nm and 280 nm
c) 1120 nm and 373 nm
d) 1120 nm and 560 nm
e) 2240 nm and 747 nm

....

Q4: Why do lenses coated with an anti-reflection layer often appear purple in colour when viewed in white light?

a) You cannot make an anti-reflection coating to cut out red or violet.
b) The coating is too thick to work properly.
c) The green light is coherent but the red and violet light is not.
d) The coating is only anti-reflecting for the green part of the visible spectrum.
e) Light only undergoes a phase change upon reflection at that wavelength.

....

Q5: What is the minimum thickness of magnesium fluoride ($n = 1.38$) that can form an anti-reflection coating on a glass lens for light with wavelength 500 nm?

a) 9.06×10^{-8} m
b) 1.36×10^{-7} m
c) 1.81×10^{-7} m
d) 2.72×10^{-7} m
e) 5.00×10^{-7} m

There is one final point to add before we leave thin film interference. In the previous Topic, we discussed coherence, and stated that we required coherent radiation to observe interference effects. But we have seen that we can produce interference with an incoherent source such as sunlight. How is this possible? The answer is that the process of interference by division of amplitude means that we are taking one wave, splitting it up and then recombining it with itself. We do not have one wave interfering with another, and so we do not need a source of coherent waves. Division of amplitude means we can use an extended source to produce interference.

7.3 Wedge fringes

Interference fringes can also be produced by a thin wedge of air. If we place a thin piece of foil between two microscope slides at one end, we form a thin air wedge. This is shown in Figure 7.3, where the size of the wedge angle has been exaggerated for clarity.

Figure 7.3: The side and top views of a wedge fringes experiment

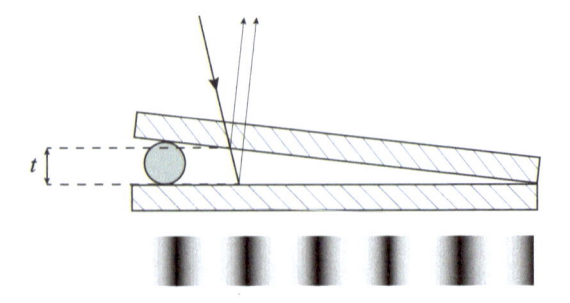

TOPIC 7. DIVISION OF AMPLITUDE

When the wedge is illuminated with a monochromatic source, bright and dark bands are seen in the reflected beam. Interference is taking place between rays reflected from the lower surface of the top slide and the upper surface of the bottom slide. Because of the piece of foil, the thickness of the air wedge is increasing from right to left, so the optical path difference between the reflected rays is increasing. A bright fringe is seen when the optical path difference leads to constructive interference, and a dark fringe occurs where destructive interference is taking place.

We can calculate the wedge thickness required to produce a bright or dark fringe. We will be considering light falling normally onto the glass slides. Remember that the angle between the slides is extremely small, so it can be assumed to be approximately zero in this analysis. The path difference between the two rays in Figure 7.3 is the extra distance travelled **in air** by the ray reflected from the lower slide, which is equal to $2t$.

Do either of the rays undergo a phase change upon reflection?

The answer is yes - the ray reflected by the lower slide is travelling in a low refractive index medium (air) and being reflected at a boundary with a higher n medium (glass) so this ray does undergo a $\lambda/2$ phase change. The other ray does not, as it is travelling in the higher n medium. So the total optical path difference between the two rays is

$$opd = 2t + \frac{\lambda}{2}$$

For constructive interference and a bright fringe,

$$opd = m\lambda$$
$$2t + \frac{\lambda}{2} = m\lambda$$
$$\therefore 2t = \left(m - \frac{1}{2}\right)\lambda$$
$$\therefore t = \frac{\left(m - \frac{1}{2}\right)\lambda}{2}$$

(7.5)

In this case, m = 1, 2, 3...

UNIT 2. QUANTA AND WAVES

For destructive interference, leading to a dark fringe

$$opd = \left(m + \frac{1}{2}\right)\lambda$$
$$2t + \frac{\lambda}{2} = \left(m + \frac{1}{2}\right)\lambda$$
$$\therefore 2t = m\lambda$$
$$\therefore t = \frac{m\lambda}{2}$$

(7.6)

Here, m is a whole number again, but we can also put $m = 0$, which corresponds to $t = 0$. Check back to Figure 7.3, and you can see that the $m = 0$ case gives us a dark fringe where the two slides are touching and $t = 0$.

Finally, the fringe separation can be determined if we know the size and separation of the glass slides.

Figure 7.4: Wedge separation analysis

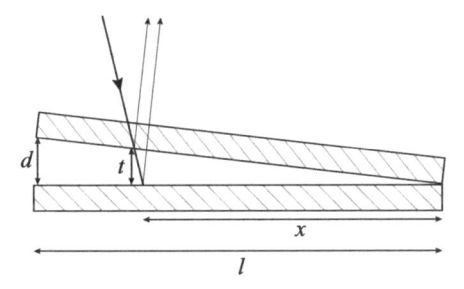

The length of the glass slides shown in Figure 7.4 is l m, and the slides are separated by d m at one end. The mth dark fringe is formed a distance x m from the end where the slides are in contact. By looking at similar triangles in Figure 7.4, we can see that

$$\frac{t}{x} = d$$
$$\therefore x = \frac{tl}{d}$$

Substituting for t using Equation 7.6

TOPIC 7. DIVISION OF AMPLITUDE

$$x = \frac{m\lambda l}{2d}$$

The distance Δx between the mth and $(m+1)^{\text{th}}$ dark fringes (the fringe separation) is therefore

$$\Delta x = \frac{(m+1)\lambda l}{2d} - \frac{m\lambda l}{2d}$$
$$\therefore \Delta x = \frac{\lambda l}{2d}$$

(7.7)

Wedge fringes demonstration   Go online

There is an online activity which demonstrates how varying the thickness of the gap, colour of the light and the index of refraction of the material between the slides (normally air).

Wedge fringes

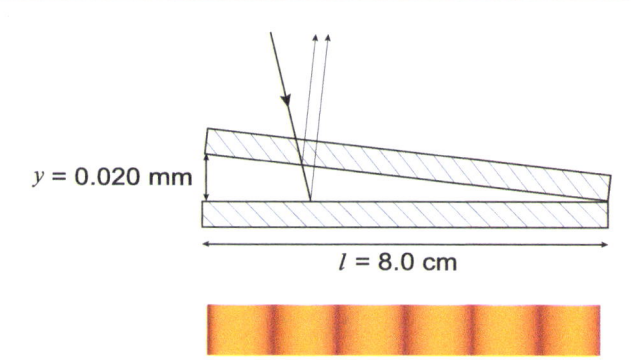

A wedge interference experiment is being carried out. The wedge is formed using two microscope slides, each of length 8.0 cm, touching at one end. At the other end, the slides are separated by a 0.020 mm thick piece of foil. What is the fringe spacing when the experiment is carried out using:

1. A helium-neon laser, at wavelength 633 nm?

2. An argon laser, at wavelength 512 nm?

© HERIOT-WATT UNIVERSITY

Quiz: Wedge fringes

Q6: In a wedge interference experiment carried out in air using monochromatic light, the 8th bright fringe occurs when the wedge thickness is 1.80×10^{-6} m.

What is the wavelength of the light?

a) 225 nm
b) 240 nm
c) 450 nm
d) 480 nm
e) 960 nm

....

Q7: An air-wedge interference experiment is being carried out using a mixture of blue light (λ = 420 nm) and red light (λ = 640 nm).

What is the colour of the bright fringe that is seen closest to where the glass slides touch?

a) Blue
b) Red
c) Purple - both blue and red appear at the same place
d) It is dark, as the two colours cancel each other out
e) Impossible to say without knowing the thickness of the wedge

....

Q8: A wedge interference experiment is carried out using two glass slides 10.0 cm long, separated at one end by 1.00×10^{-5} m.

What is the fringe separation when the slides are illuminated by light of wavelength 630 nm?

a) 1.00×10^{-3} m
b) 1.25×10^{-3} m
c) 1.58×10^{-3} m
d) 3.15×10^{-3} m
e) 6.30×10^{-3} m

....

Q9: What is the minimum air-wedge thickness that would produce a bright fringe for red light of wavelength 648 nm?

a) 81.0 nm
b) 162 nm
c) 324 nm
d) 648 nm
e) 972 nm

....

TOPIC 7. DIVISION OF AMPLITUDE

Q10: During a wedge fringes experiment with monochromatic light, air between the slides is replaced by water.

What would happen to the fringes?

a) The bright fringes would appear brighter.
b) The fringes would move closer together.
c) The fringes would move further apart.
d) There would be no difference in their appearance.
e) The colour of the fringes would change.

7.4 Extended information

Web links — Go online

These web links should serve as an insight to the wealth of information available online and allow you to explore the subject further.

- http://www.schoolphysics.co.uk/age16-19/Wave%20properties/Interference/text/Wedge_fringes/index.html
- http://hyperphysics.phy-astr.gsu.edu/hbase/phyopt/thinfilm.html
- http://physics.bu.edu/~duffy/PY106/Diffraction.html

7.5 Summary

Interference of light waves can be observed when two (or more) coherent beams are superposed. This usually requires a source of coherent light waves. Division of amplitude - splitting a wave into two parts which are later re-combined - can be used to produce interference effects without requiring a coherent source.

Two experimental arrangements for viewing interference by division of amplitude have been presented in this topic. In thin film interference, one ray travels an extra distance through a different medium. The thickness and refractive index of the film must be known in order to predict whether constructive or destructive interference takes place. In a thin wedge interference experiment, only the thickness of the wedge is required.

© HERIOT-WATT UNIVERSITY

Summary

You should now be able to:

- state what is meant by the principle of division of amplitude, and describe how the division of amplitude allows interference to be observed using an extended source;
- state the conditions under which a light wave will undergo a phase change upon reflection;
- derive expressions for maxima and minima to be formed in a "thin film" reflection, and perform calculations using these expressions;
- explain the formation of coloured fringes when a thin film is illuminated by white light;
- explain how a lens can be made non-reflecting for a particular wavelength of light;
- derive the expression for the minimum thickness of a non-reflecting coating, and carry out calculations using this expression;
- explain why a coated ("bloomed") lens appears coloured when viewed in daylight;
- derive the expression for the distance between fringes formed by "thin wedge" reflection, and carry out calculations using this expression.

7.6 End of topic 7 test

End of topic 7 test   Go online

The following data should be used when required:

Speed of light in a vacuum c	3.00×10^8 m s^{-1}
Speed of sound	340 m s^{-1}
Acceleration due to gravity g	9.8 m s^{-2}

Q11: A soap bubble consists of a thin film of soapy water ($n = 1.32$) surrounded on both sides by air.

Calculate the minimum film thickness if light of wavelength 602 nm is strongly reflected by the film.

_____ m

TOPIC 7. DIVISION OF AMPLITUDE

Q12: White light is incident on an oil film (thickness 2.66 × 10⁻⁷ m, refractive index 1.35) floating on water (refractive index 1.33). The white light has a wavelength range of 350 - 750 nm.

Calculate the two wavelengths in this range which undergo destructive interference upon reflection.

1. Calculate the longer wavelength.
   _____ nm
2. Calculate the shorter wavelength.
   _____ nm

....

Q13: A 1.02 × 10⁻⁷ m coating of a transparent polymer ($n = 1.33$) is deposited on a glass lens to make an anti-reflection coating.

If the lens has refractive index 1.52, calculate the wavelength of light in the range 350 - 750 nm for which the coating is anti-reflecting.

_____ nm

....

Q14: In a standard air wedge interference experiment, calculate the thickness of the wedge which gives the 10th bright fringe for light of wavelength 474 nm.

_____ nm

....

Q15: Wedge fringes are formed in the air gap between two glass slides of length 16.5 cm, separated at one end by a 10.0 μm piece of paper. The wedge is illuminated by monochromatic light of wavelength 510 nm.

Calculate the distance between adjacent dark fringes. Give your answer in metres.

_____ m

....

Q16: Two glass slides are laid together, separated at one end by a 15.0 μm sliver of foil. The slides each have length 10.0 cm.

When illuminated by monochromatic light, a series of light and dark reflection fringes appear, with fringe separation 1.75 mm.

Calculate the wavelength of the light.

_____ nm

Unit 2 Topic 8

Division of wavefront

Contents

- 8.1 Introduction ... 332
- 8.2 Interference by division of wavefront ... 332
- 8.3 Young's slits experiment ... 333
- 8.4 Extended information ... 338
- 8.5 Summary ... 338
- 8.6 End of topic 8 test ... 339

Prerequisites

- A prior learning for this topic is an understanding of coherent light waves.

Learning objective

By the end of this topic you should be able to:

- show an understanding of the difference between interference by division of wavefront and division of amplitude;
- describe the Young's slits experiment;
- derive the expression for fringe spacing in a Young's slits experiment, and use this expression to determine the wavelength of a monochromatic source.

8.1 Introduction

The work in this topic will concentrate on the experiment known as Young's slits, which is an example of interference by division of wavefront. We will see the difference between this process and interference by division of amplitude.

The experimental arrangement for producing interference will be described, followed by an analysis of the interference pattern. This analysis will reveal that the spacing between interference fringes is in direct proportion to the wavelength of the light.

8.2 Interference by division of wavefront

Light emitted from a point source radiates uniformly in all directions. This is often illustrated by showing the wavefronts perpendicular to the direction of travel, as in Figure 8.1.

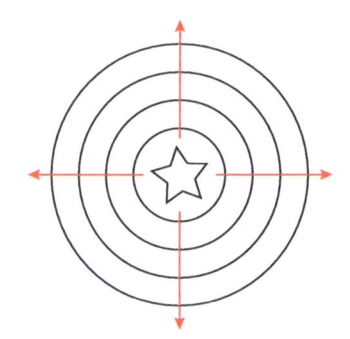

Figure 8.1: Wavefronts emitted by a point source

Each wavefront joins points in phase, for example the crest or trough of a wave. If the light source is monochromatic then the points on the wavefront are coherent, as they have the same wavelength and are in phase. If we can take two parts of a wavefront and combine them, then we will see interference effects. This is known as **interference by division of wavefront**. This is a different process to division of amplitude, in which a single wave was divided and re-combined. Here we are combining two separate waves. The two waves must be coherent to produce stable interference when they are combined.

An extended source acts like a collection of point sources, and cannot be used for a division of wavefront experiment. To overcome this problem, an extended source is often used behind a small aperture in a screen (see Figure 8.2). The size of the aperture must be of the same order of magnitude as the wavelength of the light. In this way the light appears to come from a point source.

TOPIC 8. DIVISION OF WAVEFRONT

Figure 8.2: Coherent beam produced by an extended (monochromatic) source

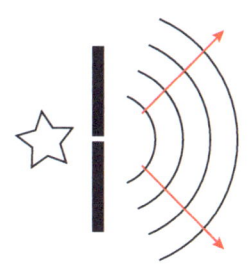

In the next Section, we will see how coherent waves produced in this manner can be combined to show interference effects. Note that if a coherent source such as a laser is used, we do not need to pass the beam through an aperture.

8.3 Young's slits experiment

This experiment is one of the earliest examples of interference by division of wavefront, first carried out by Thomas Young in 1801. The experimental arrangement is shown in Figure 8.3.

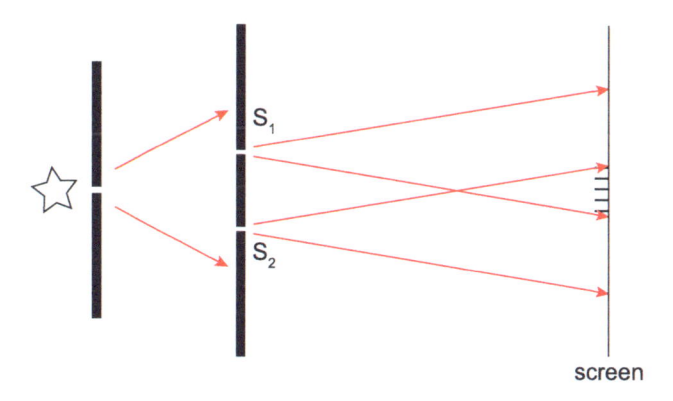

Figure 8.3: Young's slits experiment

Monochromatic light is passed through one narrow slit to give a coherent source, as described earlier. The division of wavefront takes place at the two slits S_1 and S_2. These slits are typically less than 1 mm apart, and act as point sources. A screen is placed about 1 m from the slits. Where the two beams overlap, a symmetrical pattern of fringes is formed, with a bright fringe at the centre.

334　UNIT 2. QUANTA AND WAVES

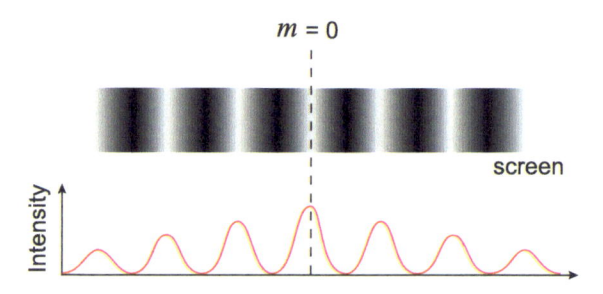

Figure 8.4: Young's slits interference fringes

Figure 8.4 shows the fringe pattern seen on the screen. The lower part of Figure 8.4 shows a plot of irradiance across the screen. The brightest fringe occurs at the centre of the interference pattern as this point is equidistant from the slits, and so the waves arrive exactly in phase. The first dark fringes occur on either side of this when the optical path difference between the beams is exactly half a wavelength. This is followed by the next bright fringe, due to a path difference of exactly one wavelength, and so on.

In general, a bright fringe occurs when the path difference between the two beams is $m\lambda$, where m = 0, 1, 2... The central bright fringe corresponds to $m = 0$.

The fringe spacing can be analysed to determine the wavelength of the light.

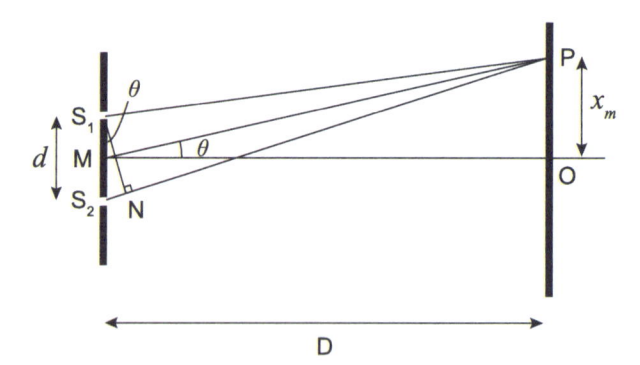

Figure 8.5: Analysis of Young's slits experiment

In Figure 8.5, the slits are separated by a distance d, and M marks the midpoint between the slits S$_1$ and S$_2$. The screen is placed a distance D ($>> d$) from the slits, with O being the point directly opposite M where the $m = 0$ bright fringe occurs. The mth bright fringe is located at the point P, a distance x_m from O.

The path difference between the two beams is $S_2P - S_1P = m\lambda$. If the length PN equal to S_1P is marked on S_2P, then the path difference is the distance $S_2N = m\lambda$.

© HERIOT-WATT UNIVERSITY

TOPIC 8. DIVISION OF WAVEFRONT

Since PM is very much larger than S_1S_2, the line S_1N meets S_2P at approximately a right angle and S_1S_2N is a right-angled triangle. In this triangle

$$\sin\theta = \frac{S_2N}{S_1S_2} = \frac{m\lambda}{d}$$

We can also look at the right-angled triangle formed by $M\ P\ O$

$$\tan\theta = \frac{OP}{MO} = \frac{x_m}{D}$$

Because θ is very small, $\sin\theta \approx \tan\theta \approx \theta$. Therefore

$$\frac{x_m}{D} = \frac{\lambda}{d}$$
$$\therefore x_m = \frac{m\lambda D}{d}$$

To find the separation Δx between fringes, we need to find the distance $x_{m+1} - x_m$ between the $(m+1)^{\text{th}}$ and m^{th} bright fringes

$$\Delta x = x_{m+1} - x_m$$
$$\therefore \Delta x = \frac{(m+1)\lambda D}{d} - \frac{m\lambda D}{d}$$
$$\therefore \Delta x = \frac{\lambda D}{d}$$

(8.1)

Rearranging Equation 8.1 in terms of λ

$$\lambda = \frac{\Delta x d}{D}$$

(8.2)

So a Young's slits experiment can be used to determine the wavelength of a monochromatic light source.

Carry out the following activity to see how the experimental parameters affect the appearance of the interference fringes.

Young's slits experiment

Go online

At this stage there is an online activity which demonstrate Young's slits experiment where the wavelength, slit separation and screen location can be changed.

Example

A Young's slits experiment is set up with a slit separation of 0.400 mm. The fringes are viewed on a screen placed 1.00 m from the slits. The separation between the $m = 0$ and $m = 10$ bright fringes is 1.40 cm. What is the wavelength of the monochromatic light used?

We are given the separation for 10 fringes as 1.40 cm, so the fringe separation $\Delta x = 0.140$ cm. Converting all the distances involved into metres, we have $\Delta x = 1.40 \times 10^{-3}$ m, $d = 4.00 \times 10^{-4}$ m and $D = 1.00$ m. Using Equation 8.2

$$\lambda = \frac{\Delta x d}{D}$$
$$\therefore \lambda = \frac{1.40 \times 10^{-3} \times 4.00 \times 10^{-4}}{1.00}$$
$$\therefore \lambda = 5.60 \times 10^{-7} \text{m}$$
$$\therefore \lambda = 560 \text{ nm}$$

There is one more problem to consider - what happens if a Young's slits experiment is performed with white light? What does the interference pattern look like then?

If white light is used, there is no difference to the $m = 0$ bright fringe. Since this appears at an equal distance from both slits, then constructive interference will occur whatever the wavelength. For the $m = 1, 2, 3...$ fringes, Equation 8.1 tells us that the fringe separation $\Delta x \propto \lambda$, so the fringe separation is smallest for short wavelengths. Thus the violet end of the spectrum produces the $m = 1$ bright fringe closest to $m = 0$. The red end of the spectrum produces a bright fringe at a larger separation. At higher orders, the fringe patterns of different colours will overlap.

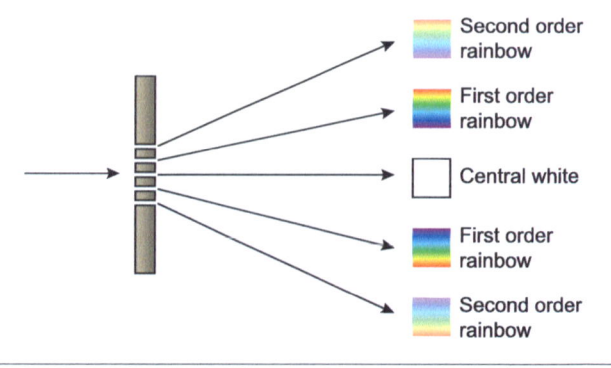

TOPIC 8. DIVISION OF WAVEFRONT

Quiz: Young's slits

Go online

Q1: The Young's slits experiment is an example of

a) interference by division of amplitude.
b) interference by division of wavefront.
c) the Doppler effect.
d) phase change.
e) refractive index.

....

Q2: A Young's slits experiment is set up using slits 0.50 mm apart. The fringes are detected 1.2 m from the slits, and are found to have a fringe spacing of 0.80 mm.
What is the wavelength of the radiation used?

a) 113 nm
b) 128 nm
c) 200 nm
d) 333 nm
e) 720 nm

....

Q3: What happens to the fringes in a Young's slits experiment if a red light source (λ = 640 nm) is replaced by a green light source (λ = 510 nm)?

a) The fringes disappear.
b) The fringes move closer together.
c) The fringes move further apart.
d) There is no change to the position of the fringes.
e) No fringes are formed in either case.

....

Q4: A Young's slits experiment is carried out by passing coherent light of wavelength 480 nm through slits 0.50 mm apart, with a screen placed 2.5 m from the slits.
What is the spacing between bright fringes viewed on the screen?

a) 2.4 mm
b) 3.8 mm
c) 4.8 mm
d) 5.8 mm
e) 9.6 mm

....

© HERIOT-WATT UNIVERSITY

Q5: Using a similar set-up to the previous question, with a red light source at 650 nm, how far from the central bright fringe is the 5th bright fringe formed?

a) 0.00065 m
b) 0.0033 m
c) 0.0041 m
d) 0.0065 m
e) 0.016 m

8.4 Extended information

Web links   Go online

These web links should serve as an insight to the wealth of information available online and allow you to explore the subject further.

- http://www.physicsclassroom.com/class/light/Lesson-3/Young-s-Experiment
- http://www.studyphysics.ca/newnotes/20/unit04_light/chp1719_light/lesson58.htm
- https://www.youtube.com/watch?v=Q1YqgPAtzho

8.5 Summary

Summary

You should now be able to:

- show an understanding of the difference between interference by division of wavefront and division of amplitude;
- describe the Young's slits experiment;
- derive the expression for fringe spacing in a Young's slits experiment, and use this expression to determine the wavelength of a monochromatic source.

8.6 End of topic 8 test

End of topic 8 test — Go online

The following data should be used when required:

Acceleration due to gravity g	9.8 m s^{-2}
Speed of sound	340 m s^{-1}
Speed of light in a vacuum c	9.8 m s^{-2}

Q6: A Young's slits experiment is carried out using monochromatic light of wavelength 525 nm. The slit separation is 4.35×10^{-4} m, and the screen is placed 1.00 m from the slits.

Calculate the spacing between adjacent bright fringes.

_____ m

....

Q7: The wavelength of a monochromatic light source is determined by a Young's slits experiment. The slits are separated by 4.95×10^{-4} m and located 2.50 m from the viewing screen.

If the fringe spacing is 2.48 mm, calculate the wavelength of the light.

_____ nm

....

Q8: In a Young's slits experiment, the fringes are viewed on a screen 1.45 m from the slits. The 8th bright fringe is observed 12.8 mm from the central maximum. The slit separation is 0.400 mm.

Calculate the wavelength of the monochromatic light source.

_____ nm

....

Q9: Coherent light at wavelength 490 nm is passed through a pair of slits placed 1.65 mm apart. The resulting fringes are viewed on a screen placed 1.50 m from the slits.

Calculate how far from the central bright fringe the 5th bright fringe appears.

_____ mm

....

Q10: A Young's slits experiment was carried out using two monochromatic sources A and B. The longer wavelength source (A) had wavelength 642 nm. It was found that the 5th bright fringe from A occurred at the same distance from the central bright fringe as the 6th bright fringe from B.

Calculate the wavelength of B.

_____ nm

Unit 2 Topic 9

Polarisation

Contents

9.1 Introduction 342
9.2 Polarised and unpolarised waves 342
9.3 Polaroid 343
    9.3.1 Polarisation of light waves using Polaroid 344
9.4 Polarisation by reflection 346
    9.4.1 Brewster's law 346
9.5 Applications of polarisation 348
    9.5.1 Other methods of producing polarised light 348
    9.5.2 Applications of polarisation 348
9.6 Extended information 353
9.7 Summary 353
9.8 End of topic 9 test 354

Prerequisites

- Introduction to waves.
- Division of amplitude.
- Reflection and Refraction of Light and Snell's Law.

Learning objective

By the end of this topic you should be able to:

- understand the difference between plane polarised and unpolarised waves;
- to understand Brewster's Law of Polarisation;
- to be able to derive the equation for Brewster's angle $n = \tan i_p$;
- to be able to explain some applications for polarisation.

9.1 Introduction

One of the properties of a transverse wave is that it can be polarised. This means that all the oscillations of the wave are in the same plane. In this topic we will investigate the production and properties of polarised waves. Most of this topic will deal with light waves, and some of the applications of polarised light will be described at the end of the topic.

Light waves are electromagnetic waves, made up of orthogonal (perpendicular) oscillating electric and magnetic fields. When we talk about the oscillations of a light wave, we will be describing the oscillating electric field. For clarity, the magnetic fields will not be shown on any of the diagrams in this chapter - this is the normal practice when describing electromagnetic waves.

Top tip

Suggested practicals:

- Investigate polarisation of microwaves and light.
- Investigate reflected laser (polarised) light from a glass surface through a polarising filter as the angle of incidence is varied.
- Investigate reflected white light through a polarising filter.

9.2 Polarised and unpolarised waves

Let us consider a transverse wave travelling in the x-direction. Although we will be concentrating on light waves in this topic, it is useful to picture transverse waves travelling along a rope. Figure 9.1 shows transverse waves oscillating in the y-direction.

Figure 9.1: Transverse waves with oscillations in the y-direction

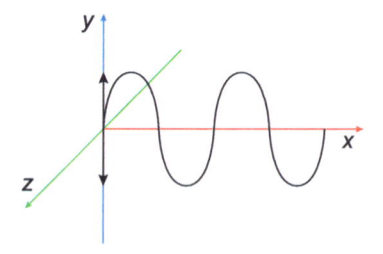

The oscillations are not constrained to the y-direction (the vertical plane). The wave can make horizontal oscillations in the z-direction, or at any angle ϕ in the y-z plane, so long as the oscillations are at right angles to the direction in which the wave is travelling (see Figure 9.2).

Figure 9.2: Transverse waves oscillating (a) in the z-direction, and (b) at an angle in the y-z plane.

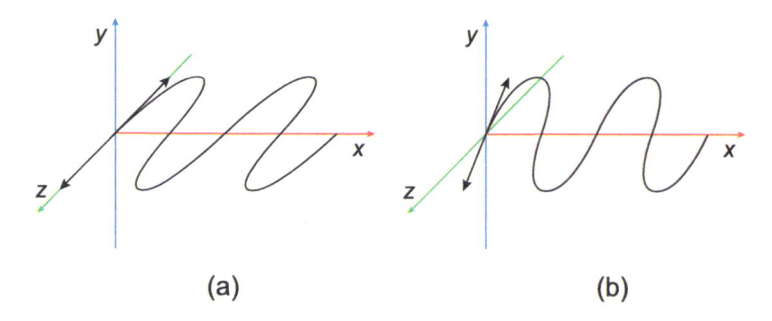

(a)   (b)

When all the oscillations occur in one plane, as shown in Figure 9.1 and Figure 9.2, the wave is said to be **polarised**. If oscillations are occurring in many or random directions, the wave is **unpolarised**. The difference between polarised and unpolarised waves is shown in Figure 9.3.

Figure 9.3: (a) Polarised, and (b) unpolarised waves

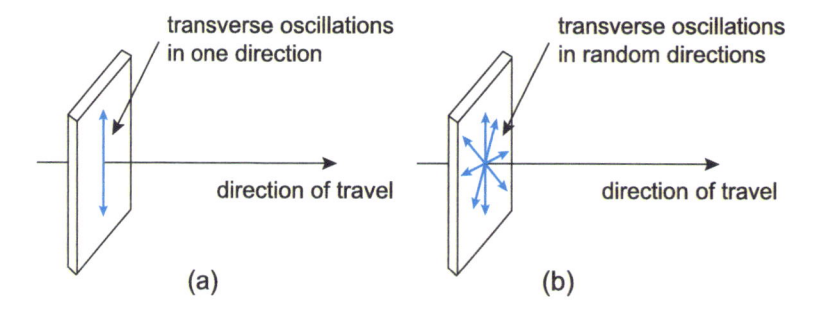

(a)   (b)

Light waves produced by a filament bulb or strip light are **unpolarised**. In the next two Sections of this topic different methods of producing **polarised light** will be described. You should note that longitudinal waves cannot be polarised since the oscillations occur in the direction in which the wave is travelling. This means that sound waves, for example, cannot be **polarised**.

9.3 Polaroid

Before we look at how to polarise light waves, let us think again about a (polarised) transverse wave travelling along a rope in the x-direction with its transverse oscillations in the y-direction. In Figure 9.4 the rope passes through a board with a slit cut into it. Figure 9.4 (a) shows what happens if the slit is aligned parallel to the y-axis. The waves pass through, since the oscillations of the rope are parallel to the slit. In Figure 9.4 (b), the slit is aligned along the z-axis, perpendicular to the oscillations. As a result, the waves cannot be transmitted through the slit.

Figure 9.4: (a) Transverse waves passing through a slit parallel to its oscillations, and (b) a slit perpendicular to the oscillations blocking the transmission of the waves

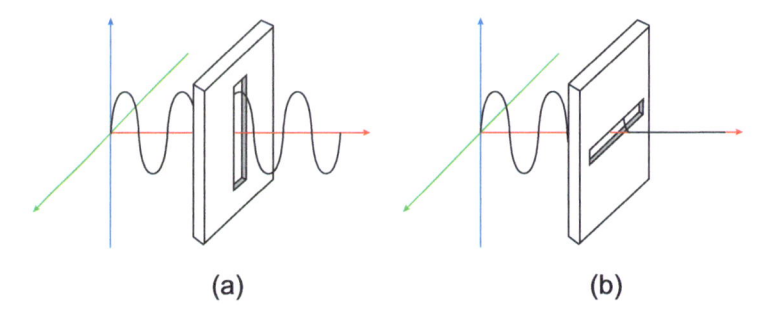

What would happen if the wave incident on the slit was oscillating in the y-z plane, making an angle of 45° with the y-axis, for instance? Under those conditions, the amplitude of the incident wave would need to be resolved into components parallel and perpendicular to the slit. The component parallel to the slit is transmitted, whilst the component perpendicular to the slit is blocked. The transmitted wave emerges polarised parallel to the slit. Later in this Section we will derive the equation used for calculating the irradiance of light transmitted through a polariser whose transmission axis is not parallel to the plane of polarisation of the light waves.

9.3.1 Polarisation of light waves using Polaroid

Although the mechanical analogy is helpful it cannot be carried over directly to the comparable situation involving light waves. Consider a sheet of Polaroid, a material consisting of long, thin polymer molecules (doped with iodine) that are aligned with each other. Because of the way a polarised light wave interacts with the molecules, the sheet of Polaroid only transmits the components of the light with the electric field vector perpendicular to the molecular alignment. The direction which passes the polarised light waves is called the **transmission axis**. The Polaroid sheet blocks the electric field component that is parallel to the molecular alignment.

Figure 9.5: Action of a sheet of Polaroid on unpolarised light

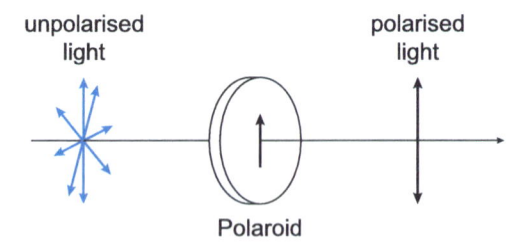

In Figure 9.5, light from a filament bulb is unpolarised. This light is incident on a sheet of Polaroid whose transmission axis is vertical. The beam that emerges on the right of the diagram is polarised in the same direction as the transmission axis of the Polaroid. Remember, this means that the electric field vector of the electromagnetic wave is oscillating in the direction shown.

TOPIC 9. POLARISATION

Polarised light

There is an animation available online showing the electric and magnetic fields of a polarised light wave.

Quiz: Polarisation

Q1: Light waves can be polarised. This provides evidence that light waves are

a) coherent.
b) stationary waves.
c) monochromatic.
d) longitudinal waves.
e) transverse waves.

....

Q2: Unpolarised light passes through a sheet of Polaroid whose transmission axis is parallel to the y-axis. It then passes through a second Polaroid whose transmission axis is at 20° to the y-axis. At what angle is the plane of polarisation of the emergent beam?

a) Parallel to the y-axis.
b) At 10° to the y-axis.
c) At 20° to the y-axis.
d) At 70° to the y-axis.
e) Perpendicular to the y-axis.

....

Q3: Can sound waves be polarised?

a) Yes, any wave can be polarised.
b) No, because the oscillations are parallel to the direction of travel.
c) Yes, because the oscillations are perpendicular to the direction of travel.
d) No, because sound waves are not coherent.
e) Yes, because they are periodic waves.

9.4 Polarisation by reflection

Light reflected by the surface of an electrical insulator is partially, and sometimes fully, polarised. The degree of polarisation is determined by the angle of incidence of the beam and the refractive index of the reflecting material.

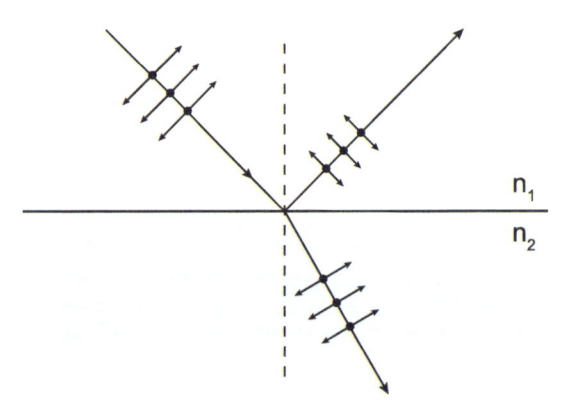

Figure 9.6: Polarisation by reflection

In Figure 9.6 the solid circles represent the components of the incident beam that are polarised parallel to the surface of the reflecting material. The double-headed arrows represent the components at right angles to those shown by the circles. The refracted (transmitted) beam contains both of these components, although the component in the plane of incidence is reduced. The refracted beam is therefore partially polarised, but the reflected beam can be completely polarised parallel to the reflecting surface and perpendicular to the direction in which the beam is travelling.

Usually the reflected beam is not completely polarised, and contains some of the 'arrows' components. We shall look now at the special case in which the reflected beam does become completely polarised.

9.4.1 Brewster's law

The Scottish physicist Sir David Brewster discovered that for a certain angle of incidence, monochromatic light was 100% polarised upon reflection. The refracted beam was partially polarised, but the reflected beam was completely polarised parallel to the reflecting surface. Furthermore, he noticed that at this angle of incidence, the reflected and refracted beams were perpendicular, as shown in Figure 9.7.

Figure 9.7: Brewster's law

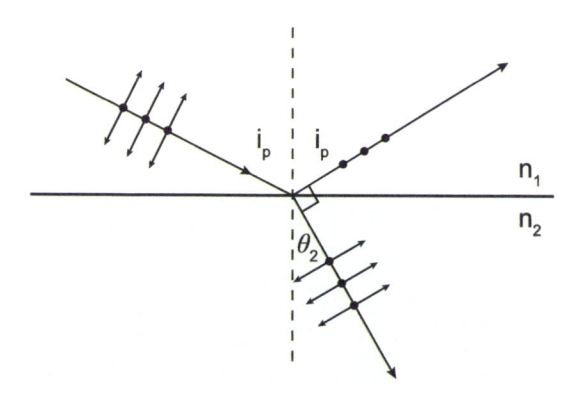

Light is travelling in a medium with refractive index n_1, and being partially reflected at the boundary with a medium of refractive index n_2. The angles of incidence and reflection are i_p, the polarising angle. The angle of refraction is θ_2. Snell's law for the incident and refracted beams is $n_1 \sin i_p = n_2 \sin \theta_2$.

According to Brewster

$$i_p + \theta_2 = 90°$$
$$\therefore \theta_2 = 90° - i_p$$

We can substitute for sin θ_2 in the Snell's law equation

$$n_1 \sin i_p = n_2 \sin \theta_2$$
$$\therefore n_1 \sin i_p = n_2 \sin(90 - i_p)$$
$$\therefore n_1 \sin i_p = n_2 \cos i_p$$
$$\therefore \frac{\sin i_p}{\cos i_p} = \frac{n_2}{n_1}$$
$$\therefore \tan i_p = \frac{n_2}{n_1}$$

(9.1)

This equation is known as **Brewster's law**. Usually the incident beam is travelling in air, so $n_1 \approx 1.00$, and the equation becomes $\tan i_p = n_2$. The polarising angle is sometimes referred to as the Brewster angle of the material.

> **Example**
>
> What is the polarising angle for a beam of light travelling in air when it is reflected by a pool of water ($n = 1.33$)?
>
> Using Brewster's law
>
> $$\tan i_p = n_2$$
> $$\therefore \tan i_p = 1.33$$
> $$\therefore i_p = 53.1°$$

The refractive index of a material varies slightly with the wavelength of incident light. The polarising angle therefore also depends on wavelength, so a beam of white light does not have a unique polarising angle.

> **Brewster's law**   Go online
>
> At this stage there is an online activity which allow you to investigate polarisation by reflection.

9.5 Applications of polarisation

9.5.1 Other methods of producing polarised light

Two methods for polarising a beam of light have been discussed. You should be aware that other techniques can be used. **Birefringent** materials such as calcite (calcium carbonate) have different refractive indices for perpendicular polarisation components. An unpolarised beam incident on a calcite crystal will be split into two beams polarised at right angles to each other. **Dichroic** crystals act in the same way as Polaroid. Their crystal structure allows only light with electric field components parallel to the crystal axis to be transmitted.

9.5.2 Applications of polarisation

Polarised light can be used to measure strain in **photoelastic** materials, such as glass and celluloid. These are materials that become birefringent when placed under mechanical stress. One application of this effect is in stress analysis. A celluloid model of a machine part, for example, is placed between a crossed polariser and analyser. The model is then placed under stress to simulate working conditions. Bright and dark fringes appear, with the fringe concentration highest where the stress is greatest. This sort of analysis gives important information in the design of mechanical parts and structures.

© HERIOT-WATT UNIVERSITY

Figure 9.8: Shows this technique used with plastic forks

Polaroid sunglasses and camera lens filters are often used to reduce glare. In the ideal case, light reflected from a horizontal surface will be polarised in a horizontal plane as described earlier, so a Polaroid with a vertical transmission axis should prevent transmission completely. In practice, reflected light is only partially polarised, and is not always being reflected from a horizontal surface, so glare is only partially reduced by Polaroid sunglasses and filters.

Optically active materials can change the plane of polarisation of a beam of light. This process comes about because of the molecular structure of these materials, and has been observed in crystalline materials such as quartz and organic (liquid) compounds such as sugar solutions. The degree of optical activity can be used to help determine the molecular structure of these compounds. In the technique known as **saccharimetry**, the angle of rotation of the plane of polarisation is used as a measure of the concentration of a sugar solution. Polarised light is passed through an empty tube, and an analyser on the other side of the tube is adjusted until no light is transmitted through it. The tube is then filled with the solution, and the analyser is adjusted until the transmission through it is again zero. The adjustment needed to return to zero transmission is the angle of rotation.

Perhaps the most common everyday use of optical activity is in **liquid crystal displays** (LCDs). A typical LCD on a digital watch or electronic calculator consists of a small cell of aligned crystals sandwiched between two transparent plates between a crossed polariser and analyser. This arrangement is shown schematically in Figure 9.9.

Figure 9.9: LCD display with no field applied (a) and with a field applied (b)

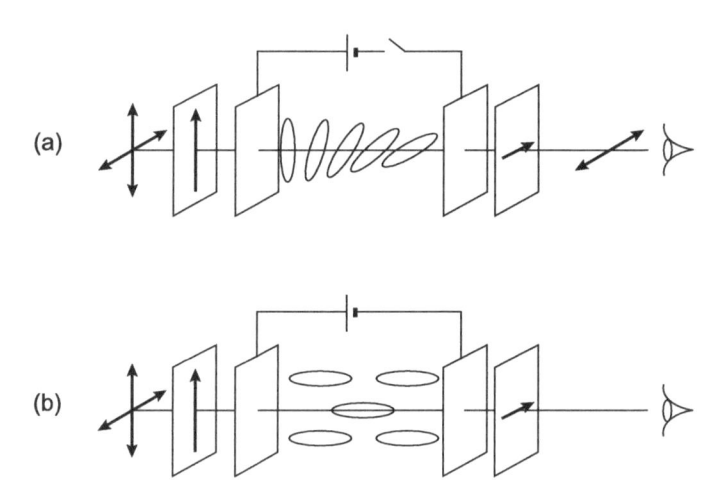

When no electric field is applied across the cell, the liquid crystal molecules are arranged in a helical twist. The polarised light entering the cell has its polarisation angle changed as it travels through the cell, and emerges polarised parallel to the transmission axis of the analyser. The cell appears light in colour, and is thus indistinguishable from the background. When a field is applied, the liquid crystal molecules align in the same direction (the direction of the electric field), and do not change the polarisation of the light. The emerging light remains polarised perpendicular to the analyser transmission axis. This light is not transmitted by the analyser, and so the cell looks dark. In this case, a black segment is seen against a lighter background.

Polarising filters are also used lots in photography, especially for intensifying the colour in images. Blue skies in many photos tend to look washed out and polarising filters are used to darken the sky and remove distracting reflections.

TOPIC 9. POLARISATION

The lenses of some 3D polaroid glasses are both polarising lenses but they are aligned at right angles to each other, meaning that light which can pass through one is blocked by the other and vice versa. Similar to the older red/green or red/blue 3D glasses, they work by delivering slightly different views to each eye to produce a 3D floating image. This is commonly seen in 3D cinemas and has the major advantage that they allow full colour displays which the red/green ones always lacked.

The newest type of 3D glasses is the RealD 3D system which uses circularly polarised light instead of the linear polarisation described above. The advantage of this is that the person wearing the glasses can tilt their head without seeing double or darkened images.

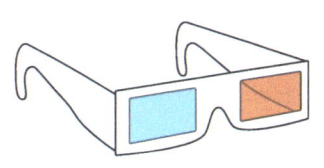

© HERIOT-WATT UNIVERSITY

Quiz: Brewster's law and applications of polarisation

Go online

Q4: What is the polarisation angle for monochromatic light travelling in air, incident on a sheet of glass of refractive index 1.52?

a) 0.026°
b) 33.3°
c) 41.1°
d) 48.8°
e) 56.7°

....

Q5: Light reflected at the polarising angle is

a) polarised parallel to the reflecting surface.
b) polarised perpendicular to the reflecting surface.
c) polarised parallel and perpendicular to the reflecting surface.
d) polarised in the direction in which the wave is travelling.
e) completely unpolarised.

....

Q6: It is found that a beam of light is 100% polarised when reflected from a smooth plastic table-top. If the angle of incidence is 61.0°, what is the refractive index of the plastic?

a) 1.00
b) 1.14
c) 1.31
d) 1.80
e) 2.06

....

Q7: Which of the following sentences best describe a material that exhibits 'optical activity'?

a) The material changes its polarisation when under mechanical stress.
b) The material can rotate the plane of polarisation of a beam of light.
c) The material does not transmit polarised light.
d) The material reflects polarised light.
e) The material has different refractive indices for perpendicular polarisation components.

....

Q8: Polaroid sunglasses are most effective at reducing glare when the transmission axis of the Polaroid is

a) vertical.
b) horizontal.
c) at the polarising angle.
d) at 45° to the horizontal.
e) at 30° to the horizontal.

9.6 Extended information

Web links — Go online

These web links should serve as an insight to the wealth of information available online and allow you to explore the subject further.

- https://www.youtube.com/watch?v=BiRiUOwNdEs
- http://www.chemguide.co.uk/basicorg/isomerism/polarised.html
- https://www.youtube.com/watch?v=E-nOXpVPAPs
- http://www.sixtysymbols.com/videos/polarisation.htm

9.7 Summary

Summary

You should now be able to:

- understand the difference between plane polarised and unpolarised waves;
- to understand Brewster's Law of Polarisation;
- to be able to derive the equation for Brewster's angle $n = \tan i_p$;
- to be able to explain some applications for polarisation.

9.8 End of topic 9 test

End of topic 9 test Go online

Q9: An unpolarised beam of light is incident on a Polaroid filter. The emerging beam then travels through a second Polaroid filter. The first filter has its transmission axis at 22° to the vertical, the second is at 60° to the vertical.

State the angle of polarisation of the beam emerging from the second Polaroid, giving your answer in degrees to the vertical.

....

Q10: Three Polaroid filters are lined up along the x-axis. The first has its transmission axis aligned in the y-direction, the second has its axis at 45° in the y - z plane, and the third has its axis in the z-direction. In which direction is the emerging beam polarised?

a) In the z-direction.
b) At 45° in the y - z plane.
c) In the y-direction.
d) In the x-direction.

....

Q11: The polarising angle for a particular glass is 58.2°.

Calculate the refractive index of the glass.

....

Q12: Calculate the polarising angle, in degrees, for a beam of light travelling in water (n_w = 1.33) incident on a block of leaded glass (n_g = 1.54).

....

Q13: Light travelling in air is partially reflected from a glass block. The reflected light is found to be 100% polarised when the angle of incidence is 57.7°.

1. State the magnitude of the angle between the reflected and refracted beams.
2. Calculate the refractive index of the block.

....

Q14: Some mobile phones polarise the light emitted by their screens. A user is wearing polarising sunglasses when trying to use their phone. The screen is switched on but the image appears black to the user.

a) What is the angle between the polarising filter on the screen and the polaroid in the sunglasses?
b) If the user tilts their head, what happens to the amount of light they see?

© HERIOT-WATT UNIVERSITY

Unit 2 Topic 10

End of section 2 test

End of section 2 test

The following data should be used when required:

Quantity	Symbol	Value
Gravitational acceleration on Earth	g	9.8 m s^{-2}
Mass of an electron	m_e	9.11 x 10^{-31} kg
Planck's constant	h	6.63 x 10^{-34} J s
Speed of light in a vacuum	c	3.00 x 10^8 m s^{-1}
Speed of sound	v	340 m s^{-1}
Refractive index of air		1.00
Refractive index of water		1.33

Q1: An electron thought of as a particle is orbiting a Hydrogen with a speed of 1.45×10^6 m s^{-1}.

1. Calculate the de Broglie wavelength of the electron.
   _____ m
2. Calculate the angular momentum of the electron if it is found in the n = 3 orbit.
   _____ kg m^2 s^{-1}

....

Q2: A proton in a particle accelerator is travelling with velocity 1.45×10^5 m s^{-1}.
Calculate the de Broglie wavelength of the proton.

_____ m

....

Q3: Helical motion of particles causes impressive effects such as the Northern Lights. This is because: (choose all that apply)

1. The component of the velocity perpendicular to the magnetic field causes the particle to continue its forward motion.
2. The Northern Lights are often green due to the Oxygen particles that the charged particles collide with.
3. All charged particles from space are pulled to the Earth's surface by the strength of the Earth's magnetic field.
4. The component of the velocity perpendicular to the magnetic field causes circular motion.
5. The Northern lights are often blue due to the amount of Nitrogen in the atmosphere.
6. The component of the particle's velocity parallel to the magnetic field causes circular motion.

7. The component of the particle's velocity parallel to the magnetic field causes the particle to continue its forward motion.

Q4: An object of mass 0.15 kg is performing simple harmonic oscillations with amplitude 20 mm and periodic time 1.4 s.

1. Calculate the maximum value of the object's acceleration.
   _____ m s^{-2}
2. Calculate the maximum value of the object's velocity.
   _____ m s^{-1}

Q5: A travelling wave is represented by the equation:

$$y = 0.4 \sin\left(2\pi\left(3t - \frac{x}{1.7}\right)\right)$$

1. What is the value of the amplitude of this wave?
   _____ m
2. Calculate the speed of the wave.
   _____ m s^{-1}

Q6: A guitarist plays a note of 475 Hz on their guitar and at the same time a piano note is played at 473 Hz.

How many beats would be heard in 3 seconds of listening to the two notes?
_____ beats

Q7: Telecommunications signals are transmitted in fibre optic cable using infra-red light of wavelength (in air) 1.55×10^{-6} m. The fibre has a refractive index of 1.56.

1. Calculate the wavelength of the light in the fibre.
   _____ m
2. For a light ray which travels straight down the fibre without making any reflections from its sides, calculate the optical path length of a 3.88 m section of fibre.
   _____ m

Q8: The m = 10 dark fringe is observed for a wedge thickness 1.22×10^{-6} metres in an air-wedge interference experiment using monochromatic light.
At what wedge thickness is the m = 15 dark fringe observed?
_____ metres

Q9: The wavelength of a monochromatic light source is determined by a Young's slits experiment, in which the slits are separated by 5.45×10^{-4} m and placed 2.00 m from the viewing screen. The measured fringe spacing is 2.47×10^{-3} m.

Determine the value of the wavelength of the light obtained from this experiment.

_____ nm

....

Q10: Light travelling in air is found to be 100% polarised when it is reflected from the surface of a pool of clear liquid at an angle of incidence 50.7°.

1. Calculate the refractive index of the liquid.

   _____

2. In which direction is the reflected light polarised?
    a) Horizontally.
    b) In the same direction as the reflected ray is travelling.
    c) In the same direction as the incident ray is travelling.
    d) Vertically.

Electromagnetism

1 Electric force and field **361**
    1.1 Introduction 363
    1.2 Electric charge 363
    1.3 Coulomb's law 363
    1.4 Electric field strength 368
    1.5 Summary 374
    1.6 Extended information 375
    1.7 End of topic 1 test 377

2 Electric potential **379**
    2.1 Introduction 380
    2.2 Potential and electric field 380
    2.3 Electric potential around a point charge and a system of charges 383
    2.4 Summary 388
    2.5 Extended information 388
    2.6 End of topic 2 test 389

3 Motion in an electric field **391**
    3.1 Introduction 393
    3.2 Energy transformation associated with movement of charge 393
    3.3 Motion of charged particles in uniform electric fields 396
    3.4 Applications of charged particles and electric fields 400
    3.5 The electronvolt 406
    3.6 Summary 406
    3.7 Extended information 407
    3.8 End of topic 3 test 407

4 Magnetic fields **409**
    4.1 Introduction 411
    4.2 Magnetic forces and fields 411
    4.3 Magnetic field around a current-carrying conductor 415
    4.4 Magnetic induction 419
    4.5 Force on a current-carrying conductor in a magnetic field 420

4.6	The relationship between magnetic induction and distance from a current-carrying conductor	426
4.7	Comparison of forces	427
4.8	Summary	432
4.9	Extended information	433
4.10	End of topic 4 test	433

5 Capacitors — 437

5.1	Introduction	439
5.2	Revision from Higher	439
5.3	The time constant for a CR circuit	448
5.4	Capacitors in a.c. circuits	452
5.5	Capacitive reactance	455
5.6	Summary	456
5.7	Extended information	456
5.8	End of topic 5 test	457

6 Inductors — 461

6.1	Introduction	463
6.2	Magnetic flux and induced e.m.f.	463
6.3	Eddy currents	466
6.4	Inductors and self-inductance	469
6.5	Inductors in d.c. circuits	474
6.6	Inductors in a.c. circuits	479
6.7	Summary	482
6.8	Extended information	483
6.9	End of topic 6 test	485

7 Electromagnetic radiation — 489

7.1	Introduction	490
7.2	The unification of electricity and magnetism	490
7.3	The wave properties of em radiation	490
7.4	Permittivity, permeability and the speed of light	492
7.5	Summary	493
7.6	Extended information	493
7.7	End of topic 7 test	494

8 End of section 3 test — 495

Unit 3 Topic 1

Electric force and field

Contents

1.1	Introduction	363
1.2	Electric charge	363
1.3	Coulomb's law	363
	1.3.1 Electrostatic and gravitational forces	365
	1.3.2 Force between more than two point charges	365
1.4	Electric field strength	368
	1.4.1 Electric field due to point charges	370
1.5	Summary	374
1.6	Extended information	375
	1.6.1 Electric field around a charged conducting sphere	375
	1.6.2 Web links	376
1.7	End of topic 1 test	377

Prerequisites

- Newton's laws of motion.
- Gravitational forces and fields.

Learning objective

By the end of this topic you should be able to:

- carry out calculations involving Coulomb's law for the electrostatic force between point charges $F = \frac{Q_1 Q_2}{4\pi\varepsilon_0 r^2}$;
- describe how the concept of an electric field is used to explain the forces that stationary charged particles exert on each other;
- draw the electric field pattern around a point charge, a system of charges and in a uniform electric field;
- state that the field strength at any point in an electric field is the force per unit positive charge placed at that point in the field, and is measured in units of N C^{-1} ;
- perform calculations relating electric field strength to the force on a charged particle $F = QE$;
- apply the expression for calculating the electric field strength E at a distance r from a point charge $E = \frac{Q}{4\pi\varepsilon_0 r^2}$;
- calculate the strength of the electric field due to a system of charges.

TOPIC 1. ELECTRIC FORCE AND FIELD

1.1 Introduction

In section 2 you studied the motion of charged particles in a magnetic field. The third section is called Electromagnetism and it will build on this work by studying electric, magnetic and electromagnetic fields. Basic to this work is an understanding of electric charge.

Electric forces act on static and moving electric charges. We will be using the concepts of electric field and electric potential to describe electrostatic interactions.

In this topic we will look at the force that exists between two or more charged bodies, and then introduce the concept of an electric field.

1.2 Electric charge

On an atomic scale, electrical charge is carried by protons (positive charge) and electrons (negative charge). The **Fundamental unit of charge** e is the magnitude of charge carried by one of these particles. The value of e is 1.60×10^{-19} coulombs (C). A charge of one coulomb is therefore equal to the charge on 6.25×10^{18} protons or electrons. It should be noted that one coulomb is an extremely large quantity of charge, and we are unlikely to encounter such a huge quantity of charge inside the laboratory. The sort of quantities of charge we are more likely to be dealing with are of the order of microcoulombs (1 μC = 10^{-6} C), nanocoulombs (1 nC = 10^{-9} C) or picocoulombs (1 pC = 10^{-12} C).

1.3 Coulomb's law

Let us consider two charged particles. We will consider point charges; that is to say, we will neglect the size and shape of the two particles and treat them as two points with charges Q_1 and Q_2 separated by a distance r. The force between the two charges is proportional to the magnitude of each of the charges.
That is to say

$$F \propto Q_1 \quad \text{and} \quad F \propto Q_2$$

The force between the two charges is also inversely proportional to the square of their separation. That is to say

$$F \propto {1}/{r^2}$$

These relationships are known as **Coulomb's law**.

UNIT 3. ELECTROMAGNETISM

The mathematical statement of Coulomb's law is

$$F = \frac{Q_1 Q_2}{4\pi\varepsilon_0 r^2}$$

(1.1)

The constant ε_0 is called the permittivity of free space, and has a value of 8.85 × 10^{-12} F m^{-1} or C^2 N^{-1} m^{-2}. So the constant of proportionality in Equation 1.1 is $\frac{1}{4\pi\varepsilon_0}$, which has the value 8.99 × 10^9 N m^2 C^{-2}. This force is called the electrostatic or Coulomb force. It is important to remember that force is a vector quantity, and the **direction** of the Coulomb force depends on the sign of the two charges. You should already be familiar with the rule that 'like charges repel, unlike charges attract'. Also, from Newton's third law of motion, we can see that each particle exerts a force of the same magnitude but the opposite direction on the other particle.

Example

Two point charges A and B are separated by a distance of 0.200 m. If the charge on A is +2.00 μC and the charge on B is -1.00 μC, calculate the force each charge exerts on the other.

Figure 1.1: Coulomb force acting between the two particles

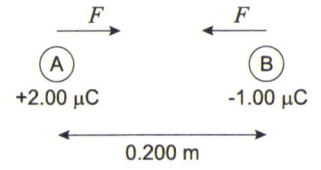

The force acting on the particles is given by Coulomb's law:

$$F = \frac{Q_1 Q_2}{4\pi\varepsilon_0 r^2}$$

$$\therefore F = \frac{2.00 \times 10^{-6} \times (-1.00 \times 10^{-6})}{4\pi \times 8.85 \times 10^{-12} \times (0.200)^2}$$

$$\therefore F = \frac{-2.00 \times 10^{-12}}{4\pi \times 8.85 \times 10^{-12} \times 0.04}$$

$$\therefore F = \frac{-2.00}{4\pi \times 8.85 \times 0.04}$$

$$\therefore F = -0.450 \text{ N}$$

The size of the force is 0.450 N. The minus sign indicates that we have two oppositely charged particles, and hence each charge exerts an attractive force on the other.

1.3.1 Electrostatic and gravitational forces

The Coulomb's law equation looks very similar to the equation used to calculate the gravitational force between two particles (Newton's law of gravitation).

$$F = \frac{Q_1 Q_2}{4\pi\varepsilon_0 r^2} \qquad\qquad F = G\frac{m_1 m_2}{r^2}$$

$$F \propto \frac{Q_1 Q_2}{r^2} \qquad\qquad F \propto \frac{m_1 m_2}{r^2}$$

In both cases, the size of the force follows an **inverse square law** dependence on the distance between the particles. One important difference between these forces is that the gravitational force is always attractive, whereas the direction of the Coulomb force depends on the charge carried by the two particles.

1.3.2 Force between more than two point charges

So far we have used Coulomb's law to calculate the force due to two charged particles, and we find equal and opposite forces exerted on each particle. We will now consider what happens if another charged particle is introduced into the system.

To calculate the force on a charged particle due to two (or more) other charged particles we perform a Coulomb's law calculation to work out each individual force. The **total** force acting on one particle is then the **vector sum** of all the forces acting on it. This makes use of the **principle of superposition of forces**, and holds for any number of charged particles.

Example

Earlier we looked at the problem of two particles A (+2.00 μC) and B (-1.00 μC) separated by 0.200 m. Let us now put a third particle X (+3.00 μC) at the midpoint of AB. What is the magnitude of the total force acting on X, and in what direction does it act?

Figure 1.2: Separate forces acting on a charge placed between two charged particles

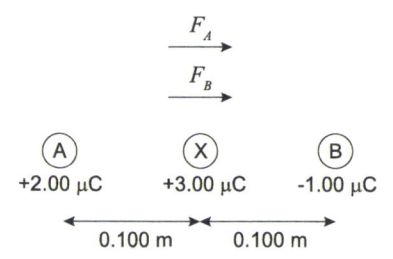

When solving a problem like this, you should always draw a sketch of all the charges, showing their signs and separations. In this case, Figure 1.2 shows the force that A exerts on X is repulsive, and the force that B exerts on X is attractive. Thus both forces act in the **same** direction.

In calculating the size of the two forces we will ignore any minus signs. What we are looking for is just the magnitude of each force. The vector diagram we have drawn shows us the direction of the two forces.

$$F_{AX} = \frac{Q_A Q_X}{4\pi\varepsilon_0 r_{AX}^2}$$
$$\therefore F = \frac{2.00 \times 10^{-6} \times 3.00 \times 10^{-6}}{4\pi \times 8.85 \times 10^{-12} \times 0.100^2}$$
$$\therefore F = \frac{6.00 \times 10^{-12}}{4\pi \times 8.85 \times 10^{-12} \times 0.01}$$
$$\therefore F = \frac{6.00}{4\pi \times 8.85 \times 0.01}$$
$$\therefore F = 5.40 \text{ N}$$

$$F_{BX} = \frac{Q_B Q_X}{4\pi\varepsilon_0 r_{BX}^2}$$
$$\therefore F = \frac{1.00 \times 10^{-6} \times 3.00 \times 10^{-6}}{4\pi \times 8.85 \times 10^{-12} \times 0.100^2}$$
$$\therefore F = \frac{3.00 \times 10^{-12}}{4\pi \times 8.85 \times 10^{-12} \times 0.01}$$
$$\therefore F = \frac{3.00}{4\pi \times 8.85 \times 0.01}$$
$$\therefore F = 2.70 \text{ N}$$

The vector sum of these two forces is 5.40 + 2.70 = 8.10 N. The direction of the force on X is towards B. The same technique of finding the vector sum would be used in the more general case where the charges were not placed in a straight line.

Three charged particles in a line

Two point charges are separated by 1.00 m. One of the charges (X) is +4.00 μC, the other (Y) is -6.00 μC. A third charge of +1.00 μC is placed between X and Y. Without performing a detailed calculation, sketch a graph to show how the force on the third charge varies as it is moved along the straight line from X to Y.

The total Coulomb force acting on a charged object is equal to the vector sum of the individual forces acting on it.

Quiz: Coulomb force

Go online

Useful data:

Fundamental charge e	1.60 × 10^{-19} C
Permittivity of free space ε_0	8.85 × 10^{-12} F m^{-1}

TOPIC 1. ELECTRIC FORCE AND FIELD

Q1: Two particles J and K, separated by a distance r, carry different positive charges, such that the charge on K is twice as large as the charge on J. If the electrostatic force exerted on J is F N, what is the magnitude of the force exerted on K?

a) $4F$ N
b) $2F$ N
c) F N
d) $F/2$ N
e) $F/4$ N

....

Q2: How many electrons are needed to carry a charge of -1 C?

a) 1.60×10^{-19}
b) 8.85×10^{-12}
c) 1.00
d) 6.25×10^{18}
e) 1.60×10^{19}

....

Q3: A point charge of +5.0 μC sits at point A, 10 cm away from a -2.0 μC charge at point B. What is the Coulomb force acting on the charge placed at point A?

a) 7.2 N away from B.
b) 7.2 N towards B.
c) 9.0 N away from B.
d) 9.0 N towards B.
e) 13.5 N away from B.

....

Q4: A small sphere charged to -4.0 μC experiences an attractive force of 12.0 N due to a nearby point charge of +2.0 μC. What is the separation between the two charged objects?

a) 7.7 cm
b) 6.7 cm
c) 1.0 cm
d) 6.0 mm
e) 4.5 mm

....

Q5: Three point charges X, Y and Z lie on a straight line with Y in the middle, 5.0 cm from both of the other charges. If the values of the charges are X = +1.0 μC, Y = -2.0 μC and Z = +3.0 μC, what is the net force exerted on Y?

a) 0.0 N
b) 14.4 N towards Z
c) 14.4 N towards X
d) 28.8 N towards Z
e) 28.8 N towards X

© HERIOT-WATT UNIVERSITY

1.4 Electric field strength

Earlier in this topic we noted the similarity between Coulomb's law and Newton's law of gravitation. In our work on gravitation we introduced the idea of a gravitational field. Similarly, an **electric field** can be defined as the space that surrounds electrically charged particles and in which a force is exerted on other electrically charged particles. Just as the gravitational field strength is the force acting per unit mass placed in a gravitational field, electric field strength E is the force F acting per unit positive charge Q placed at a point in the electric field.

$$E = \frac{F}{Q}$$

(1.2)

The units of E are N C^{-1}. E is a vector quantity, and the direction of the vector, like the direction of the force F, is the direction of the force acting on a positive charge Q. We can define the electric field strength as being the force acting on a unit positive charge (+1 C) placed in the field. This definition gives us not only the magnitude, but also the direction of the field vector. Rearranging Equation 1.2 as $F = QE$, we can see that this is of the same form as the relationship between gravitational field and force: $F = mg$.

We can calculate the field at a distance r from a point charge Q using Equation 1.1 and Equation 1.2. Starting from Equation 1.1

$$F = \frac{Q_1 \, Q_2}{4\pi\varepsilon_0 r^2}$$

Replacing Q_1 by the charge Q and Q_2 by the unit positive test charge Q_{test} gives us

$$F = \frac{Q \, Q_{test}}{4\pi\varepsilon_0 r^2}$$

Substituting for $F = EQ_{test}$ (from Equation 1.2)

$$EQ_{test} = \frac{Q \, Q_{test}}{4\pi\varepsilon_0 r^2}$$
$$\therefore E = \frac{Q}{4\pi\varepsilon_0 r^2}$$

(1.3)

Using this equation, we can calculate the field strength due to a point charge at any position in the field.

TOPIC 1. ELECTRIC FORCE AND FIELD

Example

The electric field strength 2.0 m away from a point charge is 5.0 N C⁻¹. Calculate the value of the charge.

To find the charge Q, we need to rearrange Equation 1.3

$$E = \frac{Q}{4\pi\varepsilon_0 r^2}$$
$$\therefore Q = E \times 4\pi\varepsilon_0 r^2$$

Now we can insert the values of E and r given in the question

$$Q = 5.0 \times 4\pi \times 8.85 \times 10^{-12} \times 2.0^2$$
$$\therefore Q = 2.2 \times 10^{-9} \text{ C}$$

We can sketch the electrical field lines around the charge as shown in Figure 1.3, but remember that the sign of the charge determines the direction of the field vectors.

Also remember that the closer the field lines, the stronger the electric field. So the spacing of the field lines shows us that the greater the distance from the charge, the weaker the electric field.

Figure 1.3: Field lines around (a) an isolated positive charge; (b) an isolated negative charge

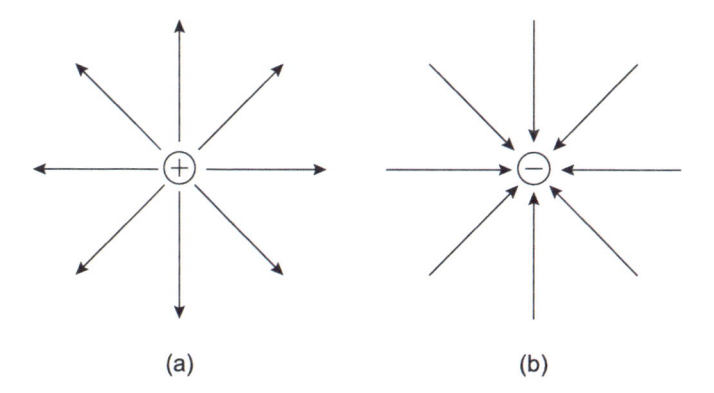

(a)   (b)

1.4.1 Electric field due to point charges

The electric field is another vector quantity. We can work out the total electric field due to more than one point charge by calculating the vector sum of the fields of each individual charge.

Example

Two point charges Q_A and Q_B are placed at points A and B, where the distance AB = 0.50 m. Q_A = +2 µC. Calculate the total electric field strength at the midpoint of AB if

1. Q_B = +3 µC
2. Q_B = -3 µC.

Solution

1. When the two charges have the same sign, the fields at the midpoint of AB are in opposite directions, so the total field strength is given by the difference between them. As shown in Figure 1.4, since Q_A and Q_B are both positive charges, a test charge placed between them will be repulsed by both.

Figure 1.4: Electric fields due to two positive charges

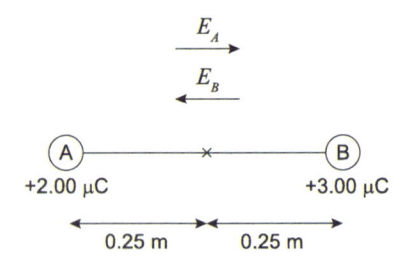

Total field $E = E_B - E_A$

$$\therefore E = \frac{Q_B}{4\pi\varepsilon_0 r^2} - \frac{Q_A}{4\pi\varepsilon_0 r^2}$$

$$\therefore E = \frac{Q_B - Q_A}{4\pi\varepsilon_0 r^2}$$

$$\therefore E = \frac{(3.0 \times 10^{-6}) - (2.0 \times 10^{-6})}{4\pi\varepsilon_0 (0.25)^2}$$

$$\therefore E = \frac{1.0 \times 10^{-6}}{4\pi\varepsilon_0 \times 0.0625}$$

$$\therefore E = 1.4 \times 10^5 \, \text{N}\,\text{C}^{-1}$$

The total field strength is 1.4×10^5 N C^{-1} directed towards A.

TOPIC 1. ELECTRIC FORCE AND FIELD

2. When Q_B is a negative charge, the two field components at the midpoint are pointing in the same direction, as shown in Figure 1.5. The total field strength then becomes the sum of the two components.

Figure 1.5: Electric fields due to two opposite charges

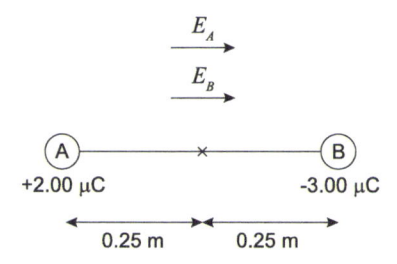

$$E = \frac{Q_A}{4\pi\varepsilon_0 r^2} + \frac{Q_B}{4\pi\varepsilon_0 r^2}$$
$$\therefore E = \frac{Q_A + Q_B}{4\pi\varepsilon_0 r^2}$$
$$\therefore E = \frac{(2.0 \times 10^{-6}) + (3.0 \times 10^{-6})}{4\pi\varepsilon_0 (0.25)^2}$$
$$\therefore E = \frac{5.0 \times 10^{-6}}{4\pi\varepsilon_0 \times 0.0625}$$
$$\therefore E = 7.2 \times 10^5 \, \text{N}\,\text{C}^{-1}$$

The total field strength is 7.2×10^5 N C^{-1} directed towards B. As usual, making a quick sketch showing the charges and their signs helps avoid mistakes.

The electric field pattern between two point charges depends upon their polarity. Remember that the field lines always point in the direction that positive charge would move. An electron would move in the opposite direction to the arrows.

UNIT 3. ELECTROMAGNETISM

Figure 1.6: Electric field pattern for (a) two opposite point charges, (b) two positive point charges and (c) two negative point charges.

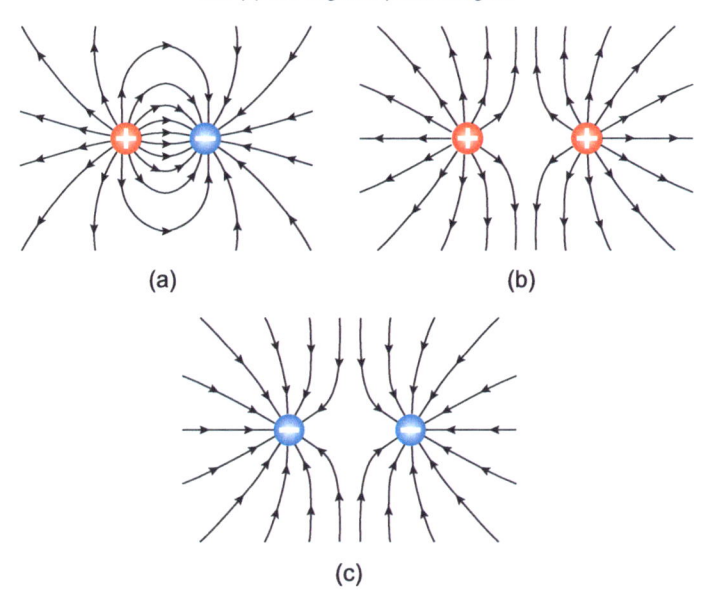

(a)

(b)

(c)

Quiz: Electric field

Useful data:

Fundamental charge e	1.60×10^{-19} C
Permittivity of free space ε_0	8.85×10^{-12} C^2 N^{-1} m^{-2}

Q6: At a distance x m from an isolated point charge, the electric field strength is E N C^{-1}. What is the strength of the electric field at a distance $2x$ m from the charge?

a) $E/4$
b) $E/2$
c) E
d) $2E$
e) $4E$

TOPIC 1. ELECTRIC FORCE AND FIELD

Q7: A +2.50 nC charged sphere is placed in an electric field of strength 5.00 N C^{-1}.
What is the magnitude of the force exerted on the sphere?

a) 5.00×10^{-10} N
b) 2.00×10^{-9} N
c) 2.50×10^{-9} N
d) 5.00×10^{-9} N
e) 1.25×10^{-8} N

....

Q8: A 50.0 N C^{-1} electric field acts in the positive x-direction.
What is the force on an electron placed in this field?

a) 0.00 N
b) 6.25×10^{-18} N in the -x-direction
c) 6.25×10^{-18} N in the +x-direction
d) 8.00×10^{-18} N in the -x-direction
e) 8.00×10^{-18} N in the +x-direction

....

Q9: Two point charges, of magnitudes +30.0 nC and +50.0 nC, are separated by a distance of 2.00 m.
What is the magnitude of the electric field strength at the midpoint between them?

a) 1.35×10^{-5} N C^{-1}
b) 30.0 N C^{-1}
c) 45.0 N C^{-1}
d) 180 N C^{-1}
e) 720 N C^{-1}

....

Q10: A +1.0 μC charge is placed at point X. A +4.0 μC charge is placed at point Y, 50 cm from X.
How far from X, on the line XY, is the point where the electric field strength is zero?

a) 10 cm
b) 17 cm
c) 25 cm
d) 33 cm
e) 40 cm

© HERIOT-WATT UNIVERSITY

1.5 Summary

In this topic we have studied electrostatic forces and fields. An electrostatic (Coulomb) force exists between any two charged particles. The magnitude of the force is proportional to the product of the two charges, and inversely proportional to the square of the distance between them. The direction of the force acting on each of the particles is determined by the sign of the charges. If more than two charges are being considered, the total force acting on a particle is the vector sum of the individual forces.

An electric field is a region in which a charged particle will be subject to the Coulomb force. The electric field strength is measured in N C^{-1}. The direction of the electric field vector at a point in a field is the direction in which a Coulomb force would act on a positive charge placed at that point.

Summary

You should now be able to:

- carry out calculations involving Coulomb's law for the electrostatic force between point charges $F = \frac{Q_1 Q_2}{4\pi\varepsilon_0 r^2}$;
- describe how the concept of an electric field is used to explain the forces that stationary charged particles exert on each other;
- draw the electric field pattern around a point charge, a system of charges and in a uniform electric field;
- state that the field strength at any point in an electric field is the force per unit positive charge placed at that point in the field, and is measured in units of N C^{-1} ;
- perform calculations relating electric field strength to the force on a charged particle $F = QE$;
- apply the expression for calculating the electric field strength E at a distance r from a point charge
$E = \frac{Q}{4\pi\varepsilon_0 r^2}$;
- calculate the strength of the electric field due to a system of charges.

1.6 Extended information

1.6.1 Electric field around a charged conducting sphere

Let's consider the electric field in the region of the a charged metal sphere. The first point to note is that the charge resides on the surface of the conductor. The electrostatic repulsion between all the individual charges means that, in equilibrium, all the excess charge (positive or negative) rests on the surface. The interior of the conductor is neutral. The distribution of charges across the surface means that **the electric field is zero at any point within a conducting material.**

The same is true of a hollow conductor. Inside the conductor, the field is zero at every point. If we were to place a test charge somewhere inside a hollow charged sphere, there would be no net force acting on it. This fact was first demonstrated by Faraday in his 'ice pail' experiment, and has important applications today in electrostatic screening.

If we plot the electric field inside and outside a hollow conducting sphere, we find it follows a $1/r^2$ dependence outside, but is equal to zero inside, as shown in Figure 1.7

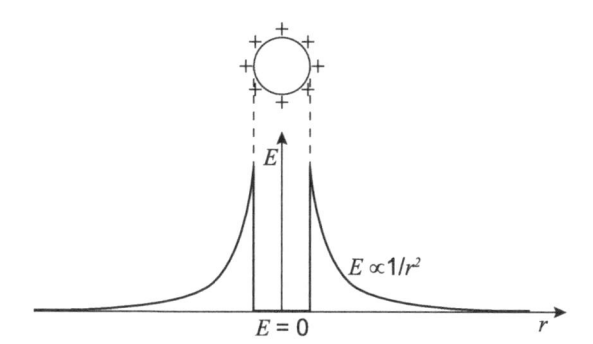

Figure 1.7: Electric field in and around a charged hollow conductor

The fact that there is zero net electric field inside a hollow charged conductor means that we can use a hollow conductor as a shield from electric fields. This is another effect that was first demonstrated by Faraday, who constructed a metallic 'cage' which he sat inside holding a sensitive electroscope. As the cage was charged up, there was no deflection of the electroscope, indicating there was no net electric field present inside the cage.

This electrostatic screening is used to protect sensitive electronic circuitry inside equipment such as computers and televisions. By enclosing these parts in a metal box they are shielded from stray electric fields from other appliances (vacuum cleaners etc.). This principle is illustrated in Figure 1.8

Figure 1.8: Electrostatic screening inside a hollow conductor placed in an electric field

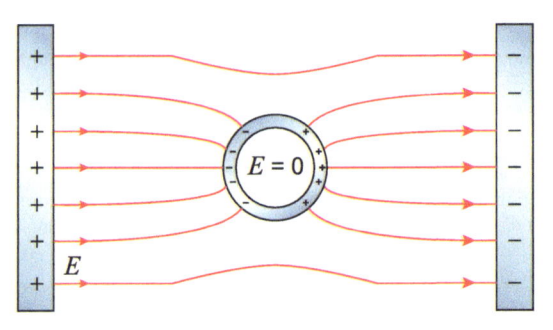

The electric field lines are perpendicular to the surface of the conductor. Induced charge lies on the surface of the sphere, and the net field inside the sphere is zero.

Electrostatic screening also means that the safest place to be when caught in a lightning storm is inside a (metal) car. If the car were to be struck by lightning, the charge would stay on the outside, leaving the occupants safe.

Faraday's ice pail experiment Go online

There is an online activity which shows how Faraday used a metal ice pail connected to an electroscope to demonstrate that charge resides on the outside of a conductor.

1.6.2 Web links

Web links Go online

These web links should serve as an insight to the wealth of information available online and allow you to explore the subject further.

- https://www.youtube.com/watch?v=Z_H156sgnbQ&list=PLiISjSDTPUjy2OKK5-BI2WrVu91LAe_cm&index=18
- https://www.youtube.com/watch?v=jIZSyMv4vpE&list=PLiISjSDTPUjy2OKK5-BI2WrVu91LAe_cm&index=19
- https://www.youtube.com/watch?v=ePxA7XiMXUs&list=PLiISjSDTPUjy2OKK5-BI2WrVu91LAe_cm&index=20
- https://sites.google.com/site/physicsflash/home/coulomb
- https://www.youtube.com/watch?v=B5LVoU_a08c&feature=youtu.be
- https://www.youtube.com/watch?v=mr-ApJ0SwpA
- https://www.youtube.com/watch?v=yUPdtFqilXo

TOPIC 1. ELECTRIC FORCE AND FIELD

1.7 End of topic 1 test

End of topic 1 test   Go online

The following data should be used when required:

Fundamental charge e	1.60×10^{-19} C
Permittivity of free space ε_0	8.85×10^{-12} F m^{-1}

Q11: How many electrons are required to charge a neutral body up to -0.032 C?

_____

....

Q12: Two identical particles, each carrying charge +7.65 µC, are placed 0.415 m apart. Calculate the magnitude of the electrical force acting on each of the particles.

_____ N

....

Q13: The Coulomb force between two point charges is 2.3 N.

Calculate the new magnitude of the Coulomb force if the distance between the two charges is halved.

_____ N

....

Q14: Three charged particles A, B and C are placed in a straight line, with AB = BC = 100 mm. The charges on each of the particles are A = -8.3 µC, B = -2.2 µC and C = +4.3 µC. What is the magnitude of the total force acting on B?

_____ N

....

Q15: Calculate the electric field strength at a distance 3.2 m from a point charge of +6.5 nC.

_____ N C^{-1}

....

Q16: Two charged particles, one with charge -18 nC and the other with charge +23 nC are placed 1.0 m apart.

Calculate the electric field strength at the point midway between the two particles.

_____ N C^{-1}

....

Q17: A small sphere of mass 0.055 kg, charged to 1.7 µC, is suspended by a thread. If the thread is cut, the sphere will fall to the ground under gravity.

© HERIOT-WATT UNIVERSITY

Calculate the magnitude of the electric field acting vertically which would hold the sphere in position when the string is cut.

_____ N C⁻¹

....

Q18: An electron is placed in an electric field of strength 0.013 N C⁻¹.

Calculate the acceleration of the electron.

_____ m s⁻²

Unit 3 Topic 2

Electric potential

Contents

2.1 Introduction 380
2.2 Potential and electric field 380
2.3 Electric potential around a point charge and a system of charges 383
    2.3.1 Calculating the potential due to one or more charges 383
2.4 Summary 388
2.5 Extended information 388
    2.5.1 Potential around a hollow conductor 388
    2.5.2 Web links 389
2.6 End of topic 2 test 389

Prerequisites

- Electric force and field.

Learning objective

By the end of this topic you should be able to:

- state that the electric potential V at a point is the work done by external forces in moving a unit positive charge from infinity to that point;
- apply the expression $E = \frac{V}{d}$ for a uniform electric field;
- explain what it meant by a conservative field;
- state that an electric field is a conservative field;
- state and apply the equation $V = \frac{Q}{4\pi\varepsilon_0 r}$ for the potential V at a distance r from a point charge Q.

2.1 Introduction

In this topic we will be considering the electric potential V. You should already have encountered V in several different contexts - the e.m.f. of a battery, the potential difference across a resistor, calculating the energy stored by a capacitor, and so on. From all these, you should be aware that the potential is a measure of work done or energy in a system. We will be investigating this idea more fully in this topic, and finding how the potential V relates to the electric field E.

We will also investigate the potential due to a point charge and a system of point charges.

2.2 Potential and electric field

In the previous topic, we found similarities between the electric field and force, and the gravitational field and force. We can again draw on this similarity to describe the electric potential.

The gravitational potential tells us how much work is done per unit mass in moving an object which has been placed in a gravitational field. The **electric potential** (or more simply the potential) tells us how much work is done in moving a unit positive charge placed in an electric field. The gravitational potential at a point in a gravitational field was defined as the work done in bringing unit mass from infinity to that point. Similarly, the electric potential V at a point in an electric field can be defined as the work done E_W in bringing unit positive charge Q from infinity to that point in the electric field. This gives us

$$V = \frac{work\,done}{Q}$$
$$\text{or } work\,done = E_W = QV$$

(2.1)

From this expression, we can define the unit of electric potential, the volt:

one volt (1 V) = one joule per coulomb (1 J C^{-1})

The **potential difference** V between two points A and B, separated by a distance d, is defined as the work done in moving one unit of positive charge from A to B. Let us consider the case of a uniform electric field E, such as that which exists between the plates of a large parallel-plate capacitor. We will assume A and B lie on the same electric field line.

TOPIC 2. ELECTRIC POTENTIAL

Figure 2.1: The electric field between the plates of a charged capacitor. The field is uniform everywhere between the plates except near the edges

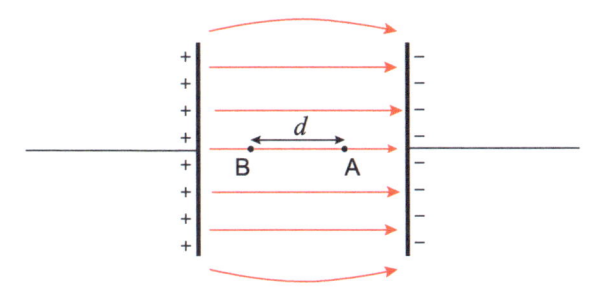

We can use the definition of *work done = force × distance* along with Equation 2.1 to find the work done in moving a unit positive charge from A to B. Remember that the electric field strength is defined as the force acting per unit charge

$$Work\,done = force \times distance$$
$$\therefore QV = (QE) \times d$$
$$\therefore V = E \times d$$

(2.2)

This equation is only valid in the special case of a uniform electric field, where the value of E is constant across the entire distance d.

Example

The potential difference between the two plates of a charged parallel-plate capacitor is 12 V. What is the electric field strength between the plates if their separation is 200 μm?

Rearranging Equation 2.2 gives us

$$E = \frac{V}{d}$$
$$\therefore E = \frac{12}{200 \times 10^{-6}}$$
$$\therefore E = 6.0 \times 10^4 \, \text{N}\,\text{C}^{-1}$$

Looking at the rearranged Equation 2.2, the electric field strength is given as a potential difference divided by a distance. This means that we can express E in the units V m^{-1}, which are equivalent to N C^{-1}.

There is one final point we should note about potential difference. Looking back to Figure 2.1, we moved a unit charge directly from A to B by the shortest possible route. The law of conservation of energy tells us that the work done in moving from A to B is independent of the route taken.

© HERIOT-WATT UNIVERSITY

Figure 2.2: Different routes from A to B

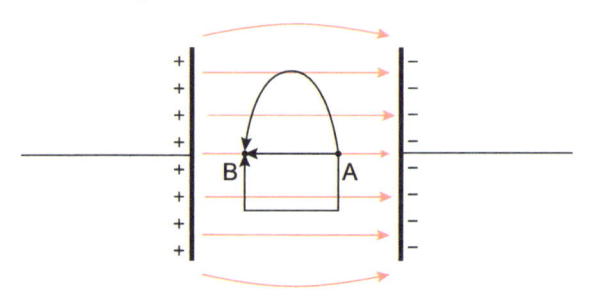

Irrespective of the route taken, the start and finish points are the same in Figure 2.2. If the potential difference is V between A and B, the same amount of work must be done in moving a unit of charge from A to B, whatever path is taken. This is because the electric field is a **conservative field**.

Quiz: Potential and electric field   Go online

Useful data:

Fundamental charge e	1.60×10^{-19} C
Permittivity of free space ε_0	8.85×10^{-12} F m^{-1}

Q1: The uniform electric field between two plates of a charged parallel-plate capacitor is 4000 N C^{-1}. If the separation of the plates is 2.00 mm, what is the potential difference between the plates?

a) 500 mV
b) 2.00 V
c) 8.00 V
d) 2000 V
e) 8000 V

....

Q2: Which of the following corresponds to the units of electric field?

a) J C^{-1}
b) N m^{-1}
c) J V^{-1}
d) V m^{-1}
e) N V^{-1}

....

© HERIOT-WATT UNIVERSITY

TOPIC 2. ELECTRIC POTENTIAL

Q3: A particle carrying charge + 20 mC is moved through a potential difference of 12 V. How much work is done on the particle?

a) 0.24 J
b) 0.60 J
c) 1.67 J
d) 240 J
e) 600 J

....

Q4: If 4.0 J of work are done in moving a 500 μC charge from point M to point N, what is the potential difference between M and N?

a) 2.0×10^{-5} V
b) 0.80 V
c) 12.5 V
d) 2000 V
e) 8000 V

2.3 Electric potential around a point charge and a system of charges

This section covers:

- Calculating the potential due to one or more charges

2.3.1 Calculating the potential due to one or more charges

When calculating the electrical potential at a point, we describe the charges as point charges. As we have seen with electric fields when a different charged shape is involved the electric field associated with it becomes more complicated.

The electric field E at a distance r from a point charge Q is given by,

$$E = \frac{Q}{4\pi\varepsilon_0 r^2}$$

The expression to determine the potential V at a distance r from the point charge is,

$$V = \frac{Q}{4\pi\varepsilon_0 r}$$

(2.3)

We need to be careful with the sign of the potential.

Remember that the potential due to a point charge is zero at an infinite distance from the charge. In moving a positive charge from infinity to r, the charge will have gained potential energy, as work is done on the charge against the electric field.

If we have defined the potential to be zero at infinity, the potential V must be positive for all r less than infinity.

Note the difference here between electric and gravitational potential. In both cases, we define zero potential at $r = \infty$. The difference is that in moving unit mass from infinity in a gravitational field, the field does work on the mass, making the potential less than at infinity, and hence a negative number.

In moving a unit *positive* charge, we must do work *against* the E-field, so the potential increases, and hence is a positive number.

Unlike the electric field, the electric potential around a point charge decays as $1/r$, not $1/r^2$. The potential is a scalar quantity, not a vector quantity, although its sign is determined by the sign of the charge Q. The field strength and potential around a positive point charge are plotted in Figure 2.3.

Figure 2.3: Plots of field strength and potential with increasing distance from a point charge

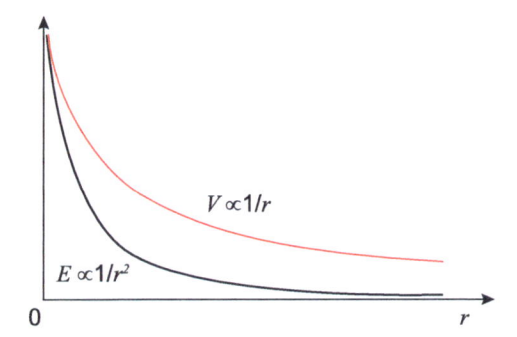

Example

At a distance 40 cm from a positive point charge, the electric field is 200 N C^{-1} and the potential is 24 V. What are the electric field strength and electric potential 20 cm from the charge?

The electric field strength E varies as $1/r^2$, so if the distance from the charge is halved, the field strength increases by 2^2.

That is to say

$$E_r \propto \frac{1}{r^2}$$

So if we replace r by $r/2$ in the above equation

TOPIC 2. ELECTRIC POTENTIAL

$$E_{r/2} \propto \frac{1}{(r/2)^2} = \frac{4}{r^2}$$

So

$$\frac{E_{r/2}}{E_r} = \frac{4/r^2}{1/r^2} = 4$$

E therefore increases by a factor of four, and the new value of E is $4 \times 200 = 800$ N C^{-1}.

V is proportional to $1/r$, so if r is halved, V will double. Hence the new value of V is 48 V.

When more than one charged particle is present, we can calculate the total potential at a point by adding the individual potentials. This is similar to the way in which we worked out the total Coulomb force and the total electric field. Once again, care must be taken with the signs of the different charges present.

Example

What is the net electric potential at the point midway between two point charges of +2.00 µC and -5.00 µC, if the two charges are 2.00 m apart?

Figure 2.4: Two charged objects 2.00 m apart

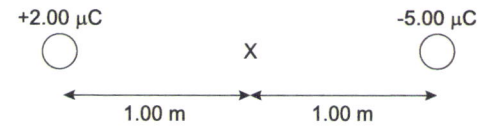

+2.00 µC   -5.00 µC
     ◯   X   ◯
        ←—— 1.00 m ——→←—— 1.00 m ——→

As usual, a sketch such as Figure 2.4 helps in solving the problem.

© HERIOT-WATT UNIVERSITY

The potential due to the positive charge on the left is

$$V_1 = \frac{Q_1}{4\pi\varepsilon_0 r_1}$$
$$\therefore V_1 = \frac{2.00 \times 10^{-6}}{4\pi\varepsilon_0 \times 1.00}$$
$$\therefore V_1 = 1.80 \times 10^4 \text{ V}$$

The potential due to the negative charge on the right is

$$V_2 = \frac{Q_2}{4\pi\varepsilon_0 r_2}$$
$$\therefore V_2 = \frac{-5.00 \times 10^{-6}}{4\pi\varepsilon_0 \times 1.00}$$
$$\therefore V_2 = -4.50 \times 10^4 \text{ V}$$

Combining these, the total potential at the mid-point is

$$V = V_1 + V_2$$
$$\therefore V = (1.80 \times 10^4) - (4.50 \times 10^4)$$
$$\therefore V = -2.70 \times 10^4 \text{ V}$$

Quiz: Electrical potential due to point charges   Go online

Useful data:

Fundamental charge e	1.60×10^{-19} C
Permittivity of free space ε_0	8.85×10^{-12} F m^{-1}

Q5: Calculate the electrical potential at a distance of 250 mm from a point charge of +4.0 μC.

a) 0.58 V
b) 2.3 V
c) 1.4×10^5 V
d) 5.8×10^5 V
e) 1.8×10^6 V

TOPIC 2. ELECTRIC POTENTIAL

Q6: An alpha particle has a charge of 3.2×10^{-19} C.

Determine the potential energy of an alpha particle at the position outlined in the previous question.

a) -4.48×10^{-14} J
b) 1.14×10^{-24} J
c) 2.29×10^{-24} J
d) 2.24×10^{-14} J
e) 4.48×10^{-14} J

....

Q7:

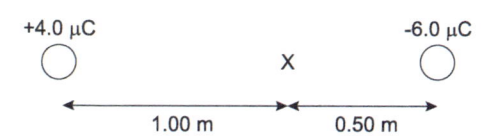

+4.0 μC   -6.0 μC
   ◯   X   ◯
   ←————— 1.00 m —————→←— 0.50 m —→

Determine the electrical potential at position X.

a) -7.2×10^{4} V
b) -1.8×10^{5} V
c) -1.4×10^{5} V
d) 7.2×10^{4} V
e) 1.4×10^{5} V

....

Q8: At a point 20 cm from a charged object, the ratio of electric field strength to electric potential (E/V) equals 100 m^{-1}.

What is the value of E/V 40 cm from the charge?

a) 25 m^{-1}
b) 50 m^{-1}
c) 100 m^{-1}
d) 200 m^{-1}
e) 400 m^{-1}

2.4 Summary

We have defined electric potential and considered the potential difference between two points. We have also shown how electric field and electric potential are related.

> **Summary**
>
> You should now be able to:
>
> - state that the electric potential V at a point is the work done by external forces in moving a unit positive charge from infinity to that point;
> - apply the expression $E = \frac{V}{d}$ for a uniform electric field;
> - explain what it meant by a conservative field;
> - state that an electric field is a conservative field;
> - state and apply the equation $V = \frac{Q}{4\pi\varepsilon_0 r}$ for the potential V at a distance r from a point charge Q.

2.5 Extended information

2.5.1 Potential around a hollow conductor

In the previous topic we plotted the electric field in and around a hollow conductor. We found that the field followed a $1/r^2$ dependence outside the conductor, but was equal to zero on the inside. What happens to the potential V inside a hollow conducting shape?

The example shown in Figure 2.5 is a hollow sphere. The definition of potential at any point is the work done in moving unit charge from infinity to that point. So outside the sphere, the potential follows the $1/r$ dependence we have just derived. Inside the sphere, the total field is zero, so there is no additional work done in moving charge about inside the sphere. The potential is therefore constant inside the sphere, and has the same value as at the edge of the sphere.

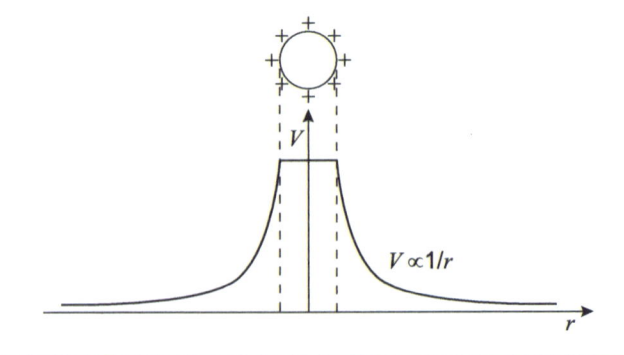

Figure 2.5: Electric potential in and around a hollow conducting sphere

2.5.2 Web links

Web links — Go online

These web links should serve as an insight to the wealth of information available online and allow you to explore the subject further.

- https://www.youtube.com/watch?v=DyUrrzZXmbo&list=PLiISjSDTPUjy2OKK5-BI2WrVu91LAe_cm&index=21
- https://www.youtube.com/watch?v=nnLf090OPNg
- http://ocw.mit.edu/resources/res-tll-004-stem-concept-videos-fall-2013/videos/derivatives-and-integrals/electric-potential/

2.6 End of topic 2 test

End of topic 2 test — Go online

The following data should be used when required:

Fundamental charge e	1.60×10^{-19} C
Permittivity of free space ε_0	8.85×10^{-12} F m^{-1}

Q9: 1.7 mJ of work are done in moving a 6.3 μC charge from point A to point B.

Calculate the potential difference between A and B.

_____ V

Q10: A 8.8 mC charged particle is moved at constant speed through a potential difference of 7.2 V.

Calculate how much work is done on the particle.

_____ J

Q11: The potential difference between points M and N is 80 V and the uniform electric field between them is 1600 N C^{-1}.

Calculate the distance between M and N.

_____ m

Q12: A charged particle is moved along a direct straight path from point C to point D, 28 mm away in a uniform electric field. 26 J of work are done on the particle in moving it along this path.

How much work must be done in order to move the particle from C to D along an indirect, curved path of distance 56 mm?

_____ J

....

Q13: Calculate the electric potential at a distance 0.62 m from a point charge of 26 μC.

_____ V

....

Q14: Two point charges, one of +2.5 nC and the other of +3.7 nC are placed 1.9 m apart. Calculate the electric potential at the point midway between the two charges.

_____ V

....

Q15: A +3.25 μC charge X is placed 2.50 m from a -2.62 μC charge Y.

Calculate the electric potential at the point 1.00 m from X, on the line joining X and Y.

_____ V

....

Q16: At a certain distance from a point charge, the electric field strength E is 2800 N C^{-1} and the electric potential V is 6300 V.

1. Calculate the distance from the charge at which E and V are being measured.
   _____ m
2. Calculate the magnitude of the charge.
   _____ C

Unit 3 Topic 3

Motion in an electric field

Contents

3.1 Introduction 393
3.2 Energy transformation associated with movement of charge 393
3.3 Motion of charged particles in uniform electric fields 396
3.4 Applications of charged particles and electric fields 400
      3.4.1 Cathode ray tubes 400
      3.4.2 Particle accelerators 402
      3.4.3 Rutherford scattering 404
3.5 The electronvolt 406
3.6 Summary 406
3.7 Extended information 407
3.8 End of topic 3 test 407

Prerequisites

- Coulomb's law.
- Electric potential and the volt.
- Kinematic relationships.
- Newton's laws of motion.

Learning objective

By the end of this topic you should be able to:

- describe the energy transformation that takes place when a charged particle is moving in an electric field;
- carry out calculations using $E_w = QV$;
- define an electronvolt;
- describe the motion of a charged particle in a uniform electric field, and use the kinematic relationships to calculate the trajectory of this motion;
- perform calculations to solve problems involving charged particles in electric fields, including the collision of a charged particle with a stationary nucleus.

TOPIC 3. MOTION IN AN ELECTRIC FIELD

3.1 Introduction

In this topic we will study the motion of charged particles in an electric field, drawing upon some of the concepts of the previous two topics. As we progress through this topic, we will also come across some other concepts you should have met before: Newton's second law of motion and Rutherford scattering of α-particles. You may find it useful to refresh your memory of these subjects before starting this topic.

As well as studying the theory, we will also be looking at some of the practical applications of applying electric fields to moving charged particles, such as the cathode ray tubes found in oscilloscopes.

3.2 Energy transformation associated with movement of charge

We have already defined one volt as being equivalent to one joule per coulomb. That is to say, if a charged particle moves through a potential difference of 1 V, it will gain or lose 1 J of energy per coulomb of its charge. Put succinctly, the energy E_W gained by a particle of charge Q being accelerated through a potential V is given by

$$E_W = QV$$

(3.1)

A charged particle placed in an electric field will be acted on by the Coulomb force. If it is free to move, the force will accelerate the particle, hence it will gain kinetic energy. So we can state, using Equation 3.1, that the gain in kinetic energy is

$$\frac{1}{2}mv^2 = QV$$

(3.2)

Hence we can calculate the velocity gained by a charged particle accelerated by a potential. Note that this is the velocity gained **in the direction of the electric field vector**. You should remember from the previous topic that the electric field points from high to low potential.

Example

An electron is accelerated from rest through a potential of 50 V. What is its final velocity?

In this case, Equation 3.2 becomes

$$\frac{1}{2}m_e v^2 = eV$$

Rearranging this equation

$$v = \sqrt{\frac{2eV}{m_e}}$$

Now we use the values of $e = 1.60 \times 10^{-19}$ C and $m_e = 9.11 \times 10^{-31}$ kg in this equation

$$v = \sqrt{\frac{2 \times 1.60 \times 10^{-19} \times 50}{9.11 \times 10^{-31}}}$$

$$\therefore v = 4.19 \times 10^6 \text{ m s}^{-1}$$

In general, we can state that for a particle of charge Q and mass m, accelerated from rest through a potential V, its final velocity will be

$$v = \sqrt{\frac{2QV}{m}}$$

(3.3)

An electron accelerated from rest can attain an extremely high velocity from acceleration through a modest electric potential. It should be noted that as the velocity of the electron increases, or the accelerating potential is increased, relativistic effects will become significant. Broadly speaking, once a charged particle is moving with a velocity greater than 10% of the speed of light ($c = 3.00 \times 10^8$ m s^{-1}) then relativistic effects need to be taken into account.

TOPIC 3. MOTION IN AN ELECTRIC FIELD

Quiz: Acceleration and energy change

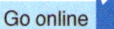

Go online

Useful data:

Fundamental charge e	1.6×10^{-19} C
Mass of an electron m_e	9.11×10^{-31} kg
Speed of light c	3.00×10^{8} m s^{-1}
Mass of an α-particle	6.65×10^{-27} kg
Charge of an α-particle	$+3.20 \times 10^{-19}$ C

Q1: A free electron is accelerated towards a fixed positive charge.
Which **one** of the following statements is true?

a) The electron gains kinetic energy.
b) The electron loses kinetic energy.
c) There are no force acting on the electron.
d) The electron's velocity is constant.
e) A repulsive force acts on the electron.

....

Q2: A -3.0 μC charge is accelerated through a potential of 40 V.
How much energy does it gain?

a) 7.5×10^{-8} J
b) 1.2×10^{-4} J
c) 0.075 J
d) 13 J
e) 120 J

....

Q3: An α-particle is accelerated from rest through a potential of 1.00 kV.
What is its final velocity?

a) 9.80×10^{3} m s^{-1}
b) 2.19×10^{5} m s^{-1}
c) 3.10×10^{5} m s^{-1}
d) 3.00×10^{8} m s^{-1}
e) 9.62×10^{10} m s^{-1}

....

© HERIOT-WATT UNIVERSITY

UNIT 3. ELECTROMAGNETISM

Q4: Two parallel plates are 50 mm apart. The electric field strength between the plates is 1.2×10^4 N C^{-1}.

An electron is accelerated between the plates. How much kinetic energy does it gain?

a) 3.8×10^{-17} J
b) 9.6×10^{-17} J
c) 1.9×10^{-15} J
d) 3.8×10^{-14} J
e) 9.6×10^{-14} J

3.3 Motion of charged particles in uniform electric fields

If a charged particle is placed in an electric field, we know that the force acting on it will be equal to QE, where Q is the charge on the particle and E is the magnitude of the electric field. If this is the only force acting on the particle, and the particle is free to move, then Newton's second law of motion tells us that it will be accelerated. If the particle has mass m, then

$$F = ma$$
$$\therefore QE = ma$$
$$\therefore a = \frac{QE}{m}$$

(3.4)

Equation 3.4 tells us the magnitude of the particle's acceleration **in the direction of the electric field**. We must be careful with the direction of the acceleration. The E-field is defined as positive in the direction of the force acting on a positive charge. An electron will be accelerated in the opposite direction to the E-field vector, whilst a positively-charged particle will be accelerated in the same direction as the E-field vector.

Let us consider what happens when an electron enters a uniform electric field at right angles to the field, as shown in Figure 3.1.

TOPIC 3. MOTION IN AN ELECTRIC FIELD

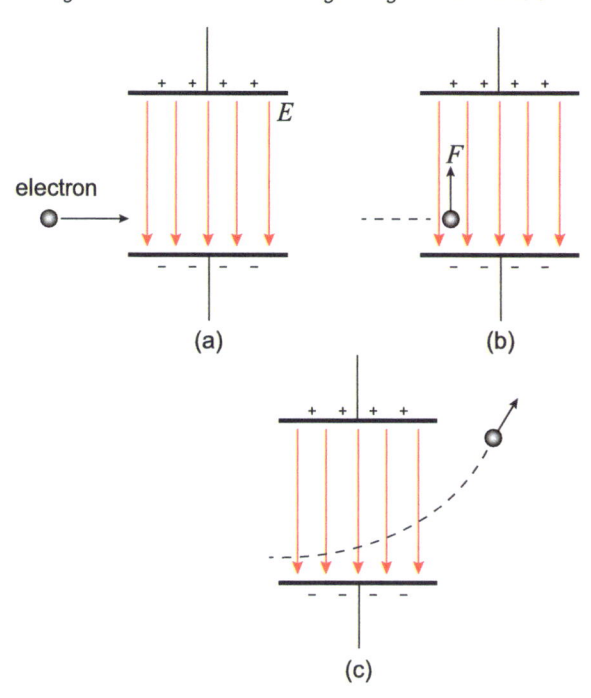

Figure 3.1: An electron travelling through an electric field

(a) (b) (c)

Note that the motion of the electron is similar to that of a body projected horizontally in the Earth's gravitational field. At right angles (orthogonal) to the field, there is no force acting and the electron moves with a uniform velocity in this direction. Parallel to the field, a constant force ($F = QE$, analogous to $F = mg$) acts causing a uniform acceleration parallel to the field lines. This means that we can solve problems involving charged particles moving in electric fields in the same way that we solved two-dimensional trajectory problems, splitting the motion into orthogonal components and applying the kinematic relationships of motion with uniform acceleration.

Example

Two horizontal plates are charged such that a uniform electric field of strength E = 200 N C^{-1} exists between them, acting upwards. An electron travelling horizontally enters the field with speed 4.00×10^6 m s^{-1}, as shown in Figure 3.2.

1. Calculate the acceleration of the electron.

2. How far (vertically) is the electron deflected from its original path when it emerges from the plates, given the length l = 0.100 m?

© HERIOT-WATT UNIVERSITY

Figure 3.2: An electron travelling in an electric field

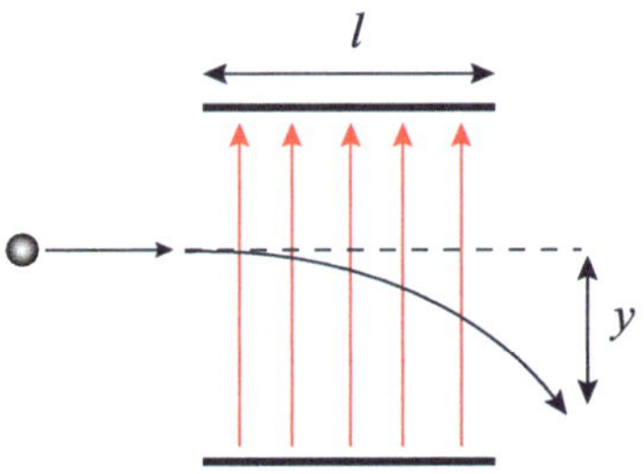

1. Considering the vertical motion, we use Equation 3.4 to calculate the downward acceleration

$$a = \frac{QE}{m}$$
$$\therefore a = \frac{eE}{m_e}$$
$$\therefore a = \frac{1.60 \times 10^{-19} \times 200}{9.11 \times 10^{-31}}$$
$$\therefore a = 3.51 \times 10^{13} \text{ m s}^{-2}$$

Note that this acceleration is many orders of magnitude greater than the acceleration due to gravity acting on the electron. We will be able to ignore the effects of gravity in all the problems we encounter concerning the motion of charged particles in electric fields since the effects of the E-fields will always be far greater.

2. To calculate the deflection y, we first need to calculate the time-of-flight of the electron between the plates, which we do by considering the horizontal motion of the electron. Since this is unaffected by the E-field, the horizontal component of the velocity is unchanged and we can use the simple relationship

$$t = \frac{l}{v_h}$$
$$\therefore t = \frac{0.100}{4.00 \times 10^6}$$
$$\therefore t = 2.50 \times 10^{-8} \text{ s}$$

Now, considering the vertical component of the motion, we know $u_v = 0$ m s^{-1}, $t = 2.50 \times 10^{-8}$ s and $a = 3.51 \times 10^{13}$ m s^{-2}. The displacement y is the unknown, so we will use the kinematic relationship $s = ut + \frac{1}{2}at^2$.

TOPIC 3. MOTION IN AN ELECTRIC FIELD

In this case

$$y = u_v t + \tfrac{1}{2}at^2$$
$$\therefore y = 0 + \left(\tfrac{1}{2} \times 3.51 \times 10^{13} \times (2.50 \times 10^{-8})^2\right)$$
$$\therefore y = 0.0110 \text{ m}$$

Note that we would normally measure the potential difference across the plates. If the plates are separated by a distance d, then the electric field $E = V/d$.

Quiz: Charged particles moving in electric fields

Go online

Useful data:

Charge on electron e	-1.60 x 10^{-19} C
Mass of an electron m_e	9.11×10^{-31} kg
Speed of light in vacuum c	3.00×10^8 m s^{-1}
Mass of an α**-particle** m_p	6.65×10^{-27} kg
Charge of an α**-particle**	$+3.20 \times 10^{-19}$ C

Q5: An electron enters a region where the electric field strength is 2500 N C^{-1}. What is the force acting on the electron?

a) 6.40×10^{-23} N
b) 1.60×10^{-19} N
c) 4.00×10^{-16} N
d) 2500 N
e) 1.56×10^{19} N

....

Q6: An electron is placed in a uniform electric field of strength 4.00×10^3 N C^{-1}. What is the acceleration of the electron?

a) 4.00×10^{-23} m s^{-2}
b) 1.42×10^{-15} m s^{-2}
c) 4.00×10^3 m s^{-2}
d) 7.03×10^{14} m s^{-2}
e) 2.50×10^{22} m s^{-2}

....

Q7: A positively-charged ion placed in a uniform electric field will be

a) accelerated in the direction of the electric field.
b) accelerated in the opposite direction to the electric field.
c) moving in a circular path.
d) moving with constant velocity.
e) stationary.

....

Q8: An α-particle enters a uniform electric field of strength 50.0 N C^{-1}, acting vertically downwards.

What is the acceleration of the particle?

a) 1.20×10^9 m s^{-2} downwards
b) 1.20×10^9 m s^{-2} upwards
c) 2.41×10^9 m s^{-2} downwards
d) 2.41×10^9 m s^{-2} upwards
e) 8.78×10^{12} m s^{-2} downwards

....

Q9: An electron moving horizontally at 1800 m s^{-1} enters a vertical electric field of field strength 1000 N C^{-1}. The electron takes 2.00×10^{-8} s to cross the field.

With what vertical component of velocity does it emerge from the field?

a) 0.00 m s^{-1}
b) 3.51×10^{-2} m s^{-1}
c) 1.80×10^3 m s^{-1}
d) 3.51×10^6 m s^{-1}
e) 7.04×10^6 m s^{-1}

3.4 Applications of charged particles and electric fields

We have discussed how electrons can be accelerated and deflected in an electric field. We now look at some practical applications of these effects.

3.4.1 Cathode ray tubes

Cathode ray tubes used to very common. Televisions, computer monitors and oscilloscopes up to about the year 2000 were nearly always made using a cathode ray tube. This meant that these devices were very large. The advent of LCD, LED and plasma screens means that cathode ray tubes have nearly all disappeared from people's homes. However the cathode ray tube is still valuable as a tool for studying electric fields and as an introduction to particle accelerators.

In a cathode ray tube, such as the one shown in Figure 3.3, electrons ('cathode rays') are freed from the heated cathode. (The electrons were originally called cathode rays because these experiments were first carried out before the electron was discovered. To the original experimenters it looked like

TOPIC 3. MOTION IN AN ELECTRIC FIELD

the cathode was emitting energy rays.) These electrons are accelerated while in the electric field set up between the cathode and the anode, gaining kinetic energy. Some electrons pass through a hole in the anode. From the anode to the *y*-plates, the electrons travel in a straight line at constant speed, obeying Newton's first law of motion.

Figure 3.3: The cathode ray tube

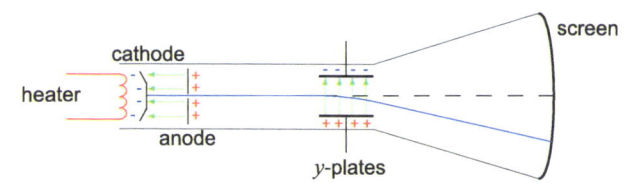

A second electric field is set up between the *y*-plates, this time at right angles to the initial direction of motion of the electrons. This electric field supplies a force to the electrons at right angles to their original direction. The resulting path of the electrons is a parabola. The motion of the electrons while between the *y*-plates is similar to the motion of a projectile thrown horizontally in a gravitational field.

When they leave the region of the *y*-plates, the electrons again travel in a straight line with constant speed (now in a different direction), eventually hitting the screen as shown.

The point on the screen where the electrons hit is determined by the strength of the electric field between the *y*-plates. This electric field strength is in turn determined by the potential difference between the *y*-plates. So the deflection of the electron beam can be used to measure a potential difference.

Example The potential between the cathode and the anode of a cathode ray tube is 200 V.

Assuming that the electrons are given off from the heated cathode with zero velocity and that all of the electrical energy given to the electrons is transformed to kinetic energy, calculate

1. the electrical energy gained by an electron between the cathode and the anode.
2. the horizontal velocity of an electron just as it leaves the anode.

(The mass of an electron is 9.11 x 10^{-31} kg)

1. The electrical energy gained by an electron is equal to the work done by the electric field between the cathode and the anode, so

    $E_W = QV$
    $E_W = 1.6 \times 10^{-19} \times 200$
    $E_W = 3.2 \times 10^{-17}$ J

2. If all of this energy is transformed to kinetic energy, then

© HERIOT-WATT UNIVERSITY

$$E_k = \frac{1}{2}mv^2$$
$$3.2 \times 10^{-17} = \frac{1}{2} \times 9.11 \times 10^{-31} \times v^2$$
$$v = 8.4 \times 10^6 \text{m s}^{-1}$$

The cathode ray tube   Go online

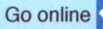

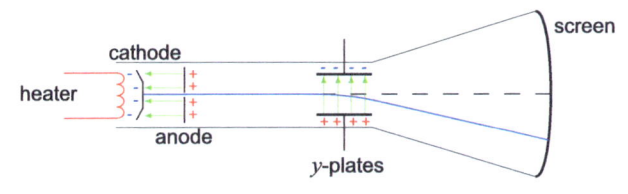

This activity allows you to see the path of electrons in the electric field set up between the cathode and the anode in a cathode ray tube, and calculate the kinetic energy gained by an electron. It also allows the path of the electrons to be changed by applying a potential difference between the y-plates.

Electrons given off from a heated cathode in a cathode ray tube are accelerated by the electric field set up between the cathode and the anode.

The path of the electrons can be changed by the electric field set up by applying a potential difference between the y-plates.

It is important to realise that increasing the potential difference between the cathode and the anode increases the speed of the electrons in the cathode ray. Altering the potential difference between the y-plates affects the position where the electrons hit the screen.

3.4.2 Particle accelerators

Particle accelerators are tools that are used to prise apart the nuclei of atoms and thereby help us increase our understanding of the nature of matter and the rules governing the particles and their interaction in the sub atomic world. Particle accelerators are massive machines that accelerate charged particles (ions) and give them enough energy to separate the constituent particles of the nucleus. They have played a significant part in the development of the standard model.

We have already met a very simple particle accelerator: the cathode ray tube. In a cathode ray tube electrons are accelerated by an electric field.

The cathode ray tube however cannot produce high enough energies to investigate the structure of matter. Larger and much more powerful particles have been developed for this purpose.

Particle accelerators are of two main types.

One type accelerates the particle in a straight line. This type is called a linear accelerator, sometimes referred to as a "linac".

The other type accelerates the particle in a circular path. The cyclotron and the more widely used

TOPIC 3. MOTION IN AN ELECTRIC FIELD

synchrotron are examples of this type.

The linear and circular accelerators both use electric fields as the means of accelerating particles and supplying them with energy.

Linear accelerator

In a linear accelerator, the particle acquires energy in a similar way to the electron in the cathode ray tube but the process is repeated a large number of times. A large alternating voltage is used to accelerate particles along in a straight line.

Figure 3.4: A linear accelerator

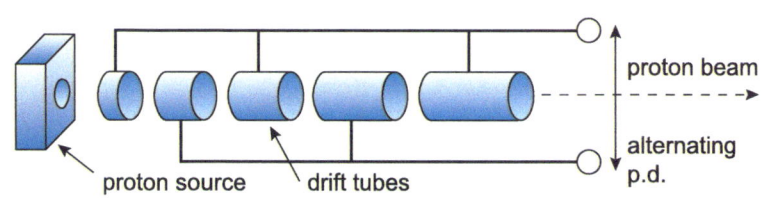

The particles pass through a line of hollow metal tubes enclosed in a long evacuated cylinder. The frequency of the alternating voltage is set so that the particle is accelerated forward each time it goes through a gap between two of the metal tubes. The metal tubes are known as drift tubes. The idea is that the particle drifts free of electric fields through these tubes at constant velocity and emerges from the end of a tube just in time for the alternating voltage to have changed polarity. The largest linac in the world, at Stanford University in the USA, is 3.2 km long.

At the end of each drift tube the charged particle is accelerated by the voltage across the gap.

- The work done on the charged particle, $E_W = QV$
- Where V = voltage across gap, Q = charge on particle being accelerated.
- The particle gains QV of energy at each gap
- This work done causes the particle to accelerate
- So the E_k increases by QV at each gap.
- The speed increases as it moves along the linear accelerator.

The length of successive drift tubes increases. This is because the speed of the charged particle is increasing and to ensure that the time taken to pass through each tube is the same, the length of the tubes must be increased. The time to pass through each drift tube is set by the frequency of the alternating voltage.

It would appear that longer linear accelerators, if they were to be built, could produce particles with yet higher speeds and energy. However special relativity sets limits on the speeds that can be achieved. At speeds comparable with the speed of light (relativistic speeds), the mass of a particle increases significantly and consequently much more energy is needed to accelerate the particle.

Linear accelerators work well but they are expensive and need a lot of space.

© HERIOT-WATT UNIVERSITY

3.4.3 Rutherford scattering

In this famous experiment, first carried out by Rutherford and his students Geiger and Marsden in 1909, a stream of alpha particles is fired at a thin sheet of gold foil. Rutherford found that although most particles travelled straight through the foil, a few were deflected, sometimes through large angles, as shown in Figure 3.5(a). Some were even deflected straight back in the direction they had come from. From this experiment, Rutherford concluded that atoms were mostly empty space, with a dense positively-charged nucleus at the centre. The scattering of α-particles was due to collisions with these nuclei.

We can now look a little closer at what happens in these 'collisions'. We can see in Figure 3.5(b) that the α-particle doesn't actually impinge upon the nucleus. Instead there is an electrostatic repulsion between the α-particle and the nucleus. It is the kinetic energy of the α-particle that determines how close it can get to the nucleus.

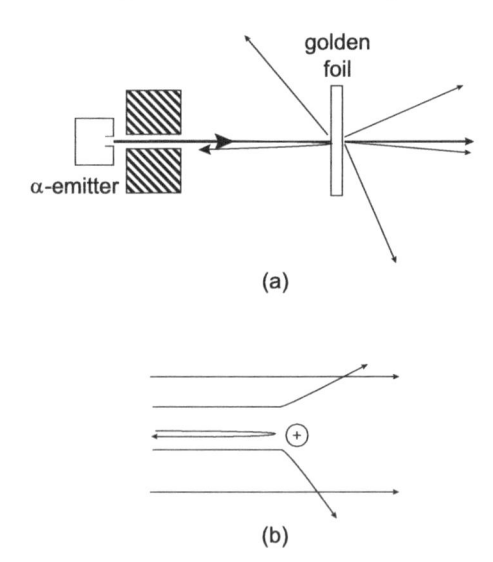

Figure 3.5: Rutherford scattering

Example

In a Rutherford scattering experiment, a beam of alpha particles is fired at a sheet of gold foil. Each α-particle has charge $2e$ and (non-relativistic) energy $E_\alpha = 1.00 \times 10^{-13}$ J. A gold nucleus has charge $79e$. What is the closest possible distance an α-particle with this energy can get to a gold nucleus?

As the α-particle approaches the nucleus, its potential energy increases since it is moving in the electric field of the nucleus. The kinetic energy of the α-particle will get less as its potential energy increases. At the point where all its kinetic energy has been converted into potential energy, the particle will be momentarily stationary, before the Coulomb repulsion force starts moving it away again.

The electric potential around the gold nucleus is calculated from

$$V = \frac{Q}{4\pi\varepsilon_0 r}$$
$$\therefore V_{gold} = \frac{79e}{4\pi\varepsilon_0 r}$$

Remember, the potential is the work done per unit charge in bringing a particle from infinity to a distance r from the object. So, by using Equation 3.1 for an α-particle of charge $2e$, the amount of work done E_W is

$$E_W = Q_\alpha V_{gold}$$
$$\therefore E_W = 2e \times \frac{79e}{4\pi\varepsilon_0 r} = 1.00 \times 10^{-13}$$
$$\text{So } r = \frac{158e^2}{4\pi\varepsilon_0 \times 1.00 \times 10^{-13}}$$
$$\therefore r = 3.64 \times 10^{-13} \text{ m}$$

At distance $r = 3.64 \times 10^{-13}$ m from the gold nucleus, all the kinetic energy of the α-particle has been turned into potential energy, and so this is the closest to the nucleus that the α-particle can get.

Rutherford scattering

Suppose a Rutherford scattering experiment was carried out firing a beam of protons at a gold foil. What would be the closest that a proton could get to a nucleus if it had a non-relativistic energy of 8.35×10^{-14} J?

($m_p = 1.67 \times 10^{-27}$ kg, $e = 1.60 \times 10^{-19}$ C, $\varepsilon_0 = 8.85 \times 10^{-12}$ F m^{-1})

3.5 The electronvolt

The electronvolt (eV) is a unit of energy commonly used in high energy particle physics. The electronvolt is equal to the kinetic energy gained by an electron when it is accelerated by a potential difference of one volt. So an electron in a high energy accelerator moving through a potential difference of 4 000 000 V will gain 4 000 000 eV of energy.

The work done when a particle of charge Q moves through a potential difference V is given by $E_W = QV$. The charge on one electron is 1.6×10^{-19} C and so one electronvolt can be expressed in joules as follows.

$$\begin{aligned} E_W &= QV \\ E_W &= 1.6 \times 10^{-19} \times 1 \\ E_W &= 1.6 \times 10^{-19} \text{ J} \end{aligned}$$

Example

The Large Hadron collider was designed to run at a maximum collision energy of 14 TeV. Express this in joules.

$$\begin{aligned} E_W &= QV \\ E_W &= 1.6 \times 10^{-19} \times 14 \times 10^{12} \\ E_W &= 2.2 \times 10^{-6} \text{ J} \end{aligned}$$

3.6 Summary

In this topic we have seen that charged particles are accelerated by electric fields, gaining kinetic energy. An electric field can be used to deflect the path of a charged particle or a beam of such particles.

Summary

You should now be able to:

- describe the energy transformation that takes place when a charged particle is moving in an electric field;
- carry out calculations using $E_w = QV$;
- define an electronvolt;
- describe the motion of a charged particle in a uniform electric field, and use the kinematic relationships to calculate the trajectory of this motion;
- perform calculations to solve problems involving charged particles in electric fields, including the collision of a charged particle with a stationary nucleus.

© HERIOT-WATT UNIVERSITY

3.7 Extended information

Web links — Go online

These web links should serve as an insight to the wealth of information available online and allow you to explore the subject further.

- https://www.youtube.com/watch?v=I-2RJajTvWc&list=PLiISjSDTPUjy2OKK5-BI2WrVu91LAe_cm&index=22
- https://www.youtube.com/watch?v=LAZ1DFA0sYA
- https://www.youtube.com/watch?v=wzALbzTdnc8

3.8 End of topic 3 test

End of topic 3 test — Go online

The following data should be used when required:

Charge on electron e	-1.60×10^{-19} C
Mass of an electron m_e	9.11×10^{-31} kg
Speed of light in vacuum c	3.00×10^{8} m s^{-1}
Mass of an α-particle	6.65×10^{-27} kg
Charge of an α-particle	$+3.20 \times 10^{-19}$ C

Q10: An electron is accelerated through a potential of 280 V.
Calculate the resultant increase in the kinetic energy of the electron.
_____ J

Q11: An electron is accelerated from rest through a potential of 75 V.
Calculate the final velocity of the electron.
_____ m s^{-1}

Q12: Consider a particle of charge 7.7 μC and mass 2.5×10^{-4} kg entering an electric field of strength 6.4×10^{5} N C^{-1}.

Calculate the acceleration of the particle.

_____ m s^{-2}

Q13: An electron travelling horizontally with velocity 1.55×10^6 m s^{-1} enters a uniform electric field, as shown below.

The electron travels a distance $l = 0.0200$ m in the field and the strength of the field is 160 N C^{-1}.

1. Calculate the vertical component of the electron's velocity when it emerges from the E-field.
   _____ m s^{-1}
2. Calculate the vertical displacement y of the electron.
   _____ m

Q14: In a Rutherford scattering experiment, an α-particle (charge +2e) is fired at a stationary gold nucleus (charge +79e).

Calculate the work done by the α-particle in moving from infinity to a distance 5.45×10^{-13} m from the gold nucleus.

_____ J

Unit 3 Topic 4

Magnetic fields

Contents

4.1	Introduction	411
4.2	Magnetic forces and fields	411
4.3	Magnetic field around a current-carrying conductor	415
4.4	Magnetic induction	419
4.5	Force on a current-carrying conductor in a magnetic field	420
	4.5.1 The electric motor	423
	4.5.2 The electromagnetic pump	424
4.6	The relationship between magnetic induction and distance from a current-carrying conductor	426
4.7	Comparison of forces	427
	4.7.1 Millikan's oil drop experiment	430
4.8	Summary	432
4.9	Extended information	433
4.10	End of topic 4 test	433

Prerequisites

- An understanding of the force on a charged particle placed in a magnetic field.
- An understanding of the concept of electrical field.
- An understanding of the concept of gravitational field.
- Basic geometrical and algebraic skills.

Learning objective

By the end of this topic you should be able to:

- state that electrons are in motion around atomic nuclei and individually produce a magnetic effect;
- state that ferromagnetism is a magnetic effect in which magnetic domains can be made to line up, resulting in the material becoming magnetised;
- state that iron, nickel, cobalt and some compounds of rare earth metals are ferromagnetic;
- sketch the magnetic field patterns around permanent magnets and the Earth;
- state that a magnetic field exists around a moving charge in addition to its electric field;
- sketch the magnetic field patterns around current carrying wires and current carrying coils;
- state that a charged particle moving across a magnetic field experiences a force;
- explain the interaction between magnetic fields and current in a wire;
- state the relative directions of current, magnetic field and force for a current-carrying conductor in a magnetic field;
- describe how to investigate the factors affecting the force on a current-carrying conductor in a magnetic field;
- use the relationship $F = IlB\sin\theta$ for the force on a current-carrying conductor in a magnetic field;
- define the unit of magnetic induction, the tesla (T);
- state and use the expression $B = \frac{\mu_0 I}{2\pi r}$ for the magnetic field B due to a straight current-carrying conductor;
- compare gravitational, electrostatic, magnetic and nuclear forces.

TOPIC 4. MAGNETIC FIELDS

4.1 Introduction

In Particles from space, you met the terms magnetic field and magnetic induction. You studied the force that acts on a charged particle moving in a magnetic field and you looked at the effect the Earth's magnetic field has on cosmic rays.

In this topic we will find out why some materials are attracted to magnets and others are not. We will look more closely at the magnetic field patterns between magnetic poles, around solenoids and around the Earth. We will describe the magnetic force on a current-carrying conductor by using a field description. We will then investigate how magnetic induction varies with distance from a current carrying wire. Finally, we will compare gravitational, electrostatic, magnetic and nuclear forces.

4.2 Magnetic forces and fields

An atom consists of a nucleus surrounded by moving electrons. Since the electrons are charged and moving, they create a magnetic field in the space around them. Some atoms have magnetic fields associated with them and behave like magnets. Iron, nickel and cobalt belong to a class of materials that are **ferromagnetic**. In these materials, the magnetic fields of atoms line up in regions called **magnetic domains**. If the magnetic domains in a piece of ferromagnetic material are arranged so that most of their magnetic fields point the same way, then the material is said to be a magnet and it will have a detectable magnetic field.

Each small arrow represents the magnetic field in a magnetic domain. A refrigerator magnet is an everyday example of ferromagnetism and this property has many applications in modern technology, such as the magnetic storage in hard disks.

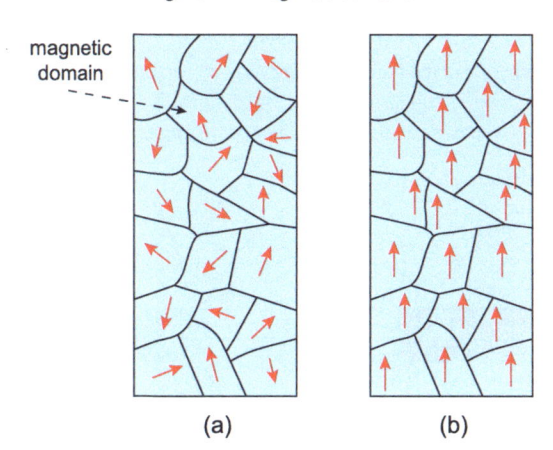

Figure 4.1: Magnetic domains

(a) Unmagnetised material where the magnetic domains cancel, (b) Magnetised material where the domains align

© HERIOT-WATT UNIVERSITY

Magnetic domains

Go online

There is an online animation showing how magnetic domains respond to an outside magnetic field.

You may recall the magnetic field pattern around a bar magnet from earlier on in the course.

Figure 4.2: The field pattern around a bar magnet

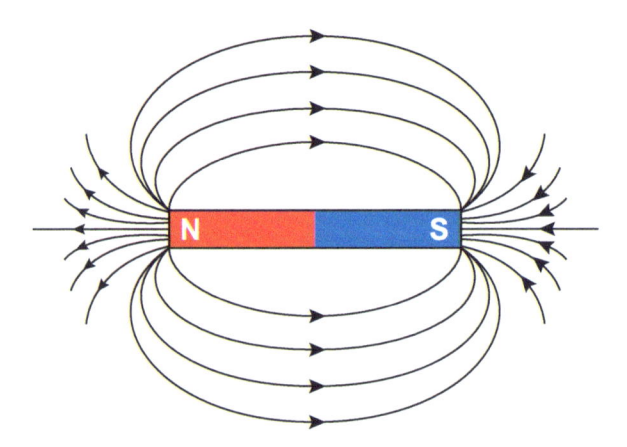

Remember that the convention is to draw the field direction as outward from the North pole and in to the South pole of the magnet. The distance between the lines increases as you move further from the magnet, since the magnetic field strength decreases.

The magnetic field pattern for a combination of magnets is shown below.

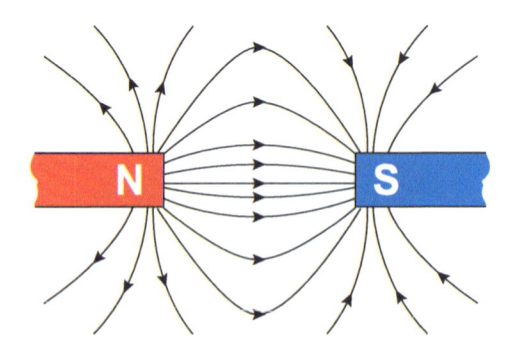

Figure 4.3: The field pattern around a pair of bar magnets - two opposite poles

TOPIC 4. MAGNETIC FIELDS

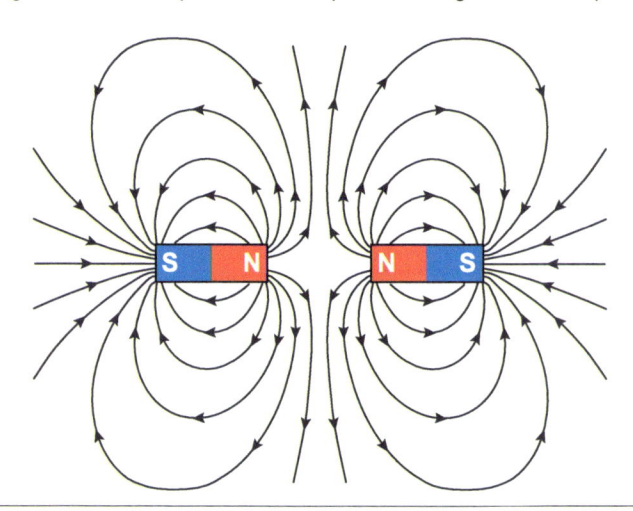

Figure 4.4: The field pattern around a pair of bar magnets - two like poles

Your teacher may provide you with equipment to confirm this to be the case.

In earlier topics, we introduced the concept of the gravitational field associated with a mass (Rotational Motion and Astrophysics - Topic 5) and the electric field associated with a charge (Electromagnetism - Topic 1).

An electric force exists between two or more charged particles whether they are moving or not.

We can explain magnetic interactions by considering that moving charges or currents create magnetic fields in the space around them, and that these magnetic fields exert forces on any other moving charges or currents present in the field.

You may recall that an interesting consequence of this is the Earth itself has a magnetic field. The flow of liquid iron within its molten core generates electric currents, which in turn produce a magnetic field. The magnetic field pattern around the Earth is similar to that of a bar magnet, but it is worth noting that the geographical north pole acts like a magnetic south pole. Therefore, the magnetic field lines actually point towards the geographical north pole.

Quiz: Magnetic fields and forces

Q1: Which one of the following statements about magnets is correct?

a) All magnets have one pole called a monopole.
b) All magnets are made of iron.
c) Ferromagnetic materials cannot be made into magnets.
d) All magnets have two poles called positive and negative.
e) All magnets have two poles called north and south.

....

Q2: Which of the following statements about magnetic field lines is/are correct?
Magnetic field lines:
(i) are directed from the north pole to the south pole of a magnet.
(ii) only intersect at right angles.
(iii) are further apart at a weaker place in the field.

a) (i) only
b) (ii) only
c) (iii) only
d) (i) and (ii) only
e) (i) and (iii) only

....

Q3: Which of the following statements about the Earth's magnetic field is/are correct?
The Earth's magnetic field:
(i) is horizontal at all points on the Earth's surface.
(ii) has a magnetic north pole at almost the same point as the geographic north pole.
(iii) is similar to the field of a bar magnet.

a) (i) only
b) (ii) only
c) (iii) only
d) (i) and (iii) only
e) (i), (ii) and (iii)

4.3 Magnetic field around a current-carrying conductor

Current is a movement of charges. We have just seen that there is a magnetic field round about moving charges, so there must be a magnetic field round a wire carrying a current. This effect was first discovered by the Danish physicist Hans Christian Oersted (1777 - 1851). Oersted was in fact the first person to link an electric current to a magnetic compass needle.

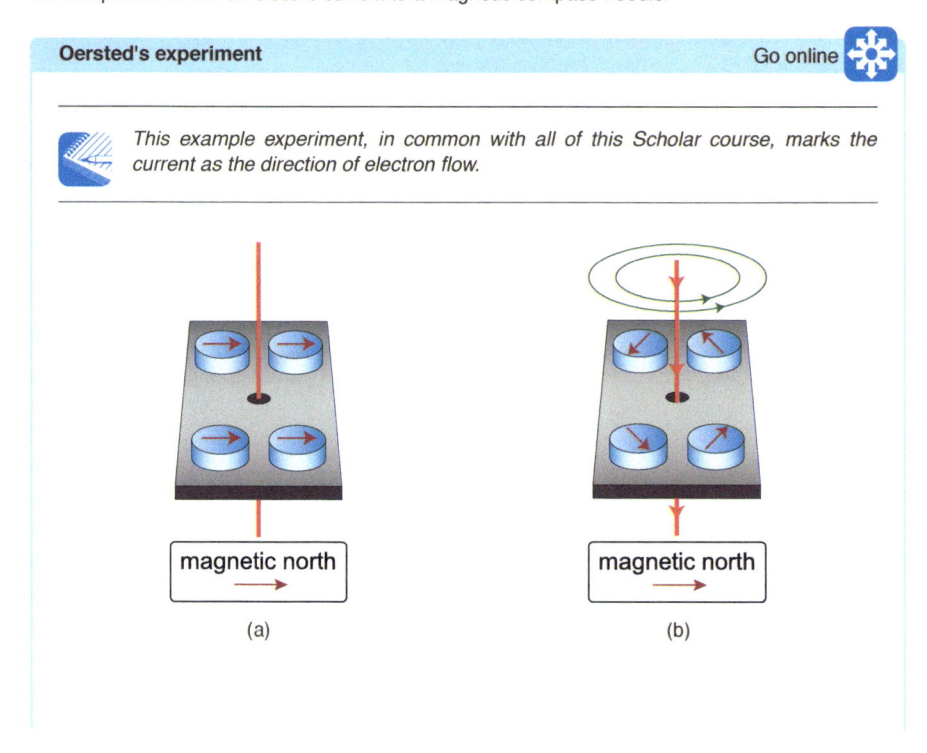

Oersted's experiment　　　　　　　　　　　　　　　　　　　　　Go online

This example experiment, in common with all of this Scholar course, marks the current as the direction of electron flow.

magnetic north (a)

magnetic north (b)

(c) magnetic north

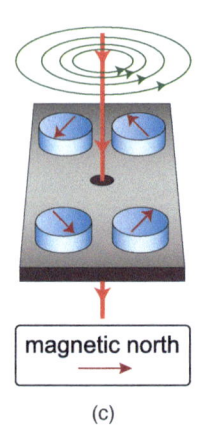

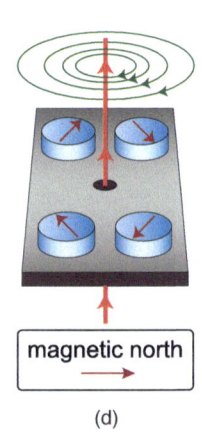

(d) magnetic north

a) When there is no current, there is no magnetic field around the wire and the compass needles react to the magnetic field around the earth.

b) A current is now passed through the wire.

1. When the current is switched on what is the shape of the magnetic field?

c) The magnitude of the current is now increased.

2. When the current is increased what happens to the strength of the magnetic field?

d) The direction of the flow of current is now reversed.

3. When the current is reversed what happens to the direction of the magnetic field?

A current through a wire produces a circular field, centred on the wire as shown in Figure 4.5. I shows the direction of electron current flow (current flows in the direction negative to positive).

Figure 4.5: The magnetic field around a straight wire

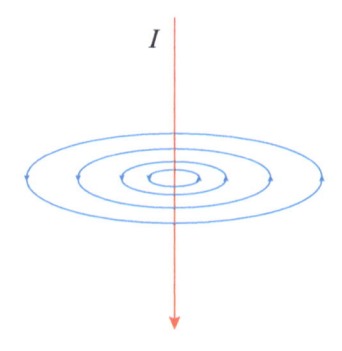

TOPIC 4. MAGNETIC FIELDS

The direction of the magnetic field can be found by using the left-hand grip rule (for electron current), as follows:

Point the thumb of the left hand in the direction of the current, that is the direction in which the electrons are moving. The way the fingers curl round the wire when making a fist is the way the magnetic field is directed. This rule is sometimes known as the left-hand grip rule.
A way to remember this is thu**M**b = Motion of electrons and **F**ingers = Field lines.

Figure 4.6: The left-hand grip rule for electron current

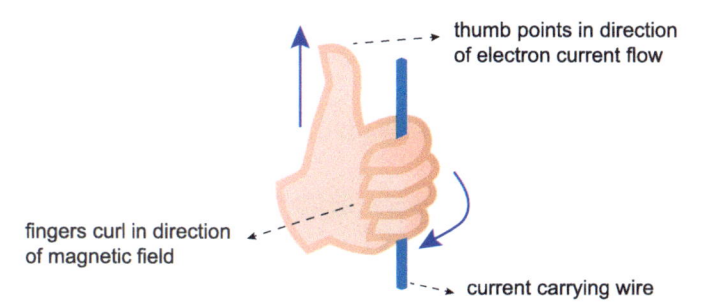

thumb points in direction of electron current flow

fingers curl in direction of magnetic field

current carrying wire

The magnetic field associated with a single straight length of wire is not very strong. If the wire is shaped into a flat circular coil, then the magnetic field inside the coil is more concentrated. The field pattern caused by a current in a flat circular coil of wire is shown in Figure 4.7.

Figure 4.7: The magnetic field pattern caused by current in a flat circular coil

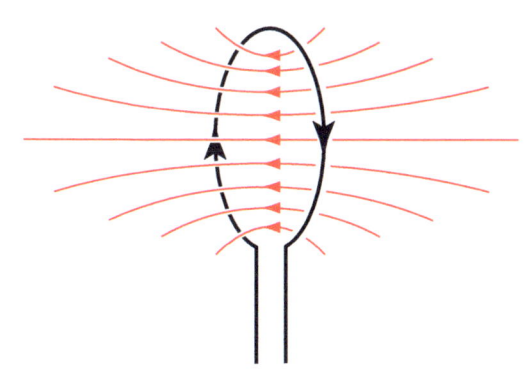

The magnetic field can be further strengthened by winding a wire into a long coil, known as a solenoid. The magnetic field pattern caused by current in a long solenoid is shown in Figure 4.8. Another version of the left-hand grip rule can be used to predict the direction of the magnetic field associated with both the flat circular coil and the long solenoid.

In this case, curl the fingers of the left hand round the coil or the solenoid in the direction of the electron current. The thumb then points towards the north end of the magnetic field produced in the solenoid. See Figure 4.7, Figure 4.8 and Figure 4.9.

© HERIOT-WATT UNIVERSITY

418 UNIT 3. ELECTROMAGNETISM

Figure 4.8: The left-hand rule for solenoids

Figure 4.9: The magnetic field pattern caused by current in a long solenoid

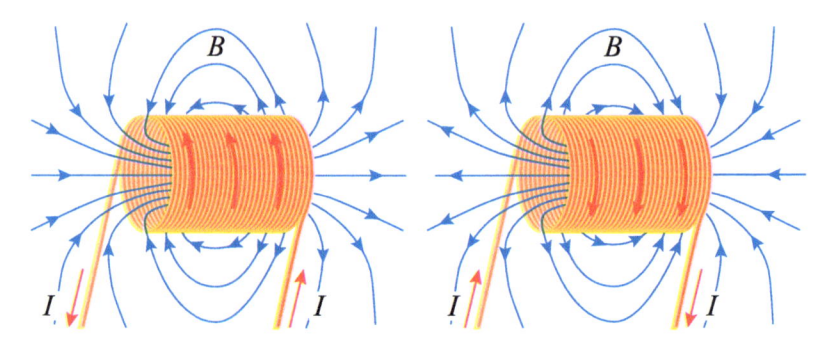

Magnetic field lines around a solenoid Go online

There is an online animation which will help you to understand the magnetic field lines around a solenoid.

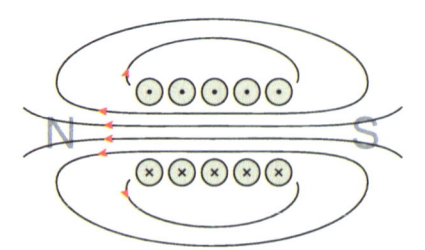

© HERIOT-WATT UNIVERSITY

4.4 Magnetic induction

So far we have used a magnetic field description without quantifying it. We will now use one of the effects of a magnetic field to do just that. The symbol that is used for the magnetic field is B. The magnetic field is a vector quantity, and so has a direction associated with it. The direction of the field at any position is defined as the way that the north pole of a compass would point in the field at that position. There are other names that are used for magnetic field - magnetic flux density, magnetic induction or magnetic B-field. They all come about from different approaches to an understanding of magnetic fields.

The unit for **magnetic induction**, the tesla (T), is obtained from the force on a conductor in a magnetic field. One tesla is the magnetic induction of a magnetic field in which a conductor of length one metre, carrying a current of one ampere perpendicular to the field is acted on by a force of one newton.

$$1 \text{ T} = 1 \text{ N A}^{-1} \text{ m}^{-1}$$

As in all areas of Physics, it is useful to have a 'feel' for the quantities that you are dealing with. The order of magnitude values shown in Table 4.1 might be of use in gaining an understanding of magnetic fields.

Table 4.1: Typical magnetic field values

Situation	Magnetic field (T)
Magnetic field of the Earth	5×10^{-5}
At the poles of a typical fridge magnet	1×10^{-3}
Between the poles of a large electromagnet	1.00
In the interior of an atom	10.0
Largest steady field produced in a laboratory	45.0
At the surface of a neutron star (estimated)	1.0×10^{8}

4.5 Force on a current-carrying conductor in a magnetic field

The forces that a magnetic field exerts on the moving charges in a conductor are transmitted to the whole of the conductor and it experiences a force that tends to make it move. Consider a **current-carrying conductor** that is in a uniform magnetic field B, as in Figure 4.10.

Figure 4.10: A current-carrying conductor in a magnetic field

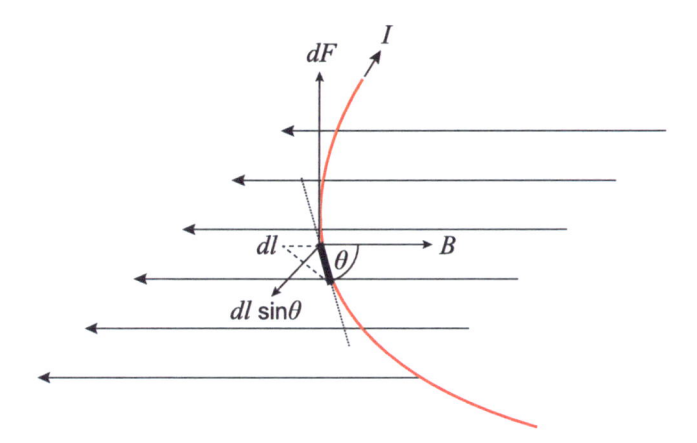

The force dF on a small length dl of the conductor is proportional to the current I, the magnetic induction B, and the component of dl perpendicular to the magnetic field, that is $dl \sin \theta$.

$$dF = BI \, dl \sin \theta$$

(4.1)

where θ is the angle between the length dl of the conductor and the magnetic field B.

For a straight conductor of length l in a uniform field B, the force on the conductor becomes

$$F = BIl \sin \theta$$

(4.2)

TOPIC 4. MAGNETIC FIELDS

If the conductor, and so also the current, is perpendicular to the field, then $\sin\theta = \sin 90° = 1$ and so the force is a maximum and is given by

$$F = BIl$$

If the conductor is parallel to the field, $\sin\theta = 0$ and the force is zero.

Example

Calculate the magnitude of the force on a horizontal conductor 10 cm long, carrying a current of 20 A from south to north, when it is placed in a horizontal magnetic field of magnitude 0.75 T, directed from east to west.

$$F = BIl\sin\theta$$
$$= 0.75 \times 20 \times 0.1 \times 1$$
$$= 1.5 \text{ N}$$

If a charge's velocity vector is not perpendicular to the magnetic field, then the component of v perpendicular to the field $v_\perp$ must be used in the equation $F = BIl\sin\theta$.

The direction of the force is at right angles to the plane containing l and B.
You may recall that the direction of this force can be established using the right hand rule.

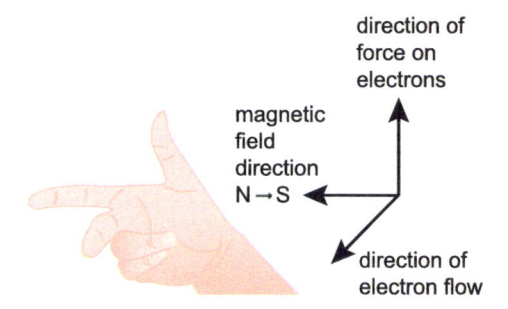

Figure 4.11: Direction of force on electrons

If the second finger points in the direction the electrons are flowing and the first finger points from north to south in the magnetic field then the thumb gives the direction of the force acting on the electrons.

Some people remember this right-hand rule as

- **T**humb for **t**hrust (force)
- **F**ore Finger for **F**ield, N → S
- **C**entre finger for **c**urrent

© HERIOT-WATT UNIVERSITY

Force on a current-carrying conductor

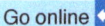

At this stage there is an online activity which demonstrates the force exerted on a current-carrying conductor placed in the field of a horseshoe magnet.

Force-on-a-conductor balance

The relationship between the magnetic induction B between two magnets and the force on a current-carrying conductor can be verified using a current balance shown in Figure 4.12

Figure 4.12: Measuring the force on a current-carrying conductor

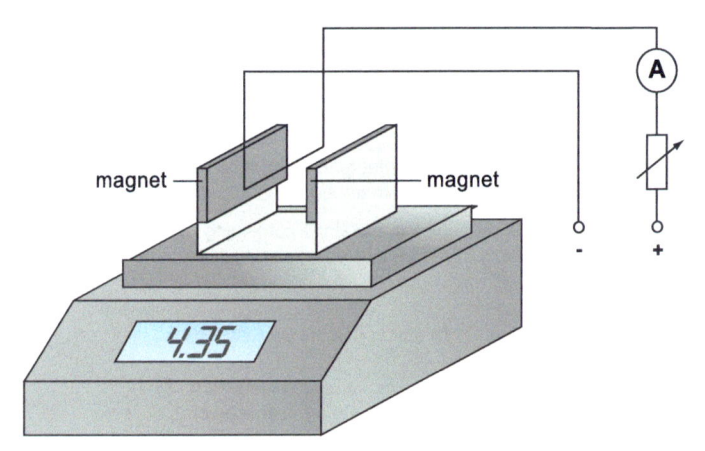

The balance is zeroed with no current in the wire. When a current is passed through the wire, the force F exerted by the wire on the magnet is seen as an apparent increase or decrease in the mass of the magnet Δm. This change in apparent mass is caused by a force of $\Delta m \, g$ newtons.

Force-on-a-conductor balance

At this stage there is an online activity which investigate the factors affecting the force on a conductor in a magnetic field.

4.5.1 The electric motor

The electric motor is device that makes use of the magnetic torque on a coil suspended in a magnetic field.

Consider the simple motor shown in Figure 4.13.

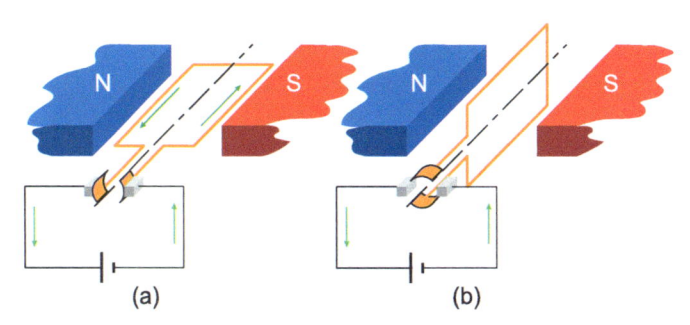

Figure 4.13: The simple electric motor

The coil, in this simple case consisting only of one turn, is called the rotor. It is free to rotate about an axis through its centre. The coil is placed in a magnetic field, which at the moment we will consider to be uniform. A current is fed into and out of the coil from an external circuit containing a source of e.m.f. through two brushes which contact with a commutator. The commutator consists of a split ring with each half connected to each end of the coil.

In Figure 4.13 (a) it can be seen that there is a force on each of the long sides of the coil. Since the current in each of these two sides is in opposite directions, these two forces supply a magnetic torque to the coil that makes it move anti-clockwise when looking in the direction shown.

This magnetic torque continues to move the coil round until it reaches the position shown in Figure 4.13 (b). At this position, if the current continued in the same direction, there would no longer be a torque on the coil (although there are still forces on each of the sides, these forces now act in opposite directions along the same line of action and so the torque has reduced to zero). Momentarily at this position, however, both sides of the commutator are in contact with both of the brushes. This stops the current in the coil. The inertia of the coil takes it slightly beyond the equilibrium position shown in Figure 4.13 (b) and this results in each brush again only connecting with one side of the commutator, restarting the current.

Although the sides of the coil have now physically changed positions, the current always enters the side of the coil that is nearest to the north pole and always leaves by the side nearest to the south pole. So the current reverses direction in the rotor every half-revolution and this current reversal, coupled with the rotation of the coil, ensures that the magnetic torque is always in the same sense.

The simple electric motor Go online

At this stage there is an online activity.

4.5.2 The electromagnetic pump

Consider a conducting fluid in a pipe with an electric current passing through it in a direction that is at right angles to the pipe. If the pipe is placed in a magnetic field that is at right angles to both the direction of the current and the pipe, then the fluid will experience a force along the length of the pipe, as shown in Figure 4.14. This will cause the fluid to flow along the pipe under the action of the magnetic force, with no external mechanical force applied to it. The twin benefits of this type of pumping action compared to a conventional mechanical pump are that the system is completely sealed and there are no moving parts other than the fluid itself.

Figure 4.14: The electromagnetic pump

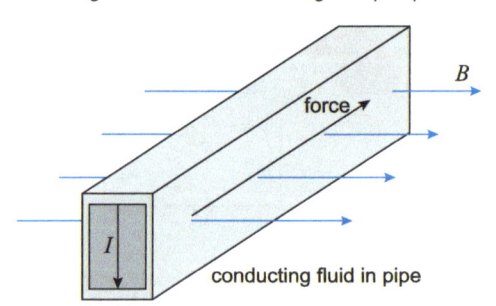

This type of pump is widely used in nuclear reactors to transport the liquid metal sodium that is used as a coolant from the reactor core to the turbine. More recently, electromagnetic pumps have been used in medical physics to transport blood in heart-lung machines and artificial kidney machines. Blood transported in this way can remain sealed and so the risk of contamination is reduced. There is also less damage to the delicate blood cells than is caused by mechanical pumps that have moving parts.

Quiz: Current-carrying conductors   Go online

Q4: The force on a conductor in a magnetic field is measured when the conductor is perpendicular to the field. Changes are made to the magnitude of the field, the current and the length of the conductor in the field.

In which one of the following situations is the force the same as the original force?

a) field halved, current the same, length the same
b) field halved, current halved, length halved
c) field doubled, current doubled, length doubled
d) field the same, current the same, length doubled
e) field the same, current doubled, length halved

....

TOPIC 4. MAGNETIC FIELDS

Q5: The force on a conductor in a magnetic field is measured when the conductor is perpendicular to the field.

Through what angle must the conductor be rotated in the direction of the magnetic field to reduce the force to half its original value?

a) 0°
b) 30°
c) 45°
d) 60°
e) 90°

....

Q6: Which is the correct description for the magnetic field around a long straight wire carrying a current?

a) radial, directed out from the wire
b) radial, directed in to the wire
c) uniform at all points
d) circular, increasing in magnitude with distance from the wire
e) circular, decreasing in magnitude with distance from the wire

....

Q7: Which of the following is equivalent to the unit of magnetic induction, the tesla?

a) $N\,A\,m^{-1}$
b) $N\,A^{-1}\,m^{-1}$
c) $N\,m^{-1}$
d) $N\,m\,A^{-1}$
e) $N\,m\,rad^{-1}$

© HERIOT-WATT UNIVERSITY

4.6 The relationship between magnetic induction and distance from a current-carrying conductor

Earlier in the topic, we saw that the magnetic field around a long straight wire carrying a current is circular, and is centred on the wire. A Hall probe, smartphone or search coil can be used to measure the magnitude of the field at various points. Such an investigation shows that the magnitude of the field, B, is directly proportional to the current, I, in the wire and is inversely proportional to the distance, r, from the wire.

$$B \propto \frac{I}{r}$$

The constant of proportionality in this relationship is written as $\mu_0/2\pi$, so the relationship becomes

$$B = \frac{\mu_0 I}{2\pi r}$$

(4.3)

The constant μ_0 in Equation 4.3 is called the **permeability of free space** and it has a value of $4\pi \times 10^{-7}$ H m^{-1} (or T m A^{-1}).

μ_0 is the magnetism counterpart to ε_0, the permittivity of free space, that appears in electrostatics. You will also have noticed that μ_0 appears in the numerator of the expression for magnetic induction, while ε_0 appears in the denominator of the expression for electric field ($E = \frac{Q}{4\pi\varepsilon_0 r^2}$).

This is partly explained by the fact that any insulating material placed in an electric field decreases the magnitude of the field, so relative permittivity appears as a divisor. On the other hand, inserting a ferromagnetic material in a magnetic field increases the magnitude of the field. Hence relative permeability appears as a multiplier.

Example

Calculate the magnitude of the magnetic field at a point in space 12 cm from a long straight wire that is carrying a current of 9.0 A.

We are given that the current I is 9.0 A. We want to calculate B at a point where r is 0.12 m.

$$B = \frac{\mu_0 I}{2\pi r}$$
$$\therefore B = \frac{4\pi \times 10^{-7} \times 9.0}{2\pi \times 0.12}$$
$$\therefore B = 1.5 \times 10^{-5} \text{ T}$$

TOPIC 4. MAGNETIC FIELDS

The hiker

A hiker is standing directly under a high voltage transmission line that is carrying a current of 500 A in a direction from north to south. The line is 10 m above the ground.

a) Calculate the magnitude of the magnetic field where the hiker is standing.

b) Calculate the minimum distance the hiker has to walk on horizontal ground to be able to rely on the reading given by his compass, assuming that any external magnetic field greater than 10% of the value of the Earth's magnetic field adversely affects the operation of a compass.

Take the magnitude of the Earth's magnetic field to be 0.5×10^{-4} T.

4.7 Comparison of forces

In our everyday lives there are two forces of nature that shape the world around us - electromagnetic and gravitational forces. Any other forces are just different manifestations of these forces. For example, you might consider the force involved when you stretch an elastic band. The tension in the band comes about because there are electrostatic attractions between the atoms in the band. As you pull on the band, these electric forces supply an opposing force. While you may think of this as a 'mechanical' force, fundamentally this is an electromagnetic interaction.

Example

Let's consider the interesting situation where both Coulomb and gravitational forces are present - which of the forces is dominant? For example, in a hydrogen atom we have two charged particles of known mass, so both forces are present. Is it the Coulomb force or the gravitational force that keeps them together as a hydrogen atom?

The proton and electron which make up a hydrogen atom have equal and opposite charges, $e = 1.60 \times 10^{-19}$ C. The mass of a proton $m_p = 1.67 \times 10^{-27}$ kg and the mass of an electron $m_e = 9.11 \times 10^{-31}$ kg. The average separation between proton and electron $r = 5.29 \times 10^{-11}$ m.

Coulomb Force

$$F = \frac{Q_1 Q_2}{4\pi\varepsilon_0 r^2}$$

$$\therefore F = \frac{1.60 \times 10^{-19} \times (-1.60 \times 10^{-19})}{4\pi\varepsilon_0 \times (5.29 \times 10^{-11})^2}$$

$$\therefore F = \frac{-2.56 \times 10^{-38}}{4\pi\varepsilon_0 \times 2.798 \times 10^{-21}}$$

$$\therefore F = -8.23 \times 10^{-8} \text{ N}$$

Gravitational Force

$$F = G\frac{m_1 m_2}{r^2}$$

$$\therefore F = G \times \frac{1.67 \times 10^{-27} \times 9.11 \times 10^{-31}}{(5.29 \times 10^{-11})^2}$$

$$\therefore F = 6.67 \times 10^{-11} \times \frac{1.521 \times 10^{-57}}{2.798 \times 10^{-21}}$$

$$\therefore F = 3.63 \times 10^{-47} \text{ N}$$

Remember the minus sign in front of the Coulomb force just indicates that we have an attractive Coulomb force (oppositely charged particles). The gravitational force is also attractive, but by convention the negative sign in omitted. Our calculations show that for atomic hydrogen, the Coulomb force is greater than the gravitational force by a factor of 10^{39}! So in this case the gravitational force is negligible compared to the electrostatic (Coulomb) force.

Now let us consider what happens inside the nucleus of an atom.

Although we can describe everyday phenomena in terms of electromagnetic or gravitational forces, you will recall from the Higher course that we need to consider other forces when describing nuclei. To understand this, let us consider a helium nucleus, consisting of two positively charged protons and two uncharged neutrons. Clearly there is an electrostatic repulsion between the protons. There is also a gravitational attraction between them. We can carry out an order-of-magnitude calculation to compare these forces.

Electrostatic and gravitational forces

Two protons in a helium nucleus are separated by around 10^{-15} m. Given the following data, perform an order-of-magnitude calculation to determine the ratio F_C/F_G of the Coulomb (electrostatic) and gravitational forces that act between the two protons. You will need to know the values of the permittivity of free space and the gravitational constant:

charge on a proton	$+1.6 \times 10^{-19}$ C
mass of a proton	1.67×10^{-27} kg
permittivity of free space ε_0	8.85×10^{-12} F m^{-1}
gravitational constant G	6.67×10^{-11} N m^2 kg^{-2}

These order-of-magnitude calculations shows that the electrostatic force is very much greater than the gravitational force. If this is the case, why doesn't this repulsive force cause the nucleus to split apart? The reason is that there is another force, called the **strong nuclear force**, that acts between any two nucleons (protons or neutrons) in a nucleus. This force, usually just called the strong force, can act between two protons, two neutrons, or a proton and a neutron, and has almost the same magnitude in each case.

How does the strong force compare to the electromagnetic and gravitational forces? The main difference is that the strong force is a short-range force. In fact its range is $< 10^{-14}$ m. This means that unless a particle is closer than about 10^{-14} m to another particle, the strong force between them is effectively zero. This is in contrast to the other two forces, both of which are long-range. For example, the gravitational force is the main interaction between planets and stars, and is effective over many millions of kilometres. The electrical force between charged objects follows a similar $1/r^2$ dependence.

In terms of strength, over a short range, the strong force is very much greater than the gravitational force, which is negligibly small between particles of such low mass. The strong force overcomes the electrostatic repulsion between protons, and prevents the nucleus from disintegrating.

Another point to note is that electrons are not affected by the strong force. A proton and an electron, or a neutron and an electron, would not interact via the strong force.

TOPIC 4. MAGNETIC FIELDS

You may also recall from the Higher course that the weak nuclear force is responsible for beta decay.

One of the biggest challenges in theoretical physics is to explain fully all four fundamental forces. Table 4.2 compares these four forces.

Table 4.2: Comparison of the forces of nature

Force	Relative magnitude	Range (m)	Example
strong nuclear	1	$< 10^{-14}$	nucleons in a nucleus
electromagnetic	$\sim 10^{-2}$	∞	majority of everyday 'contact' forces
weak nuclear	$\sim 10^{-5}$	$\sim 10^{-18}$	β-decay of a nucleus
gravitational	$\sim 10^{-38}$	∞	very large masses, e.g. planets

Gravity is clearly a much weaker force than the electric force. Furthermore, it is always attractive, whereas the electric force can be either attractive or repulsive. Despite these obvious differences, during the 1700s scientists noticed a great deal of similarity between the two forces, leading to speculation that they were perhaps really just manifestations of the same thing. Table 4.3 highlights the similarities between gravitational and electric forces. We shall further explore the unification of forces in **Topic 7**.

Table 4.3: Comparison of gravitational and electric forces

Concept	Gravitational field	Electric field
Force	Force between point masses obeys an inverse square law $$F = \frac{GMm}{r^2}$$ Only attraction	Force between point charges obeys an inverse square law $$F = \frac{Q_1 Q_2}{4\pi\varepsilon_0 r^2}$$ Repulsion or attraction
Field lines	A radial field surrounds a point mass and the field lines are drawn towards the mass.	A radial field surrounds a point charge and the field lines are drawn in the direction a positive charge would move.
Field strength	Force per unit mass $$g = F/m$$ Unit is Nkg^{-1}	Force per unit charge $$E = F/Q$$ Unit is NC^{-1}

© HERIOT-WATT UNIVERSITY

Concept	Gravitational field	Electric field
Field strength for point mass or charge	Field strength for point mass obeys inverse square law $$g = \frac{GM}{r^2}$$	Field strength for point charge obeys inverse square law $$E = \frac{Q}{4\pi\varepsilon_0 r^2}$$
Potential	$$V = \frac{GM}{r}$$ Joules per kg The zero of potential is at infinity from the planet.	$$V = \frac{Q}{4\pi\varepsilon_0 r}$$ Joules per Coulomb The zero of potential is at infinity from a charged object.
Potential Energy	$$E_p = mV$$ $$E_p = \frac{-GMm}{r}$$	$$E_p = QV$$ $$E_p = \frac{Q_1 Q_2}{4\pi\varepsilon_0 r}$$
Effect	Gravitational fields hold the Universe together. The gravitational force is always attractive.	Electric fields hold atoms and molecules together. The electric force may be attractive or repulsive.

4.7.1 Millikan's oil drop experiment

The American scientist R A Millikan conducted a experiment which involved balancing the electrostatic force on an oil drop with the gravitational force acting upon it. He designed the experiment to accurately measure the charge of an electron.

Theory

The experiment works by putting a negative electric charge on a microscopic drop of oil. The motion of the oil drop is observed as it falls between two charged horizontal plates. The magnitude of the electric field between the plates can be varied, so that the drop can be held stationary, or allowed to fall with constant velocity. Knowing the electric field strength, the charge on the oil drop can be deduced. Millikan found that every drop had a charge that was equal to an integer multiple of e, and no charged drop had a charge less than e. From these observations, he concluded firstly that charge was quantised, since the drops could only have a charge that was an integer multiple of e. Secondly, he stated that e was the fundamental unit of charge - a drop which had acquired one electron had charge e, one with two extra electrons had charge $2e$ and so on.

Experiment

The experimental arrangement is shown in Figure 4.15. The oil drops are charged either by friction as they are sprayed from a fine nozzle, or by irradiating the air near the nozzle using X-rays or a radioactive source. The drops then enter the electric field via the small aperture in the upper plate.

Figure 4.15: Experimental arrangement of Millikan's oil drop experiment

A charged drop can be held stationary in the electric field by adjusting the potential difference across the plates, or it can be allowed to fall if the pd is decreased.

Figure 4.16: Charged oil drops between charged plates

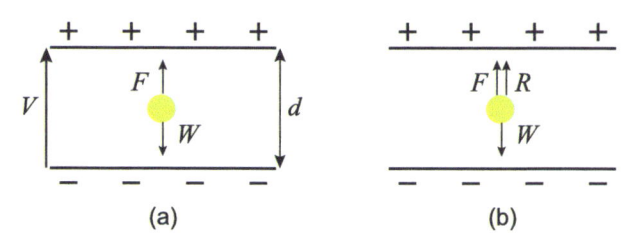

In Figure 4.16 (a) the Coulomb force F is equal to the gravitational force W acting on the drop. If the mass or radius of the drop is known, then the charge on the drop can be easily calculated. However, measuring the mass or radius of the drop is extremely difficult. Millikan devised a different method shown in Figure 4.16 (b). The drop is allowed to fall, so an additional force, air resistance R, acts on the drop. This force depends on the size of the drop and its velocity. By adjusting the pd, the drop can be made to fall at a constant velocity, so once again it is in equilibrium (Newton's First Law). If the density of the oil is known, an expression can be deduced relating the three forces acting on the drop which does not involve the mass or radius of the drop. Using this expression, the charge on the drop can be calculated.

4.8 Summary

In this topic we have seen that magnetic forces exist between moving charges. This is in addition to the electric forces that always exist between charges, moving or not.

Since a current in a wire is a movement of charges, the magnetic field around a wire was next studied. The unit for magnetic induction, the tesla, was defined. A simple way of deciding the direction of the force on a current-carrying conductor in a magnetic field was studied.

The expression for the force on a current-carrying conductor in a magnetic field was developed. It was found that the force on a conductor carrying a current in a magnetic field is proportional to both the current and the magnetic field.

We then went on to study the expression for the magnetic induction at a distance r from a current-carrying conductor. Finally, we compared the gravitational, electrostatic, magnetic and nuclear forces.

Summary

You should now be able to:

- state that electrons are in motion around atomic nuclei and individually produce a magnetic effect;
- state that ferromagnetism is a magnetic effect in which magnetic domains can be made to line up, resulting in the material becoming magnetised;
- state that iron, nickel, cobalt and some compounds of rare earth metals are ferromagnetic;
- sketch the magnetic field patterns around permanent magnets and the Earth;
- state that a magnetic field exists around a moving charge in addition to its electric field;
- sketch the magnetic field patterns around current carrying wires and current carrying coils;
- state that a charged particle moving across a magnetic field experiences a force;
- explain the interaction between magnetic fields and current in a wire;
- state the relative directions of current, magnetic field and force for a current-carrying conductor in a magnetic field;
- describe how to investigate the factors affecting the force on a current-carrying conductor in a magnetic field;
- use the relationship $F = IlB\sin\theta$ for the force on a current-carrying conductor in a magnetic field;
- define the unit of magnetic induction, the tesla (T);
- state and use the expression $B = \frac{\mu_0 I}{2\pi r}$ for the magnetic field B due to a straight current-carrying conductor;
- compare gravitational, electrostatic, magnetic and nuclear forces.

4.9 Extended information

Web links

These web links should serve as an insight to the wealth of information available online and allow you to explore the subject further.

- https://www.youtube.com/watch?v=6cNA0Yhh50U
- https://www.youtube.com/watch?v=V-M07N4a6-Y
- https://www.youtube.com/watch?v=hFAOXdXZ5TM
- https://www.youtube.com/watch?v=BqKeiiezqzc
- https://www.youtube.com/watch?v=-misJLuvHuE&list=PLiISjSDTPUjy2OKK5-BI2WrVu91LAe_cm&index=24
- https://www.youtube.com/watch?v=BvkNyXFVg40&list=PLiISjSDTPUjy2OKK5-BI2WrVu91LAe_cm&index=25
- https://www.youtube.com/watch?v=AcRCgyComEw&feature=youtu.be
- http://www.physicscentral.com/explore/action/nailpolish.cfm
- https://www.youtube.com/watch?v=MO0r930Sn_8&feature=youtu.be

4.10 End of topic 4 test

End of topic 4 test

Q8: A long straight wire is held perpendicular to the poles of a magnet of field strength 0.311 T. Assume that the field is uniform and extends for a distance of 2.03 cm.

Calculate the force on the wire when the current in it is 4.02 A.

_____ N

Q9: An electric power line carries a current of 1700 A.

Calculate the force on a 3.14 km length of this line at a position where the Earth's magnetic field has a magnitude of 5.35×10^{-5} T and makes an angle of 75.0° to the line.

_____ N

Q10: A short length of wire has a mass of 13.6 grams. It is resting on two conductors that are 3.67 cm apart, at right angles to them.

The conductors are connected to a power supply which maintains a constant current of 6.83 A in the wire.

The wire is held between the poles of a horseshoe magnet that has a uniform magnetic field of 0.237 T. The wire is perpendicular to the magnetic field.

Ignoring friction, air resistance and electrical resistance, calculate the initial acceleration of the wire.

_____ m s^{-2}

Q11: The apparatus shown is used to investigate the magnetic induction B of an electromagnet.

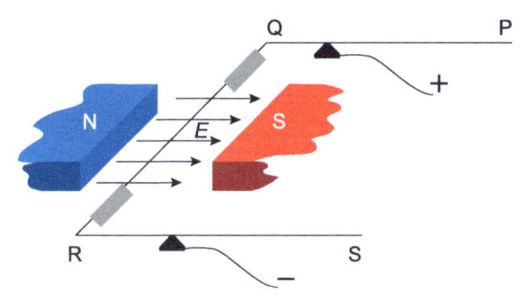

The straight wire QR has a current of 4.9 A supplied to it through the two knife edge points.

With the electromagnet switched off, the wire PQRS is balanced in a horizontal plane by hanging small masses as shown.

When the electromagnet is switched on, the mass hanging on QR has to be altered by 4.2 grams to restore PQRS to the horizontal.

The perpendicular length of QR which is in the magnetic field is 40 mm.

1. To restore PQRS to the horizontal, masses have to be
   a) removed from QR.
   b) added to QR.
2. Calculate the magnitude of the magnetic induction B.
   _____ T

Q12: A long straight conductor carries a steady current and produces a magnetic field of 4.1 × 10^{-5} T, at a distance of 11 mm.

Calculate the magnitude of the current.

_____ A

Q13: A lightning bolt can be considered as a straight current-carrying conductor.

Calculate the magnetic induction 18.7 m away from a lightning bolt in which a charge of 18.2 C is transferred in a time of 4.83 ms.

_____ T

TOPIC 4. MAGNETIC FIELDS

Q14: A device called a Hall probe can be used to find the current in a pipe carrying molten metal.

When the Hall probe is held perpendicular 0.53 m from the centre of the pipe, the maximum reading recorded by the probe is 1.7 mV.

The Hall probe has a sensitivity of 1000 mV T^{-1}.

1. Calculate the magnetic induction at this position.
   ---------- T
2. Calculate the current in the pipe.
   ---------- A

Unit 3 Topic 5

Capacitors

Contents
- 5.1 Introduction 439
- 5.2 Revision from Higher 439
    - 5.2.1 Relationship between Q and V 439
    - 5.2.2 Energy stored by a capacitor 441
    - 5.2.3 Charging and discharging capacitors in d.c. circuits 442
    - 5.2.4 Discharging a capacitor 445
- 5.3 The time constant for a CR circuit 448
- 5.4 Capacitors in a.c. circuits 452
- 5.5 Capacitive reactance 455
- 5.6 Summary 456
- 5.7 Extended information 456
- 5.8 End of topic 5 test 457

Prerequisites
- Ohm's Law and circuit rules (Higher).
- Capacitors (Higher).
- r.m.s current (Higher).
- Electric fields.

Learning objective

By the end of this topic you should be able to:

- sketch graphs of voltage and current against time for charging and discharging capacitors in series CR circuits;
- define the time constant of a circuit;
- carry out calculations relating the time constant of a circuit to the resistance and capacitance;
- use graphical data to determine the time constant of a circuit;
- define capacitive reactance;
- describe the response of a capacitive circuit to an a.c. signal;
- use the appropriate relationship to solve problems involving capacitive reactance, voltage and current;
- use the appropriate relationship to solve problems involving a.c. frequency, capacitance and capacitive reactance.

TOPIC 5. CAPACITORS

5.1 Introduction

You studied capacitors as part of the Higher course. In this topic we will now look at their behaviour in more detail, paying particular attention to the time they take to charge and discharge. To understand capacitors fully you will also need to recall the section on electric fields.

You may recall that a capacitor is a device for storing electrical energy. A capacitor consists of two conducting plates. When one plate is negatively charged and the other is positively charged, then electrical energy is stored on the capacitor. We will be reviewing how this process works, and how much energy can be stored on a capacitor.

Capacitors are important components in many electrical circuits. We will study how capacitive circuits respond to d.c. and a.c. signals. This will help us to understand some of the practical applications of capacitors.

5.2 Revision from Higher

This section will allow you to revise the content covered at CfE Higher.

5.2.1 Relationship between Q and V

A capacitor is made of two pieces of metal separated by an insulator.

When the capacitor becomes charged there will be a potential difference across the two pieces of metal.

The circuit symbol for a capacitor is

The capacitance of a capacitor is measured in farads (F) or more commonly microfarads (μF, x10^{-6}) or nanofarads (nF, x 10^{-9}). The capacitance of a capacitor depends on its construction not the charge on it or the potential difference across it.

You may recall that the charge (Q) stored by a capacitor is directly proportional to the potential difference (V) across its plates and that the constant of proportionality is the capacitance of the capacitor. The following activity will refresh your memory.

© HERIOT-WATT UNIVERSITY

Investigating V_c and Q_c

A circuit is set up to investigate the relationship between the voltage and the charge stored on the capacitor.

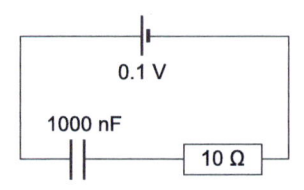

The voltage of the supply is increased (between 0.1 V and 1.0 V) and the charge on the capacitor is noted. The results obtained are used to produce the following graph.

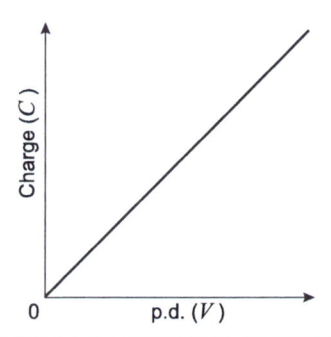

So the capacitance of a capacitor is defined by the equation

$$C = \frac{Q}{V}$$

This means that the capacitance of a capacitor is numerically equal to the amount of charge it stores when the p.d. across it is 1 volt. The unit of capacitance is the farad F, where 1F = 1 C V^{-1}. It is usually more common to express the capacitance in microfarads (1 μF = 1 × 10^{-6} F), nanofarads (1 nF = 1 × 10^{-9} F) or picofarads (1 pF = 1 × 10^{-12} F).

Example A 20 mF capacitor has a potential difference of 9.0 V across it. How much charge does it store?

$$C = \frac{Q}{V}$$
$$20 \times 10^{-3} = \frac{Q}{9.0}$$
$$Q = 0.18\ C$$

TOPIC 5. CAPACITORS

5.2.2 Energy stored by a capacitor

Let us consider the charged parallel-plate capacitor shown in Figure 5.1.

Figure 5.1: Electric field between two charged plates

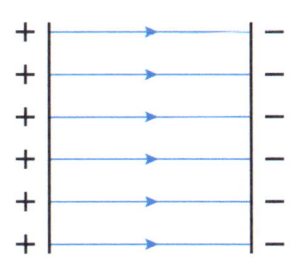

Suppose we take an electron from the left-hand plate and transfer it to the right-hand plate. We have to do work in moving the electron since the electrical force acting on it opposes this motion. The more charge that is stored on the plates, the more difficult it will be to move the electron since the electric field between the plates will be larger.

The work that is done in placing charge on the plates of a capacitor is stored as potential energy in the charged capacitor. The more charge that is stored on the capacitor, the greater the stored potential energy.

You may recall that it can be shown that the energy stored by a capacitor is given by the following equations:

$$E = \frac{1}{2}QV \qquad E = \frac{1}{2}CV^2 \qquad E = \frac{1}{2}\frac{Q^2}{C}$$

Example

A 750 nF capacitor is charged to 50 V. Calculate the charge and the energy it stores.

$$E = \frac{1}{2}CV^2$$
$$E = \frac{1}{2}750 \times 10^{-9} \times 50^2$$
$$E = 9.4 \times 10^{-4} \text{ J}$$

5.2.3 Charging and discharging capacitors in d.c. circuits

Let us now recap the behaviour of capacitors when they are connected as components in d.c. circuits. A capacitor is effectively a break in the circuit, and charge cannot flow across it. We will see now review how this influences the current in capacitive circuits.

> **Charging a capacitor** Go online
>
> There is an activity online at this stage. The activity provides a circuit with a capacitor and resistor which can be altered. The shape of the output graphs is also given.

Figure 5.2 shows a simple d.c. circuit in which a capacitor is connected in series to a battery and resistor. This is often called a series *CR* circuit.

Figure 5.2: Simple d.c. capacitive circuit

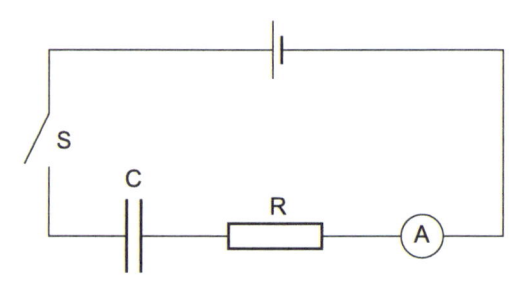

When the switch S is closed, charge can flow on to (but not across) the capacitor C. At the instant the switch is closed the capacitor is uncharged, and it requires little work to add charges to the capacitor. As we have already discussed, though, once the capacitor has some charge stored on it, it takes more work to add further charges. Figure 5.3 shows graphs of current *I* through the capacitor (measured on the ammeter) and charge *Q* on the capacitor, against time.

Increasing the R or C value increases the rise time however the final p.d. across the capacitor will remain the same. The final p.d. across the capacitor will equal the e.m.f., E, of the supply.

Figure 5.3: Plots of current and charge against time for a charging capacitor

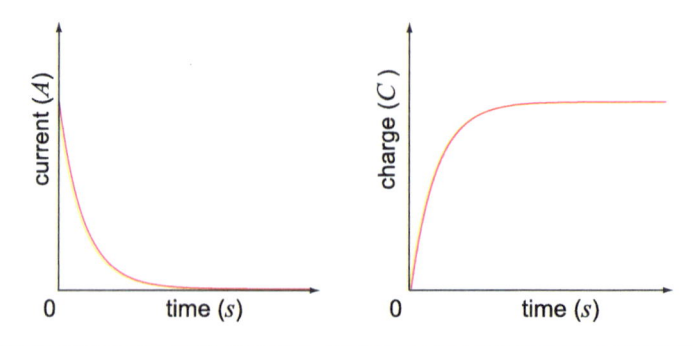

TOPIC 5. CAPACITORS

Since the potential difference across a capacitor is proportional to the charge on it, then a plot of p.d. against time will have the same shape as the plot of charge against time shown in Figure 5.3.

Suppose the battery in Figure 5.2 has e.m.f E and negligible internal resistance. The sum of the p.d.s across C and R must be equal to E at all times. That is to say,

$$V_C + V_R = E$$

where V_C is the p.d. across the capacitor and V_R is the p.d. across the resistor. At the instant switch S is closed there is no charge stored on the capacitor, so V_C is zero, hence $V_R = E$. The current in the circuit at the instant the switch is closed is given by

$$I = \frac{E}{R}$$

(5.1)

As charge builds up on the capacitor, so V_C increases and V_R decreases. This is shown in Figure 5.4.

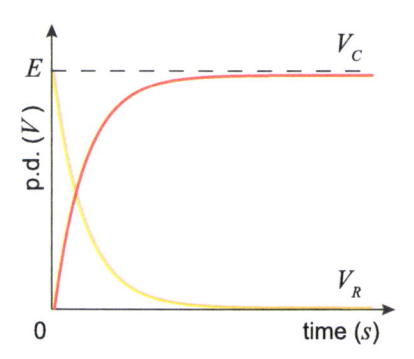

Figure 5.4: Plots of p.d. against time for a capacitive circuit

The charge and potential difference across the capacitor follow an exponential rise. (The current follows an exponential decay). The rise time (the time taken for the capacitor to become fully charged) depends on the values of the capacitance C and the resistance R. The rise time increases if either C or R increases. So, for example, replacing the resistor R in the circuit in Figure 5.2 by a resistor with a greater resistance will result in the p.d. across the capacitor C rising more slowly, and the current in the circuit dropping more slowly. We will look at this effect more closely in the next section.

© HERIOT-WATT UNIVERSITY

Example

Consider the circuit in Figure 5.5, in which a 40 kΩ resistor and an uncharged 220 μF capacitor are connected in series to a 12 V battery of negligible internal resistance.

Figure 5.5: Capacitor and resistor in series

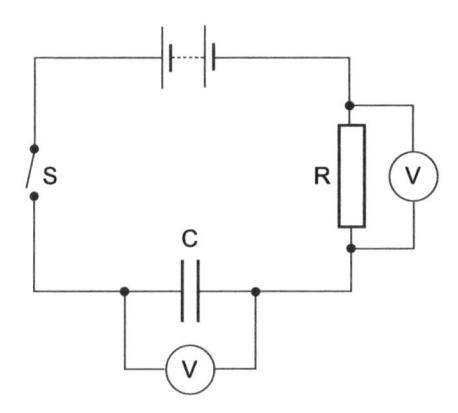

1. What is the potential difference across the capacitor at the instant the switch is closed?
2. After a certain time, the charge on the capacitor is 600 μC. Calculate the potential differences across the capacitor and the resistor at this time.

Answer:

1. At the instant the switch is closed, the charge on the capacitor is zero, so the p.d. across it is also zero.

2. We can calculate the p.d. across the capacitor using $Q = CV$ or $C = \frac{Q}{V}$:

$$V_c = \frac{Q}{C}$$

$$\therefore V_c = \frac{600 \times 10^{-6}}{220 \times 10^{-6}}$$

$$\therefore V_c = 2.7 \text{ V}$$

Since the p.d. across the capacitor is 2.7 V, the p.d. across the resistor is 12 - 2.7 = 9.3 V.

5.2.4 Discharging a capacitor

Discharging a capacitor   Go online

There is an activity online at this stage showing how the capacitor charges and discharges.

The circuit shown in Figure 5.6 can be used to investigate the charging and discharging of a capacitor.

Figure 5.6: Circuit used for charging and discharging a capacitor

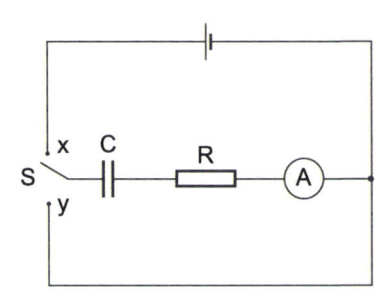

When the switch S is connected to x, the capacitor C is connected to the battery and resistor R, and will charge in the manner shown in Figure 5.3. When S is connected to y, the capacitor is disconnected from the battery, and forms a circuit with the resistor R. Charge will flow from C through R until C is uncharged. A plot of the current against time is given in Figure 5.7.

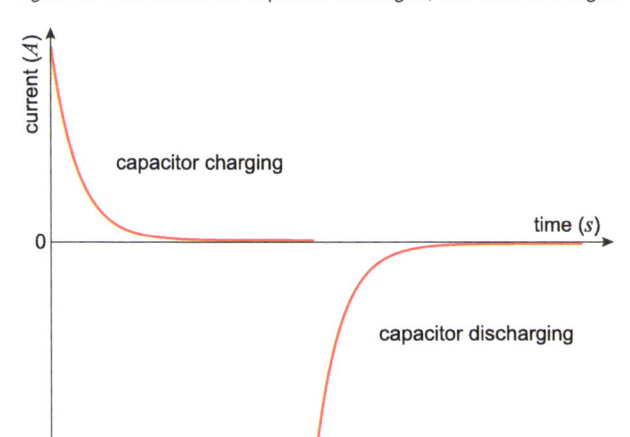

Figure 5.7: Current as the capacitor is charged, and then discharged

Remember that the capacitor acts as a break in the circuit. Charge is *not* flowing across the gap between the plates, it is flowing from one plate through the resistor to the other plate. Note that the direction of the current reverses when we change from charging to discharging the capacitor. The energy which has been stored on the capacitor is dissipated in the resistor.

© HERIOT-WATT UNIVERSITY

Charging current: The initial charging current is very large. Its value can be calculated by $I = \frac{V_{supply}}{R}$.

The current is only at this value for an instant of time. As the capacitor charges, the p.d. across the capacitor increases so the p.d. across the resistor decreases causing the current to decrease.

Discharging current: The initial discharging current is very large. Its value can be calculated by $I = \frac{V_{capacitator}}{R}$

During discharge the circuit is not connect to the supply so it is the p.d. across the capacitor, not the p.d. across the supply, which drives the current. If however the capacitor had been fully charged, the initial p.d. across the capacitor would equal the p.d. across the supply.

The current is only at this value for an instant of time. As the capacitor discharges, the p.d. across the capacitor decreases so the p.d. across the resistor also decreases causing the current to decrease. When the capacitor is fully discharged, the p.d. across it will be zero hence the current will also be zero.

Figure 5.7 shows us that at the instant when the capacitor is allowed to discharge, the size of the current is extremely large, but dies away very quickly. This leads us to one of the applications of capacitors, which is to provide a large current for a short amount of time. One example is the use of a capacitor in a camera flash unit. The capacitor is charged by the camera's batteries. At the instant the shutter is pressed, the capacitor is allowed to discharge through the flashbulb, producing a short, bright burst of light.

Using the energy stored on a capacitor   Go online

At this stage there is an online activity. If however you do not have access to the internet you should ensure that you understand the following explanation.

When a lamp is lit from a d.c. supply directly it gives a steady dim energy output.

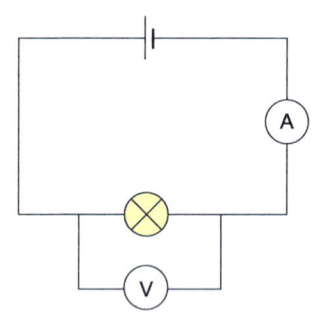

TOPIC 5. CAPACITORS

It is now connected to a capacitor and charged as shown

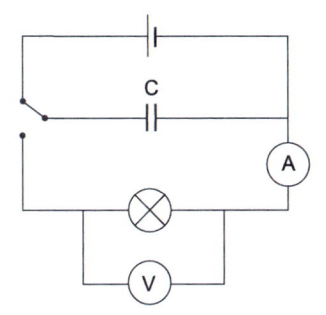

The capacitor is then discharged through the lamp by changing the switch position as shown.

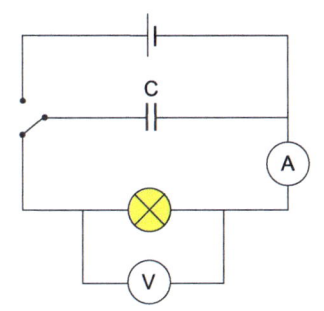

- When the capacitor powers the lamp, a large current flows for a very short period of time. This produces a bright flash of light.
- The current flows for only a short time while the capacitor discharges.
- Before the flash can be used again the capacitor must be recharged from the supply.

The following table shows how voltage and current change during charging and discharging.

	At any instant	At start	At finish
Charging	$V_S = V_C + V_R$	$V_C = 0$	$V_C = V_S$
	$I = V_R/R$	$V_R = V_S$	$V_R = 0$
		$I_0 = V_S/R$	$I = 0$
Discharging	$V_C = V_R$	$V_C = V_S$	$V_C = 0$
	$I = V_R/R$	$V_R = V_S$	$V_R = 0$
		$I_0 = V_S/R$	$I = 0$

5.3 The time constant for a CR circuit

Now that we have gone back through the main points covered at Higher, let us now look more closely at capacitor discharge.

The graphs in the last section showed that when a capacitor is discharging, the potential difference across its plates decreases exponentially with time. As $Q = CV$, this means that the charge the capacitor stores must also decrease exponentially with time.

TOPIC 5. CAPACITORS

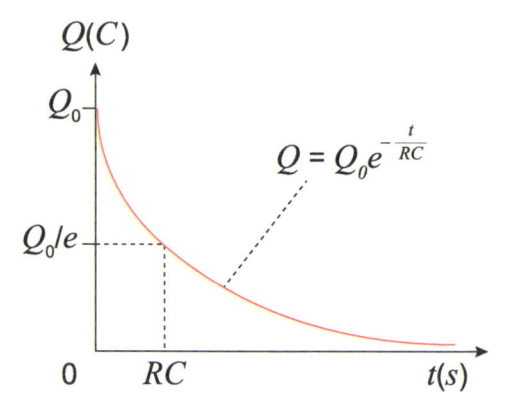

In fact, the charge left on the plates of a capacitor as it discharges is given by the equation,

$$Q = Q_0 e^{-\frac{t}{RC}}$$

where Q_0 is the charge stored by the capacitor when it is fully charged, R is the resistance of the resistor and C is the capacitance of the capacitor.

If $\tau = RC$ is put into the equation above, then $Q = Q_0 e^{-1}$.

So, when $\tau = RC$,

$$\frac{Q}{Q_0} = \frac{1}{e}$$
$$\frac{1}{e} \simeq 0.37$$

The quantity CR is called the **time constant** (τ) of the circuit. The time constant (τ) is the time taken for the charge on a discharging capacitor decrease to $\frac{1}{e}$ of its initial value, Q_0. In other words, the time constant (τ) is the time taken to discharge the capacitor to 37% of initial charge. The unit of time constant (τ) is the second.

After $t = 2\tau$, the value falls to $\frac{1}{e^2}$ of its initial value.

After $t = 3\tau$, the value falls to $\frac{1}{e^3}$ of its initial value.

The capacitor is said to be fully discharged after a time approximately equal to 5τ.

The potential difference across the capacitor and the current in the circuit also decrease exponentially in the same manner, according to,

$$V = V_0 e^{-\frac{t}{RC}}$$

Therefore, the

$$I = I_0 e^{-\frac{t}{RC}}$$

time constant can also be expressed as the time taken for the current in the circuit or potential difference across the capacitor to fall to 37% of their initial values.

So, provided the resistance of the resistor and the capacitance of the capacitor are not altered, it always takes the same length of time for the Q, I and $V_{Capacitor}$ to decrease to 37% of the original value, no matter how much charge the capacitor starts with. The value of the time constant ($\tau = RC$) can be increased by increasing the value of C or R or both C and R. If CR is large, the discharge will be slow and if it is small the discharge will be fast.

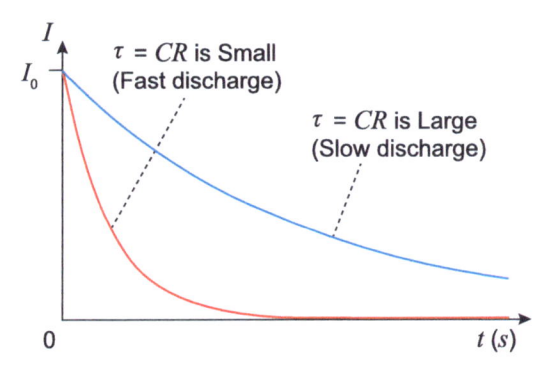

The time constant is the time taken to increase the charge stored by 63% of the difference between initial charge and maximum (full) charge. The larger the resistance in series with the capacitor and the larger the capacitance of the capacitor, the longer it takes to charge. The capacitor is said to be fully charged after a time approximately equal to 5τ.

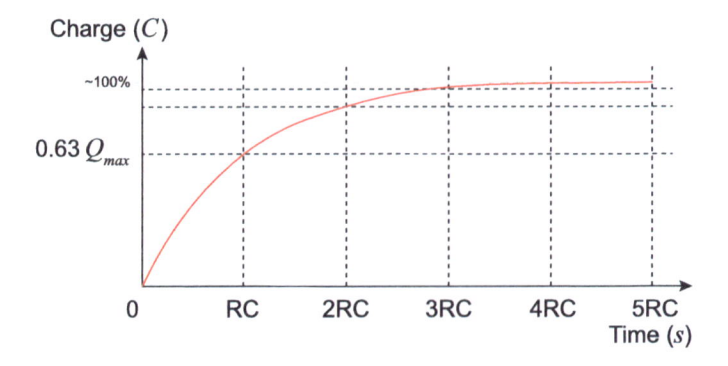

TOPIC 5. CAPACITORS

Examples

1.

A 500 μF capacitor is fully charged from a 12 V battery and is then discharged through a 3 kΩ resistor. Calculate the time taken for the charge stored by the capacitor to decrease to 37% of the initial value.

$$\tau = RC$$
$$\tau = 3000 \times 500 \times 10^{-6}$$
$$\tau = 1.5 \text{ s}$$

....

A 2.2 μF capacitor is connected in series to a resistor and a 6.0 V battery. It takes 0.055 s for the p.d. across its plates to increase to 3.78 V. Calculate the resistance of the resistor.

$$\frac{3.78}{6.0} \times 100 = 63\%$$

The capacitor has been charged to 63% of the supply voltage. Therefore, the time passed must equal the time constant.

$$\tau = RC$$
$$0.055 = R \times 2.2 \times 10^{-6}$$
$$R = 25000 \text{ } \Omega$$

Circuits in which a capacitor discharges through a resistor are often used in electronic timers. For instance, a pelican crossing uses a capacitor to activate a sequence of traffic lights for a predetermined time when a pedestrian presses a switch. The capacitor is initially charged and is then allowed to gradually discharge. Eventually the p.d. across the capacitor will fall below a set value, triggering a switching circuit.

For Interest Only

The mathematical proof for the equation $Q = Q_0 e^{-\frac{t}{RC}}$ is not examinable. It is included here for interest only.

The current at time t during capacitor discharge is given by $I = -\frac{dQ}{dt}$. The negative sign is present because the charge stored by the capacitor decreases with time.

Substituting $V = \frac{Q}{C}$ into the equation $I = \frac{V}{R}$ and gives $I = \frac{Q}{CR}$.

Therefore, we have

$$-\frac{dQ}{dt} = \frac{Q}{CR}$$

$$\int_{Q_0}^{Q} \frac{dQ}{Q} = -\frac{1}{CR} \int_0^t dt$$

$$[\ln Q]_{Q_0}^{Q} = -\frac{1}{CR}[t]_0^t$$

$$\ln \frac{Q}{Q_0} = -\frac{t}{CR}$$

$$Q = Q_0 e^{-\frac{t}{RC}}$$

5.4 Capacitors in a.c. circuits

Capacitors oppose the flow of alternating current, as do components called inductors, which we will examine in the next topic. The opposition which a capacitor offers to a.c. current flow is called its **capacitive reactance**, X_C, and is defined by

$$X_C = \frac{V}{I}$$

Capacitive reactance is measured in Ω, as is a resistor's resistance.

We are about to explore the behaviour of a capacitor in an a.c. circuit, paying particular attention to the relationship between the a.c. frequency and the capacitive reactance. However, let us first of all find out how a resistor would behave in such a circuit.

Resistors in a.c. circuits   Go online

There is an online activity which allows you to observe the effect on the alternating current through a resistor as the frequency of the supply is altered.

You have just observed that the ratio $\frac{V_{r.m.s.}}{I_{r.m.s.}}$ remains constant for a resistor no matter what the frequency of the a.c. supply is. This means the alternating r.m.s. current in a resistor does not

TOPIC 5. CAPACITORS

change with frequency. Now let's investigate the response of a capacitor to an a.c. supply.

The capacitor and a.c.   Go online

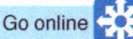

A circuit is set up to investigate the relationship between frequency of an a.c. supply and current in a capacitive circuit.

The r.m.s. voltage of the supply is kept constant.

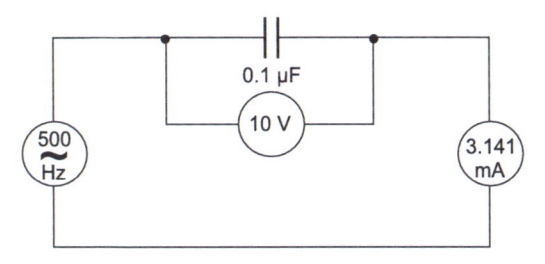

[Circuit diagram: 500 Hz a.c. source, 0.1 μF capacitor at 10 V, ammeter reading 3.141 mA]

The frequency of the supply is increased and the r.m.s. current is measured.

The results obtained are used to produce the following graph.

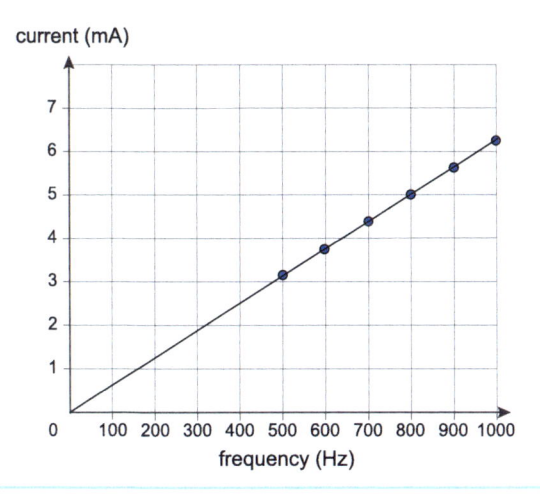

[Graph: current (mA) vs frequency (Hz), linear relationship passing through origin, reaching about 6.3 mA at 1000 Hz]

As you can see the r.m.s current in a capacitive circuit is directly proportional to the frequency of the supply. This means the capacitive reactance is inversely proportional to the frequency.

You may recall that when a d.c. supply is used in a CR circuit, the current rapidly drops to zero once the switch is closed. We have just observed that in an a.c. circuit there is a steady current through the capacitor. This is because the capacitor is charging and discharging every time a.c changes direction. A CR circuit passes high frequency a.c. much better than it does low frequency a.c. or d.c. But why is this?

© HERIOT-WATT UNIVERSITY

You should remember that charge does not flow across the plates of a capacitor. It accumulates on the plates, and the more charge that has accumulated, the more work is required to add extra charges. At all times, the *total* charge on the plates of the capacitor is zero. Charges are merely transferred from one plate to the other via the external circuit when the capacitor is charged. At low frequency, as the applied voltage oscillates, there is plenty of time for lots of charge to accumulate on the plates, which means the current drops more at low frequency (see Figure 5.3). At high frequency, there is only a short time for charge to accumulate on the plates before the direction of the current is reversed, and the capacitor discharges.

Applications

The fact a capacitor passes high frequency a.c. much better than low frequency a.c. or d.c. means it can be used as a **high-pass filter** for electrical signals. That is to say, it allows a high frequency electrical signal to pass, but blocks any low frequency signals. This is particularly useful if a small a.c. voltage is superimposed on a large d.c. voltage, and we are trying to measure the a.c. part. Figure 5.8 shows how a high pass filter is used to measure such a signal (a). The d.c. component can be filtered out using a capacitor, leaving the signal shown in Figure 5.8 (b). The sensitivity of the voltmeter can then be turned up to allow the a.c. signal to be measured accurately (c).

Figure 5.8: Filtered and amplified a.c. signal

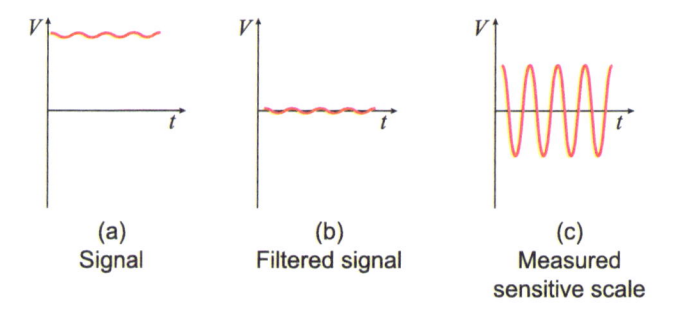

(a) Signal

(b) Filtered signal

(c) Measured sensitive scale

5.5 Capacitive reactance

The exact relationship between capacitive reactance and a.c. frequency is given by the expression $X_C = \frac{1}{2\pi fC}$

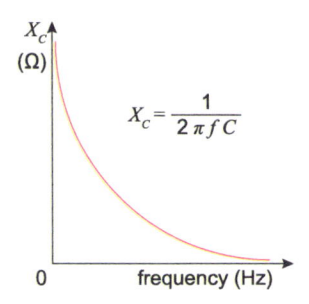

A graph of X_C against $\frac{1}{f}$ is a straight line through the origin, with gradient equal to $\frac{1}{2\pi C}$.

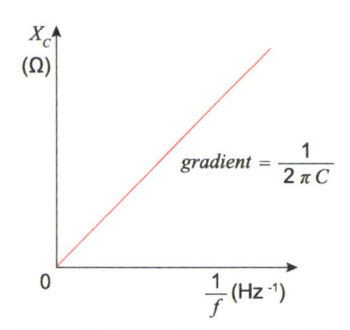

Example

A 4700 µF capacitor is connected to an a.c. supply of frequency 12 Hz. The r.m.s voltage is 6.0 V. Calculate the r.m.s. current.

$$X_C = \frac{1}{2\pi fC}$$
$$X_C = \frac{1}{2\pi \times 12 \times 4700 \times 10^{-6}}$$
$$X_C = 2.82\ \Omega$$

$$X_C = \frac{V}{I}$$
$$2.82 = \frac{6.0}{I}$$
$$I = 2.1\ \text{A}$$

5.6 Summary

> **Summary**
>
> You should now be able to:
>
> - sketch graphs of voltage and current against time for charging and discharging capacitors in series CR circuits;
> - define the time constant of a circuit;
> - carry out calculations relating the time constant of a circuit to the resistance and capacitance;
> - use graphical data to determine the time constant of a circuit;
> - define capacitive reactance;
> - describe the response of a capacitive circuit to an a.c. signal;
> - use the appropriate relationship to solve problems involving capacitive reactance, voltage and current;
> - use the appropriate relationship to solve problems involving a.c. frequency, capacitance and capacitive reactance.

5.7 Extended information

> **Web links**   Go online
>
> These web links should serve as an insight to the wealth of information available online and allow you to explore the subject further.
>
> - http://tutor-homework.com/Physics_Help/rc_circuit_simulation.html (be aware electron flow is in the opposite direction to that shown in this simulation)
> - http://physics.bu.edu/~duffy/semester2/c11_RC.html

TOPIC 5. CAPACITORS

5.8 End of topic 5 test

End of topic 5 test Go online

Q1: The following graph shows how the charge stored by a capacitor varies with time as it charges in a CR circuit.

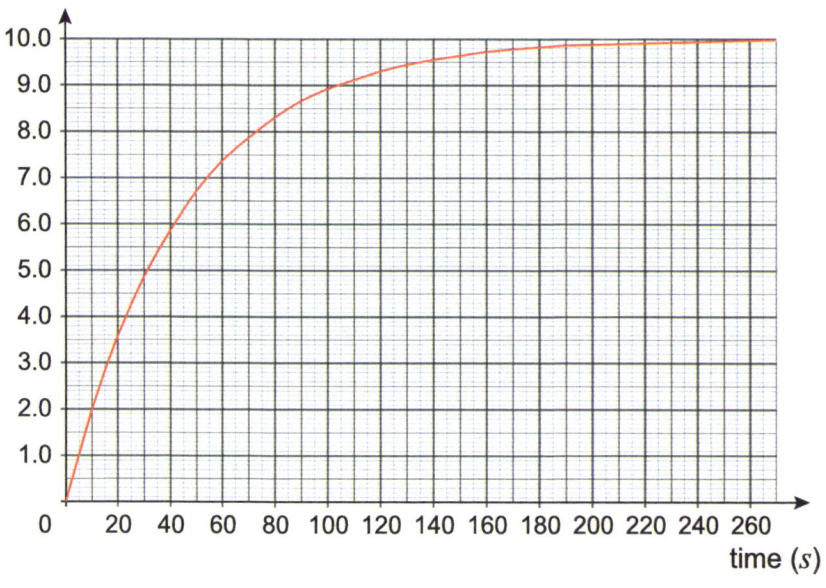

Use the graph to determine the time constant of the discharge circuit.

_____ s

Q2: The following graph shows how the potential difference across a capacitor varies with time as it discharges in a CR circuit.

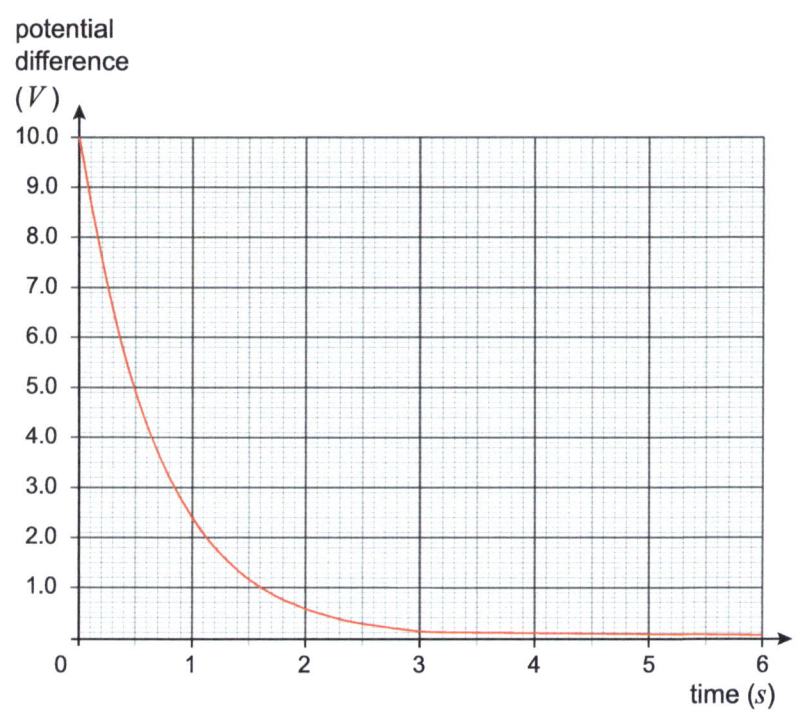

The capacitor has a capacitance of 4700 μF. Determine the resistance of the circuit.

_____ Ω

....

Q3: In a series CR circuit, a 580 nF capacitor and a 400 Ω resistor are connected to an a.c. power supply. When the frequency of the supply is 50 Hz, the r.m.s current in the circuit is 17 mA.

Calculate the r.m.s. current when the frequency of the supply is increased to 150 Hz, if the r.m.s. voltage of the power supply is kept constant.

_____ mA

....

Q4: The frequency of the output from an a.c. supply is increased.
Which graph shows how the reactance of a capacitor varies with the frequency of the supply?

TOPIC 5. CAPACITORS

(a)

(b)

(c)

(d)

(e)

....

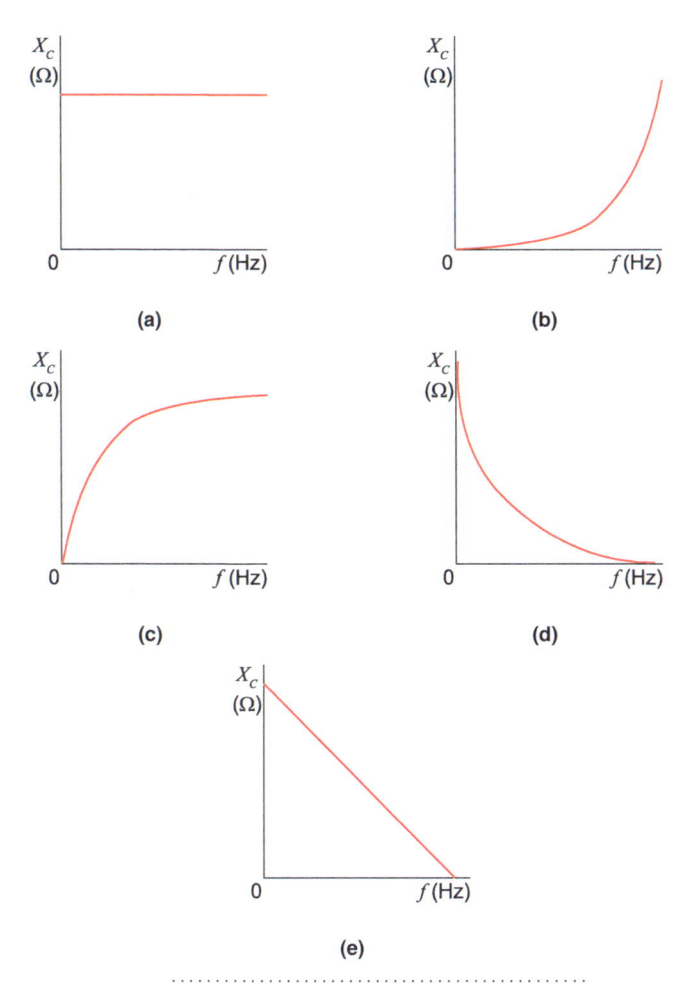

Q5: A 180 nF capacitor is connected to 14.0 V a.c. power supply. The frequency of the a.c. supply is 5800 Hz.

1. Calculate the capacitive reactance of the capacitor.
   _____ Ω
2. Calculate the current in the circuit.
   _____ A

Unit 3 Topic 6

Inductors

Contents

6.1 Introduction ... 463
6.2 Magnetic flux and induced e.m.f. ... 463
    6.2.1 Magnetic flux and solenoids ... 463
    6.2.2 Induced e.m.f. in a moving conductor ... 464
    6.2.3 Cassette players ... 465
    6.2.4 Faraday's law and Lenz's law ... 466
6.3 Eddy currents ... 466
    6.3.1 Implications of Lenz's law ... 467
    6.3.2 Induction heating ... 467
    6.3.3 Metal detectors ... 468
6.4 Inductors and self-inductance ... 469
    6.4.1 Self-inductance ... 469
    6.4.2 Energy stored in an inductor ... 470
6.5 Inductors in d.c. circuits ... 474
    6.5.1 Growth and decay of current ... 474
    6.5.2 Back e.m.f. ... 477
6.6 Inductors in a.c. circuits ... 479
6.7 Summary ... 482
6.8 Extended information ... 483
    6.8.1 Levitation of superconductors ... 483
    6.8.2 Web links ... 484
6.9 End of topic 6 test ... 485

Prerequisites

- Magnetic fields.
- Energy and power in an electric circuits (Higher).
- Current and voltage in series and parallel circuits (Higher).
- Capacitors in a.c. circuits.

Learning objective

By the end of this topic you should be able to:

- sketch graphs showing the growth and decay of current in a simple d.c. circuit containing an inductor;
- describe the principles of a method to illustrate the growth of current in a d.c. circuit;
- state that an e.m.f. is induced across a coil when the current through the coil is varying;
- explain the production of the induced e.m.f across a coil;
- explain the direction of the induced e.m.f in terms of energy;
- state that the inductance of an inductor is one henry if an e.m.f. of one volt is induced when the current is changing at a rate of one ampere per second;
- use the equation $\varepsilon = -L\frac{dI}{dt}$ and explain why a minus sign appears in this equation;
- state that the work done in building up the current in an inductor is stored in the magnetic field of the inductor, and that this energy is given by the equation $E = \frac{1}{2}LI^2$;
- calculate the maximum values of current and induced e.m.f. in a d.c. LR circuit;
- use the equations for inductive reactance $X_L = \frac{V}{I}$ and $X_L = 2\pi f L$;
- describe the response of an a.c. inductive circuit to low and high frequency signals.

6.1 Introduction

Electromagnetic induction is the production of an induced e.m.f. in a conductor when it is present in a changing magnetic field. An airport metal detector is just one example of a modern appliance that relies on electromagnetic induction for its operation. In this topic we will investigate different ways of producing an induced current and we will look at various other applications of this effect.

We will then focus on the behaviour of inductors, which are basically coils of wire designed for use in electronic circuits. We will pay particular attention to their opposition to current flow, allowing us to contrast their behaviour to that of the capacitors we studied in the last topic.

6.2 Magnetic flux and induced e.m.f.

6.2.1 Magnetic flux and solenoids

Before we look at induction, we will first review some essential points concerning magnets and magnetic fields. An important concept is **magnetic flux**. We can visualise the magnetic flux lines to indicate the strength and direction of a magnetic field, just as we have used field lines to represent electrical or gravitational fields. The magnetic flux ϕ passing through an area A perpendicular to a uniform magnetic field of strength B is given by the equation

$$\phi = BA$$

(6.1)

where ϕ is the flux density in T m^2 (or weber, Wb). While you do not need to remember this equation, the idea of magnetic flux is a useful one in understanding inductance.

Another idea that you should have met before is the magnetic field associated with a current-carrying coil, otherwise known as a solenoid. The magnetic field strength inside an air-filled cylindrical solenoid depends on the radius and length of the coil, and the number of turns of the coil. The direction of the magnetic field depends on the direction of the current, as shown in Figure 6.1.

Figure 6.1: Solenoids - the direction of the current (electron flow) tells us the direction of the magnetic field

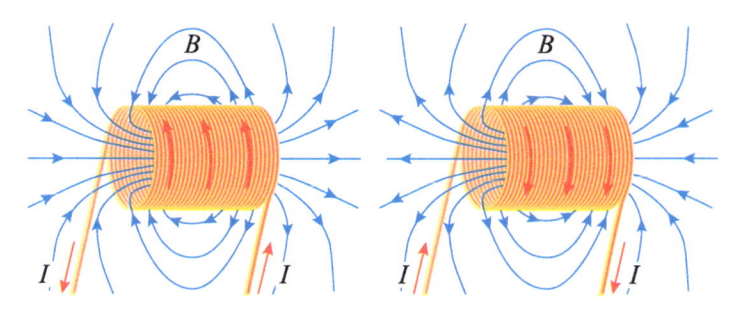

6.2.2 Induced e.m.f. in a moving conductor

In previous topics we have studied the force exerted on a charge moving in a magnetic field, such as the charged particles making up the solar wind. We begin this topic by looking at how this force can induce an e.m.f. in a conductor.

Any metallic conductor contains 'free' electrons that are not strongly bonded to any particular atom. When an e.m.f. is applied, these electrons drift along in the conductor, this movement of charges being what we call an electric current. We have used the equation $F = IlB\sin\theta$ to calculate the force on a conductor placed in a magnetic field when a current is present.

Consider now what happens when a rod of metal is made to move in a magnetic field.

Figure 6.2: Metal rod moving at right angles to a magnetic field

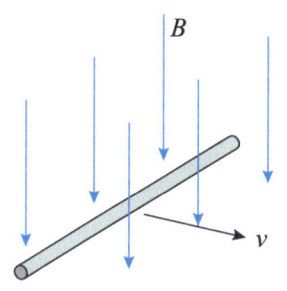

In Figure 6.2, a magnetic field acts vertically downwards in the diagram. As the conductor is moved from left to right, each free electron is a charged particle moving at right angles to a magnetic field. The force on each electron acts out of the page in the diagram, so electrons will drift that way, leaving a net positive charge behind. Thus there is a net positive charge at the far away end of the end of the rod and a net negative charge at the other end. This means there will be a potential difference between the ends of the rod.

A force will act on the charges in a conductor whenever a conductor moves across a magnetic field. We usually state that this occurs whenever a conductor crosses magnetic flux lines, as there is no induced voltage when the conductor moves parallel to the magnetic field. If the conductor is connected to a stationary circuit, as shown in Figure 6.3, then a current I is induced in the circuit.

Figure 6.3: Current due to the induced e.m.f

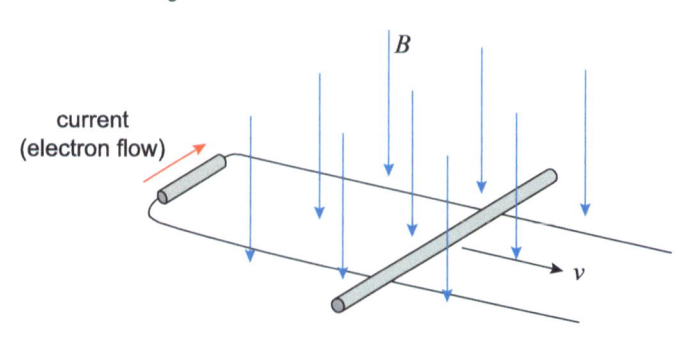

TOPIC 6. INDUCTORS

You should note that the **induced e.m.f.** occurs when there is relative motion between the magnetic field and a conductor, so we can have a stationary conductor and a moving magnetic field. An example of this is the e.m.f. induced when a magnet is moved in and out of a coil, as shown in Figure 6.4.

Figure 6.4: Induced e.m.f. causing a current to appear in a coil when the magnet is moved up and down

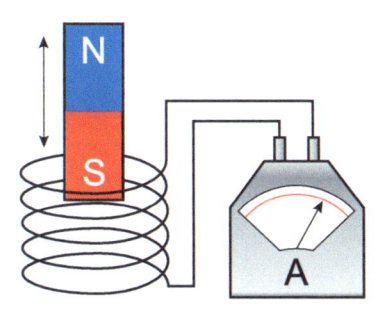

Induced e.m.f. Go online

There is an activity available online, which allows you to investigate induced current caused by the change in magnetic field.

As the magnet is moved in and out of the coil, the induced e.m.f. causes a current. The direction of the current changes as the direction of the magnet's movement changes. If the magnet is stationary, whether inside or outside the coil, no current is detected. The induced e.m.f. only appears when there is relative movement between the coil and the magnet.

In fact, it is the change of magnetic flux that causes the induced e.m.f., and the magnitude of the induced e.m.f. is proportional to the rate of change of magnetic flux. This means that if the magnetic field strength is changing, an e.m.f. is induced in a conductor placed in the field. So an e.m.f. can be induced by changing the strength of a magnetic field without needing to physically move a magnet or a conductor. This is the effect we will be studying in the remainder of this topic.

6.2.3 Cassette players

We have just described how a moving magnet can induce a current in a coil. Exactly the same principle is used in the playback head of a cassette player, the device people used to listen to music before mp3 players were invented. The tape in a pre-recorded cassette is magnetised, and is effectively a collection of very short bar magnets spaced along the tape. The head consists of an iron ring with a small gap, under which the tape passes. As the tape passes under the ring, the ring becomes magnetised, the direction and strength of the field in the ring constantly changes as the tape passes under it.

A coil of wire wound around the top of the ring is connected to an amplifier circuit. As the magnetic field in the ring changes, a current is induced in the coil. It is this electrical signal that is amplified and played through the speakers.

© HERIOT-WATT UNIVERSITY

6.2.4 Faraday's law and Lenz's law

Electromagnetic induction was investigated independently by the English physicist Michael Faraday and the German physicist Heinrich Lenz in the mid-19th century. The laws which bear their names tell us the magnitude and direction of the induced e.m.f. produced by electromagnetic induction.

Faraday's law of electromagnetic induction states that the magnitude of the induced e.m.f. is proportional to the rate of change of magnetic flux through the coil or circuit.

Lenz's law states that the induced current is always in such a direction as to oppose the change that is causing it.

These two laws are summed up in the relationship

$$\varepsilon \propto -\frac{d\phi}{dt}$$

(6.2)

where ε is the induced e.m.f. Lenz's law is essentially a statement of conservation of energy: to induce a current, we have to put work into a system.

Looking back at Figure 6.4, Faraday's law tells us that the faster we move the magnet up and down, the larger the induced e.m.f. will be. A current around a coil produces its own magnetic field (see Figure 6.1), and Lenz's law tells us that this field will cause a force that will oppose the motion of the bar magnet towards the coil. Similarly, when the bar magnet is being removed from the coil, the induced current causes an attractive force on the bar magnet, again opposing its motion.

6.3 Eddy currents

Consider a metal disc rotating about its centre, as shown in Figure 6.5.

Figure 6.5: Rotating metal disc with a magnetic field acting on a small part

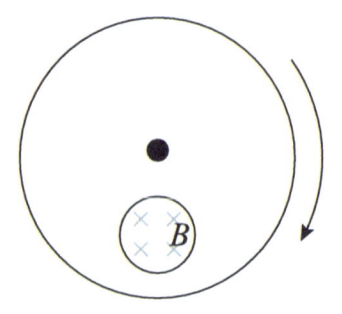

TOPIC 6. INDUCTORS

We will consider a magnetic field acting at right angles to the disc, but only acting over a small area. If the direction of the flux lines is into the disc, and the disc is rotating clockwise, then there will be an induced current in the region of the magnetic field. This induced current is shown in Figure 6.6.

Figure 6.6: Eddy currents inside (solid line) and outside (dashed line) the magnetic field.

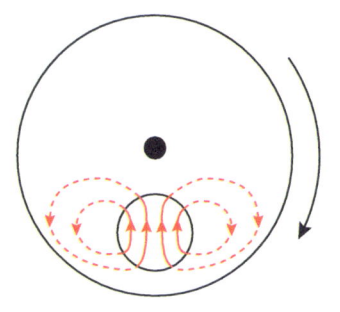

Because the field is only acting on part of the disc, charge will be able to flow back in the regions of the disc that are outside the field (shown as dashed lines in Figure 6.6). Thus **eddy currents** are induced in the disc. Note that in the region of the field, the charge is all flowing in one direction (solid lines), and the force that acts because of this is in the opposite direction to the rotation of the disc.

6.3.1 Implications of Lenz's law

Lenz's law tells us that an induced current always opposes the change that is causing it. This means that eddy currents can be used to apply **electromagnetic braking**. Consider the induced current in a localised field acting on part of a freely spinning disc, as shown in the solid lines in Figure 6.6.

The eddy currents in the part of the disc within the magnetic field cause a force to act on the disc in the opposite direction to the rotation of the disc. The currents in the opposite direction (dashed lines) are outside the field, so do not contribute a force. Thus a net force opposing the motion acts on the disc, slowing it down. This effect is used in circular saws, to bring the saw blade to rest quickly after the power is turned off. The same effect is used as the braking system in electric rapid-transit trains.

6.3.2 Induction heating

Eddy currents can lead to a large amount of energy being lost in electric motors through **induction heating**. The power dissipated when there is a current I through a resistor R is equal to $I^2 R$, so large currents can lead to a lot of energy being transferred as heat energy. In large dynamos in power stations, for example, this can make the generation of useful energy very inefficient. The laminated dynamos described earlier reduce eddy currents and hence reduce induction heating.

Induction heating is not always undesirable. In fact, it is used in circumstances where other forms of heating are impractical, such as the heat treatment of metals - welding and soldering. A piece of metal held in an electrical insulator can be heated to a very high temperature by the eddy currents. No eddy current is induced in the insulator, so it will remain cool.

Modern cookers called induction hobs work by electrical induction rather than by thermal conduction from a flame or a heating element. Therefore, such cookers require the use of a pot made of a ferromagnetic metal such as iron or stainless steel. They don't work with copper and aluminium.

© HERIOT-WATT UNIVERSITY

The cooking pot is placed above a coil of copper wire that has an alternating current passing through it. This results in a changing magnetic field, which induces eddy currents in the pot, causing it to heat up. Since nothing outside the vessel is affected by the field, it is a very efficient process, only heating the pot itself. Furthermore, it is much safer since the induction cooking surface is only heated by the pot rather than by a heating element, making people less likely to receive a burn.

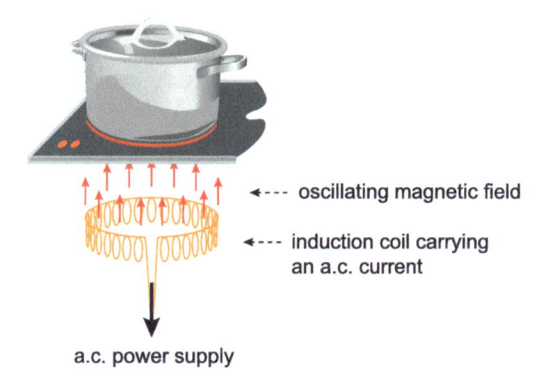

6.3.3 Metal detectors

An airport metal detector uses eddy currents to generate a magnetic field, and it is this field that is actually detected.

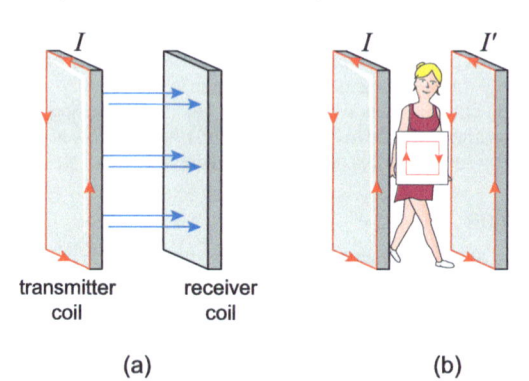

Figure 6.7: Schematic of an airport metal detector

Each passenger passes between two coils. The steady current I in the transmitter coil creates a magnetic field B (Figure 6.7 (a)). If a passenger walking between the coils is carrying a metal object, then eddy currents are induced in the object, and these currents in turn produce their own (moving) magnetic field. This new magnetic field induces a current I' in the receiver coil (Figure 6.7 (b)), triggering the alarm.

6.4 Inductors and self-inductance

We are now going to explore the function of inductors, which are coils of wire designed for use in electronic circuits.

6.4.1 Self-inductance

A coil (or inductor, as we shall see) in an electrical circuit can be represented by either of the symbols shown in Figure 6.8.

Figure 6.8: Circuit symbols for (a) an air-cored inductor; (b) an iron-cored inductor

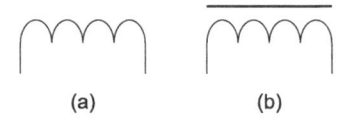

Let us consider a simple circuit in which a coil of negligible resistance is connected in series to a d.c. power supply and a resistor (Figure 6.9).

Figure 6.9: Coil connected to a d.c. power supply

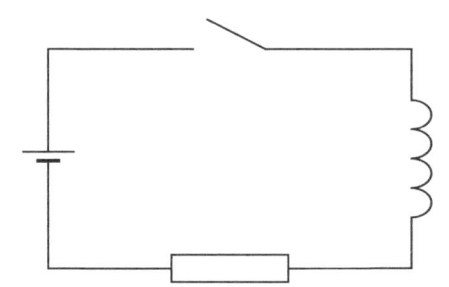

When a steady current is present, the magnetic field in and around the coil is stable. When the current changes (when the switch is opened or closed), the magnetic field changes and an e.m.f. is induced in the coil. This e.m.f. is called a self-induced e.m.f., since it is an e.m.f. induced in the coil that is caused by a change in its own magnetic field. The effect is known as **self-inductance**.

We know from Equation 6.2 that the induced e.m.f. ε is proportional to the rate of change of magnetic flux. Since the rate of change of the magnetic flux in a coil is proportional to the rate of change of current, we can state that

$$\varepsilon \propto -\frac{d\phi}{dt}$$
$$\therefore \varepsilon \propto -\frac{dI}{dt}$$

(6.3)

The constant of proportionality in Equation 6.3 is the inductance L of the coil. The inductance depends on the coil's size and shape, the number of turns of the coil, and whether there is any material in the centre of the coil. A coil in a circuit is called a self-inductor (or more usually just an **inductor**). The self-induced e.m.f. ε in an inductor of inductance L is given by the equation

$$\varepsilon = -L\frac{dI}{dt}$$

(6.4)

In Equation 6.4, dI/dt is the rate of change of current in the inductor. The SI unit of inductance is the henry (H). From Equation 6.4 we can see that an inductor has an inductance L of 1 H if an e.m.f. of 1 V is induced in it when the current is changing at a rate of 1 A s^{-1}. Note that there is a minus sign in Equation 6.4, consistent with Lenz's law. The self-induced e.m.f. always opposes the change in current in the inductor, and for this reason is also known as the **back e.m.f.**.

6.4.2 Energy stored in an inductor

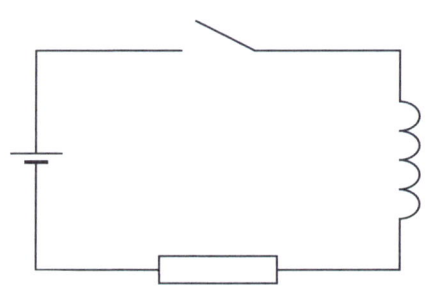

When we study the energy stored in an inductor, we look at an ideal conductor - one with negligible resistance. Otherwise, energy will be lost as work would be done overcoming the resistance.

When the switch is closed, the current in the ideal inductor increases from zero to some final value I. Work is done by the power supply against the back e.m.f.; this work is stored in the magnetic field of the inductor.

The expression to calculate how much energy E is stored in the magnetic field is:

$$E = \tfrac{1}{2}LI^2$$

The derivation of the equation is not a required learning outcome, but is shown after the example for those who wish to study it.

Example

A 2.0 H inductor is connected into a simple circuit. If a steady current of 0.80 A is present in the circuit, how much energy is stored in the magnetic field of the inductor?

TOPIC 6. INDUCTORS

Using Equation 6.5

$$E = \frac{1}{2}LI^2$$
$$\therefore E = \frac{1}{2} \times 2 \times 0.80^2$$
$$\therefore E = 0.64\,\text{J}$$

Let us return to Figure 6.9 and consider an ideal inductor - one with negligible resistance. When the switch is closed, the current in the inductor increases from zero to some final value I. Work is done by the power supply against the back e.m.f., and this work is stored in the magnetic field of the inductor. We will now find an expression to enable us to calculate how much energy E is stored in the magnetic field.

You should already be aware that if a potential difference V exists across a component in a circuit when a current I is present, then the rate P at which energy is being supplied to that component is given by the equation $P = IV$ (where P is measured in W, equivalent to J s^{-1}). If the current is varying across an inductor, then we can use Equation 6.4 and substitute for the potential difference

$$P = IV$$
$$\therefore P = I \times L\frac{dI}{dt}$$

(Since we are only concerned with the magnitude of the potential difference, we have ignored the minus sign when making this substitution.) P is the rate at which energy is being supplied, so we can substitute for $P = dE/dt$

$$\frac{dE}{dt} = LI\frac{dI}{dt}$$
$$\therefore dE = LI\,dI$$

Integrating over the limits from zero to the final current I

$$\int_0^E dE = \int_0^I LI\,dI$$
$$\therefore \int_0^E dE = L\int_0^I I\,dI$$
$$\therefore E = \frac{1}{2}LI^2$$

(6.5)

© HERIOT-WATT UNIVERSITY

The energy stored in the magnetic field of an inductor can itself be a source of e.m.f. When the current is switched off, there is a change in current so a self-induced e.m.f. will appear across the inductor opposing the change in current. The energy used to create this e.m.f. comes from the energy that has been stored in the magnetic field.

Quiz: Self-inductance Go online

Q1: A potential difference can be induced between the ends of a metal wire when it is

a) moved parallel to a magnetic field.
b) moved across a magnetic field.
c) stationary in a magnetic field.
d) stationary outside a solenoid.
e) stationary inside a solenoid.

....

TOPIC 6. INDUCTORS

Q2: Lenz's law states that

a) the induced e.m.f. in a circuit is proportional to the rate of change of magnetic flux through the circuit.
b) the magnetic field in a solenoid is proportional to the current through it.
c) magnetic flux is equal to the field strength times the area through which the flux lines are passing.
d) the induced current is always in such a direction as to oppose the change that is causing it.
e) the induced current is proportional to the magnetic field strength.

....

Q3: The current in an inductor is changing at a rate of 0.072 A s^{-1}, producing a back e.m.f. of 0.021 V.

What is the inductance of the inductor?

a) 0.0015 H
b) 0.29 H
c) 3.4 H
d) 4.1 H
e) 670 H

....

Q4: The steady current through a 0.050 H inductor is 200 mA.

What is the self-induced e.m.f. in the inductor?

a) 0 V
b) -0.010 V
c) -0.25 V
d) -4.0 V
e) -100 V

....

Q5: Which **one** of the following statements is true?

a) When the current through an inductor is constant, there is no energy stored in the inductor.
b) Faraday's law does not apply to self-inductance.
c) A back e.m.f. is produced whenever there is a current through an inductor.
d) The self-induced e.m.f. in an inductor always opposes the change in current that is causing it.
e) The principle of conservation of energy does not apply to inductors.

....

© HERIOT-WATT UNIVERSITY

Q6: How much energy is stored in the magnetic field of a 4.0 H inductor when the current through the inductor is 300 mA?

a) 0.18 J
b) 0.36 J
c) 0.60 J
d) 0.72 J
e) 2.4 J

....

Q7: An inductor stores 0.24 J of energy in its magnetic field when a steady current of 0.75 A is present. If the resistance of the inductor can be ignored, calculate the inductance of the inductor.

a) 0.10 H
b) 0.41 H
c) 0.43 H
d) 0.85 H
e) 1.2 H

6.5 Inductors in d.c. circuits

We can connect an inductor into a circuit in the same way as we would connect a resistor or a capacitor. We will now investigate how an inductor behaves when it is used as a component in a circuit.

We will begin by looking at a d.c. circuit, where we have an inductor connected in series to a resistor and a power supply such as a battery. After that we will replace the battery by an a.c. supply to investigate the response of an inductive circuit to an alternating current. We will compare the responses of inductive and capacitative circuits to an a.c. signal.

6.5.1 Growth and decay of current

Consider a simple circuit with an inductor of inductance L and negligible resistance connected in series to a resistor of resistance R, an ammeter of negligible resistance, and a d.c. power supply of e.m.f. E with negligible internal resistance.

The circuit (often called simply an LR circuit) is shown in Figure 6.10.

TOPIC 6. INDUCTORS

Figure 6.10: d.c. circuit with resistor and inductor in series

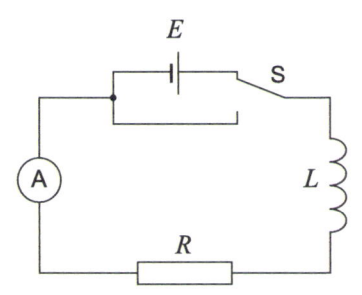

When the switch S is connected to the power supply, charge flows through the resistor and inductor, with the ammeter measuring the current. In the time taken for the current to rise from zero to its final value, the current through the inductor is changing, so a back e.m.f. is induced, which (by Lenz's law) opposes the increase in current. The rise time for the current to reach its final value in an inductive circuit will therefore be longer than it is in a non-inductive circuit.

A student could use a stopwatch to measure the time and the current could be noted from the ammeter at regular time intervals. A graph of current against time would be obtained as shown in Figure 6.11

The final, steady value of the current is given by Ohm's law, $I = E/R$, and so does not depend on the value of L. This should not be surprising, since when the current is at a steady value, there will be no induced back e.m.f.

Figure 6.11: Growth of current in a simple inductive circuit

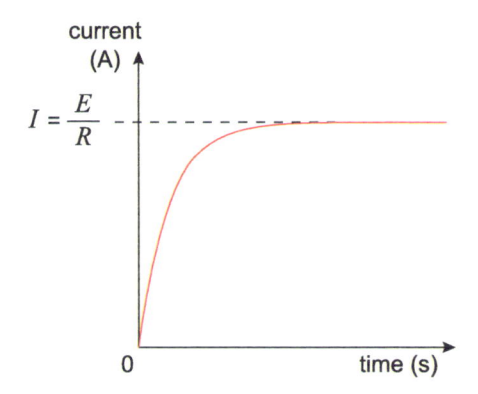

When the switch S in Figure 6.10 is switched to the down position, the power supply is no longer connected in the circuit, and the current drops from a value I to zero. Once again, the change in current produces a back e.m.f. that opposes the change. The upshot of this is that the current takes longer to decay than it would in a non-inductive circuit. This is shown in Figure 6.12.

© HERIOT-WATT UNIVERSITY

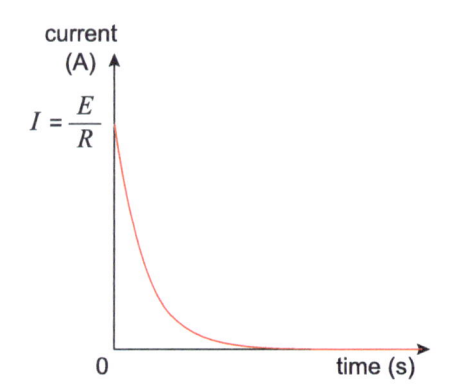

Figure 6.12: Decay of current in a simple inductive circuit

$$I = \frac{E}{R}$$

An example of the way current varies in an inductive circuit is the fact that a neon bulb connected to a battery can be lit, even although such a bulb requires a large p.d. across it. Consider the circuit shown in Figure 6.13.

Figure 6.13: Neon bulb connected to an inductive circuit

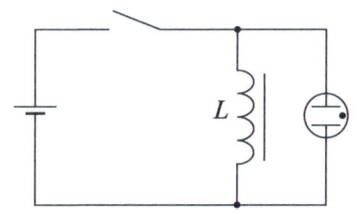

The power supply is a 1.5 V battery. We will consider an inductor that has an inductance L and a resistance R, which is connected in parallel to a neon bulb. The bulb acts like a capacitor in the circuit. Unless a sufficiently high p.d. is applied across it, the bulb acts like a break in the circuit. If the p.d. is high enough, the 'capacitor' breaks down, and charge flows between its terminals, causing the bulb to light up.

If the switch in Figure 6.13 is closed, current appears through the inductor but not through the bulb, since the p.d. across it is too low. The current rises as shown in Figure 6.11, reaching a steady value of $I = E/R$, where E is the e.m.f. of the battery (1.5 V) and R is the resistance of the inductor. The energy stored in the inductor is equal to $\frac{1}{2}LI^2$.

Opening the switch means there is a change in current through the inductor, and hence a back e.m.f. Charge cannot now flow around the left hand side of the circuit. It can only flow across the neon bulb, causing the bulb to flash.

6.5.2 Back e.m.f.

In the previous topic, we stated that the back e.m.f.

$$\varepsilon$$

induced in an inductor of inductance L is given by the equation

$$\varepsilon = -L\frac{dI}{dt}$$

Figure 6.11 shows that the rate of change of current in an LR circuit such as in Figure 6.10 is greatest just after the switch is moved to the battery side, so this is when the back e.m.f. will also be at its largest value. As the current increases, the rate of change decreases, and hence the back e.m.f. also decreases.

The sum of e.m.f.s around a circuit loop is equal to the sum of potential differences around the loop. So, at the instant when the switch is moved to the battery side, the current in the resistor is zero. This means the back e.m.f. must be equal in magnitude to E. As the current grows (and the *rate of change* of current *decreases*), the p.d. across the resistor increases and the back e.m.f. must decrease. Thus we have a maximum value for the back e.m.f., which is

$$\varepsilon_{\max} = -E$$

(6.6)

The minus sign appears as the back e.m.f. is, by its very nature, in the opposite direction to E.

Example

The circuit in Figure 6.14 contains an ideal (zero resistance) inductor of inductance 2.5 H, a 500 Ω resistor and a battery of e.m.f. 1.5 V and negligible internal resistance.

Figure 6.14: Inductor, resistor and battery in series

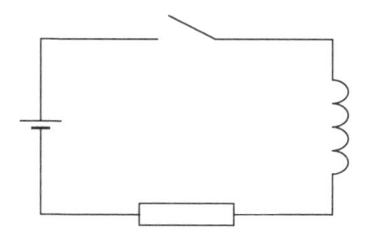

(a) What is the maximum self-induced e.m.f. in the inductor?

(b) At the instant when the back e.m.f. is 0.64 V, at what rate is the current changing in the circuit?

Solution

(a) The maximum self-induced e.m.f. is equal and opposite to the e.m.f. of the battery, which is -1.5 V.

(b) Using the equation for back e.m.f.

$$\varepsilon = -L\frac{dI}{dt}$$
$$\therefore \frac{dI}{dt} = -\frac{\varepsilon}{L}$$
$$\therefore \frac{dI}{dt} = -\frac{-0.64}{2.5}$$
$$\therefore \frac{dI}{dt} = 0.26 \text{ A s}^{-1}$$

Quiz: Inductors in d.c. circuits   Go online

Q8: A 120 mH inductor is connected in series to a battery of e.m.f. 1.5 V and negligible internal resistance, and a 60 Ω resistor. What is the maximum current in this circuit?

a) 12.5 mA
b) 25 mA
c) 180 mA
d) 3 A
e) 12.5 A

....

Q9: In the circuit in the previous question, what is the maximum potential difference across the inductor?

a) 0 V
b) 25 mV
c) 120 mV
d) 1.5 V
e) 12.5 V

....

TOPIC 6. INDUCTORS

Q10: A series circuit consists of a 9.0 V d.c. power supply, a 2.5 H inductor and a 1.0 kΩ resistor. What is the magnitude of the back e.m.f. when a steady current of 9.0 mA is present?

a) 0 V
b) 22.5 mV
c) 2.5 V
d) 3.6 V
e) 9.0 V

....

Q11: The circuit shown is used to measure the growth of current in an inductor.

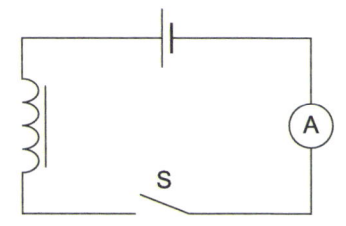

What other piece of apparatus is needed as well as the circuit above?

a) data capture device
b) digital ammeter
c) low value inductor
d) stopwatch
e) analogue voltmeter

6.6 Inductors in a.c. circuits

You will recall from the last topic that when an a.c. current is present a resistor behaves in exactly the same way as it does for a d.c. current. Meanwhile, a capacitor was found to oppose high frequency a.c. less than low frequency. That is, capacitive reactance was found to be inversely proportional to the frequency of an alternating current.

Inductors also oppose a.c. current, and we can define an inductor's reactance as

$$X_L = \frac{V}{I}$$

Inductive reactance, like capacitive reactance, is measure in ohms (Ω).

We have mentioned Lenz's law several times in this topic. The induced e.m.f. always opposes the change that is causing it. So an ideal inductor does not oppose d.c. current. So long as the current does not vary with time, an ideal inductor offers no opposition to current. However, in a.c. circuits the current and the associated magnetic field are continually changing. As the a.c. supply's frequency

© HERIOT-WATT UNIVERSITY

increases, the rate of change of current increases. So the self-induced back e.m.f. increases and therefore the inductive reactance increases. This makes the current decrease.

Let us now explore the exact relationship between inductive reactance and an a.c. supply's frequency. Assume the inductor has negligible resistance.

The inductor and a.c. Go online

There is an online activity where you can find out how the frequency of the a.c. supply affects the current.

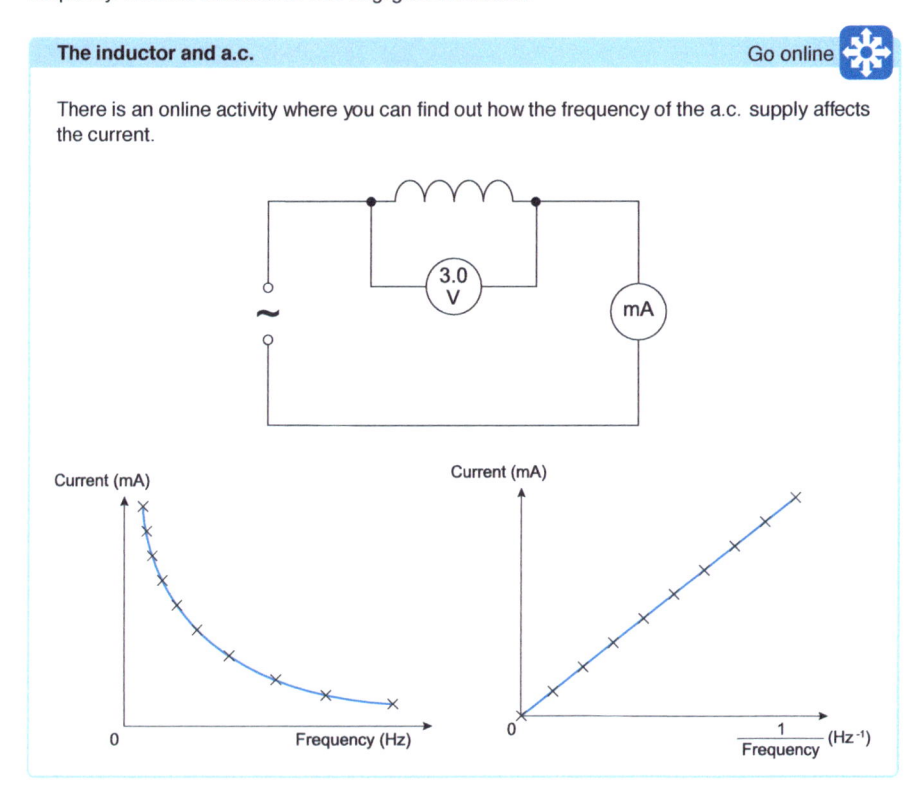

Figure 6.15: Simple a.c. inductive circuit

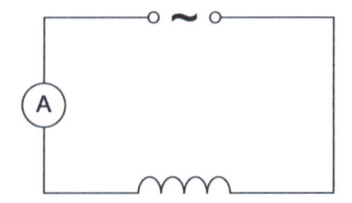

We have found that for an inductor the current is inversely proportional to the frequency. Since $X_L = \frac{V}{I}$, this means the inductive reactance must be directly proportional to the frequency of the supply.

TOPIC 6. INDUCTORS

The relationship is given by the following expression:

$$X_L = 2\pi f L$$

Since inductive reactance increases with frequency, an inductor is a good 'low-pass' filter. An inductor allows low-frequency and d.c. signals to pass but offers high 'resistance' to high-frequency signals. An inductor can be used to smooth a signal by removing high-frequency noise and spikes in the signal.

The frequency response of a resistor, capacitor and inductor can be summarised by the following graphs.

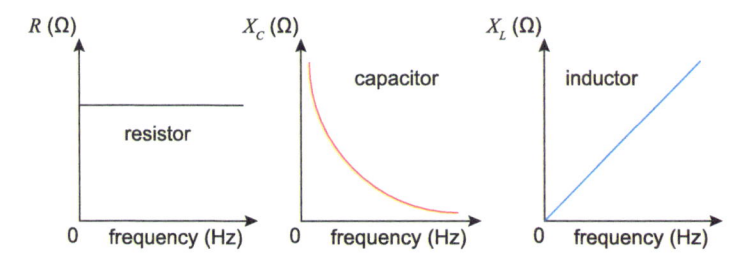

Quiz: a.c. circuits

Go online

Q12: Which of the following describes the relationship between reactance X and frequency f in an a.c. inductive circuit?

a) $X \propto 1/f$
b) $X \propto 1/f^2$
c) $X \propto f$
d) $X \propto f^2$
e) $X \propto \sqrt{f}$

....

Q13: Which *one* of the following statements is true?

a) An inductor can be used to filter out the d.c. component of a signal.
b) The inductance of an inductor is inversely proportional to the frequency of the supply.
c) An inductor is often used to filter out low frequency signals and allow only high frequency signals to pass through.
d) The reactance of a capacitor is proportional to the frequency of the a.c. current.
e) An inductor can smooth a signal by filtering out high frequency noise and spikes.

....

Q14: An experiment is carried out to investigate how the current varies with frequency in an inductive circuit. The results of such an experiment show that

a) $I \propto 1/f$
b) $I \propto 1/f^2$

© HERIOT-WATT UNIVERSITY

c) $I \propto f$
d) $I \propto f^2$
e) $I \propto \sqrt{f}$

6.7 Summary

In this topic we have seen that an e.m.f. can be induced in a conductor in a magnetic field when the magnetic flux changes. Thus an e.m.f. can be induced when a conductor moves across a magnetic field; when a magnet moves near to a stationary conductor; or when the strength of a magnetic field changes.

We have found out that a coil in a circuit is called a self-inductor or just an inductor. Since work is done against the back e.m.f. in establishing a current in an inductor, there is energy stored in its magnetic field whilst a current is present in an inductor.

We then studied the behaviour of inductors in simple d.c. and a.c. circuits. We have seen that an ideal (non-resistive) inductor does not have any effect on a steady d.c. current. When the current through an inductor is changing, the induced e.m.f. acts to oppose the change. Inductive reactance was shown to be proportional to the frequency of an a.c. supply, meaning that inductors are good at filtering out high frequency signals and allowing only low frequency signals to pass through.

Summary

You should now be able to:

- sketch graphs showing the growth and decay of current in a simple d.c. circuit containing an inductor;
- describe the principles of a method to illustrate the growth of current in a d.c. circuit;
- state that an e.m.f. is induced across a coil when the current through the coil is varying;
- explain the production of the induced e.m.f across a coil;
- explain the direction of the induced e.m.f in terms of energy;
- state that the inductance of an inductor is one henry if an e.m.f. of one volt is induced when the current is changing at a rate of one ampere per second;
- use the equation $\varepsilon = -L\frac{dI}{dt}$ and explain why a minus sign appears in this equation;
- state that the work done in building up the current in an inductor is stored in the magnetic field of the inductor, and that this energy is given by the equation $E = \frac{1}{2}LI^2$;
- calculate the maximum values of current and induced e.m.f. in a d.c. LR circuit;
- use the equations for inductive reactance $X_L = \frac{V}{I}$ and $X_L = 2\pi f L$;
- describe the response of an a.c. inductive circuit to low and high frequency signals.

6.8 Extended information

6.8.1 Levitation of superconductors

Another effect of eddy currents is the levitation of a superconductor in a magnetic field.

To explain this effect, we first need to understand a little about superconductivity. This effect, first observed by the Dutch physicist H. K. Onnes in 1911, is one that occurs at very low temperatures, at which certain metals and compounds have effectively zero electrical resistance. For many years, superconductivity could only be observed in materials cooled below the boiling point of helium, which is 4 K. In recent years, huge worldwide research activity has resulted in the development of compounds that can remain superconducting up to the boiling point of nitrogen (77 K).

As well as exhibiting zero electrical resistance, a superconducting material also has interesting magnetic properties. A piece of superconductor placed in a magnetic field will distort the field lines, so that the magnetic field inside the superconductor is zero. Figure 6.16 shows a metal sphere and a superconducting sphere in a uniform magnetic field.

Figure 6.16: (a) Uniform magnetic field; (b) iron sphere placed in the field; (c) superconducting sphere placed in the field

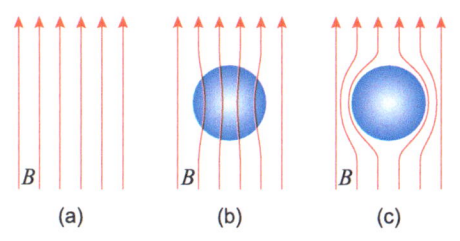

We are considering a uniform magnetic field B, shown in Figure 6.16 (a). The magnetic field strength inside an iron sphere, for example, (Figure 6.16 (b)) is enhanced. On the other hand, a superconductor placed in the field (Figure 6.16 (c)) distorts the field so that no field lines can enter it. The magnetic field inside the superconductor is therefore zero.

We can explain this phenomenon in terms of eddy currents. When the superconductor is moved into the magnetic field, eddy currents are induced on its surface. Lenz's law states that these currents will create a magnetic field opposing the external field. Since there is no electrical resistance in a superconductor, the eddy currents will continue even when the superconductor is stationary in the field. The magnetic field due to the eddy currents is in the opposite direction to the external magnetic field, with the result that the external magnetic field cannot penetrate into the superconductor. Since magnetic field lines cannot be broken, the lines must continue outside the superconductor.

We can see this effect by placing a superconductor in the field of a permanent magnet. Let us first think about what happens if we place a piece of iron near a permanent magnet. The magnetic domains within the piece of iron arrange themselves in the direction of the magnetic field lines, and the resulting attractive force draws the piece of iron towards the magnet.

The opposite happens to a piece of superconductor. Lenz's law means that the magnetic field due to the eddy currents in the superconductor opposes the field due to the permanent magnet, and a repulsive force exists between the two. We can observe magnetic levitation if we position the superconductor above the magnet, as the force of gravity acting down on the magnet can be balanced by the magnetic repulsion acting upwards.

© HERIOT-WATT UNIVERSITY

Levitating superconductor   Go online

At this stage there is a video clip which shows a demonstration of magnetic levitation. A piece of superconductor, cooled with liquid nitrogen, is suspended above a permanent magnet.

6.8.2 Web links

Web links   Go online

These web links should serve as an insight to the wealth of information available online and allow you to explore the subject further.

- https://www.youtube.com/watch?v=bZ45lTTm5_U&list=PLiISjSDTPUjy2OKK5-Bl2WrVu91LAe_cm&index=26
- https://www.youtube.com/watch?v=B62DIJs2WNI&list=PLiISjSDTPUjy2OKK5-Bl2WrVu91LAe_cm&index=27
- https://www.youtube.com/watch?v=Z5blesm2eH8&list=PLiISjSDTPUjy2OKK5-Bl2WrVu91LAe_cm&index=28
- https://www.youtube.com/watch?v=flKORY5zAJ8
- https://www.youtube.com/watch?v=BLtq4_3UFkU
- https://www.youtube.com/watch?v=T8O8aTO3ea8
- https://www.youtube.com/watch?v=aSmMFog10D0&feature=youtu.be
- https://www.youtube.com/watch?v=tC6E9J925pY
- https://www.youtube.com/watch?v=txmKr69jGBk
- https://www.youtube.com/watch?v=NgwXkUt3XxQ

6.9 End of topic 6 test

Q15: The current through a 0.55 H inductor is changing at a rate of 15 A s^{-1}.

Calculate the magnitude of the e.m.f. induced in the inductor. (Do NOT include a minus sign in your answer.)

_____ V

Q16: A 4.5 H inductor of negligible resistance is connected to a circuit in which the steady current is 460 mA.

Calculate the energy stored in the magnetic field of the inductor.

_____ J

Q17: Which of the following is equivalent to one henry?

a) 1 A V s^{-1}
b) 1 V A^{-1} s^{-1}
c) 1 V s A^{-1}
d) 1 A V^{-1} s^{-1}

Q18: Consider the circuit below, in which a variable resistor R and an inductor L of inductance 1.5 H are connected in series to a 3.0 V battery of zero internal resistance.

The variable resistor is changed from 76 Ω to 25 Ω over a time period of 2.5 s.

Calculate the average back e.m.f. across the inductor whilst the resistance is being changed. (Do NOT include a minus sign in your answer.)

_____ V

Q19: A resistor R = 14 Ω and an inductor L = 580 mH are connected to a power supply as shown below.

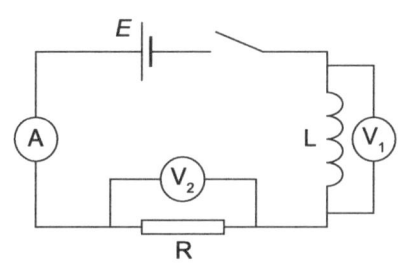

A short time after the switch is closed, the current in the circuit has reached a steady value, and the energy stored in the inductor is 0.10 J.

Calculate the e.m.f. of the power supply.

_____ V

Q20: The circuit below shows a 12 V power supply connected to a 1.7 H inductor and a 32 Ω resistor.

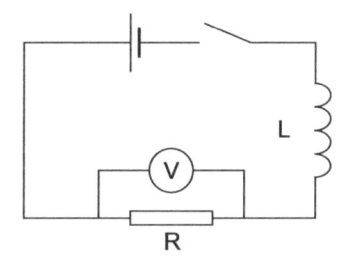

Calculate the potential difference measured by the voltmeter when the current through the resistor is changing at a rate of 2.1 A s^{-1}.

_____ V

Q21: In the circuit shown below, the voltmeters V_1 and V_2 measure the potential difference across an inductor L and a resistor R respectively.

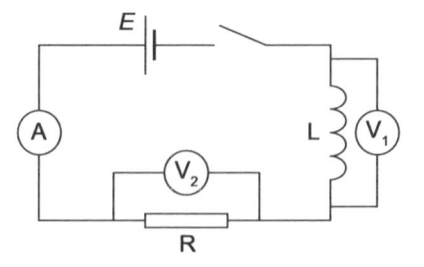

TOPIC 6. INDUCTORS

The battery has e.m.f. E.
L = 620 mH and R = 14 Ω.

1. The maximum potential difference in V recorded on the voltmeter V_2 after the switch is closed is 2.6 V.
   State the e.m.f. E of the battery.
   ---------- V
2. After the switch has been closed for several seconds, state the value of the potential difference measured by voltmeter V_1.
   ---------- V
3. Calculate the maximum current recorded on the ammeter A after the switch is closed.
   ---------- A

....

Q22: Consider the circuit shown below, in which an inductor L and resistor R are connected in series to a 1.5 V battery of negligible internal resistance.

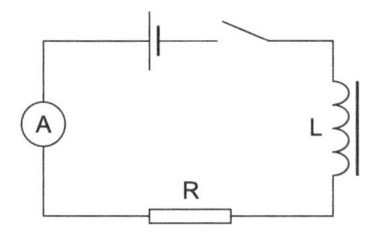

L has value 340 mH and the resistance R is 35 Ω.

1. At one instant after the switch is closed, the current in the circuit is changing at a rate of 1.9 A s^{-1}.
   Calculate the back e.m.f. at this instant. (Do NOT include a minus sign in your answer.)
   ---------- V
2. Calculate the maximum current through the inductor.
   ---------- A
3. Calculate how much energy is stored in the magnetic field of the inductor when the current reaches a steady value.
   ---------- J

....

Q23: Consider the circuit below, in which an inductor is connected to a 7.0 V battery of negligible internal resistance.

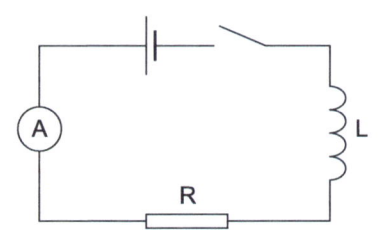

The resistance R is 40 Ω.

1. At the instant the switch is closed, the current in the circuit is changing at a rate of 60 A s^{-1}.
   Calculate the inductance L.
   _____ H
2. Calculate the maximum current in the circuit.
   _____ A
3. Calculate the energy store
   _____ J
   d in the inductor when the current has reached its maximum value.

....

Q24: Consider the circuit below.

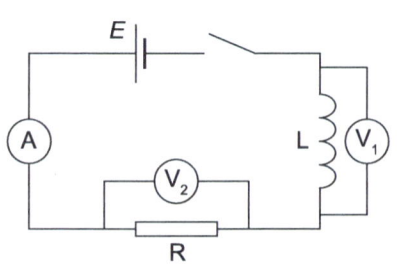

The battery has e.m.f. 2.8 V, and is connected to an inductor L = 320 mH and a resistor R = 16 Ω. The voltmeters V_1 and V_2 measure the potential difference across the inductor and the resistor respectively.

1. Calculate the maximum potential difference recorded on the voltmeter V_1 after the switch is closed.
   _____ V
2. Calculate the maximum potential difference recorded on the voltmeter V_2 after the switch is closed.
   _____ V
3. Calculate the energy stored in the magnetic field of the inductor when the current has reached a steady value.
   _____ J

Unit 3 Topic 7

Electromagnetic radiation

Contents

7.1 Introduction 490
7.2 The unification of electricity and magnetism 490
7.3 The wave properties of em radiation 490
7.4 Permittivity, permeability and the speed of light 492
7.5 Summary 493
7.6 Extended information 493
7.7 End of topic 7 test 494

Prerequisites

- Wave properties.
- Electromagnetic waves.

Learning objective

By the end of this topic you should be able to:

- state that the similarities between electricity and magnetism led to their unification i.e. the discovery that they are really manifestations of a single electromagnetic force;
- state that electromagnetic radiation exhibits wave properties i.e. electromagnetic radiation reflects, refracts, diffracts and undergoes interference;
- describe electromagnetic radiation as a transverse wave which has both electric and magnetic field components which oscillate in phase perpendicular to each other and the direction of energy propagation;
- carry out calculations using $c = \frac{1}{\sqrt{\varepsilon_0 \mu_0}}$.

7.1 Introduction

We have seen that electric currents exert forces on magnets and that time-varying magnetic fields can induce electric currents. Until the 1860s, they were thought to be unrelated. We are now going to look at the work of James Clerk Maxwell, who recognised the similarities between electricity and magnetism and developed his theory of a single electromagnetic force.

7.2 The unification of electricity and magnetism

You will be aware that the four fundamental forces of nature are gravitational, electromagnetic, and the strong and the weak nuclear forces. Theoretical physicists currently favour the idea that these four forces are actually just different manifestations of the same force. That is to say, there is only one fundamental force, and we perceive it to be acting in four different ways. One of the biggest challenges in theoretical physics is to find a Grand Unified Theory (GUT) which will unite these forces, showing that at extremely short distances, for extremely high energy particles, the four forces become one.

The Scottish physicist James Clerk Maxwell was the first to successfully unify two of these forces. His theory on electromagnetism showed that electricity and magnetism could be unified. Theoretical physicists have subsequently shown that the electromagnetic and the weak forces can be combined as an 'electroweak' force when acting over very short distances. However, this is only the case at the high energies explored in particle collisions at CERN and other laboratories. Unfortunately, it is impossible at present to study high enough energies to directly explore the unification of the other forces, but it is thought that such conditions would have existed in the early universe, almost immediately after the big bang. Instead, physicists must look for the consequences of grand unification at lower energies. Such consequences include supersymmetry, which is a theory that predicts a partner particle for each particle in the Standard Model.

7.3 The wave properties of em radiation

You will recall that electromagnetic waves such as light are made up of oscillating electric and magnetic fields. For simplicity, diagrams often only show the oscillating electric field, but it is important to remember that an electromagnetic wave has both electric and magnetic field components which oscillate in phase, perpendicular to each other and to the direction of energy propagation.

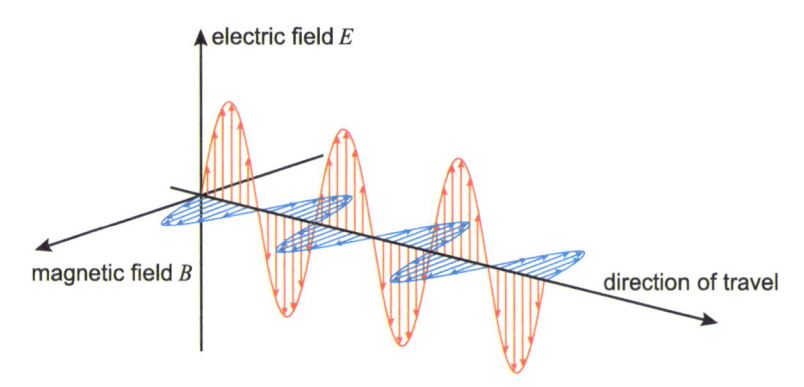

Figure 7.1: Electromagnetic wave

Electromagnetic wave — Go online

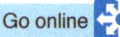

At this stage there is an online activity which shows a polarised electromagnetic wave that propagates in a positive x direction and explores the electric and magnetic fields.

Maxwell's theory of electromagnetism was particularly remarkable since he predicted electromagnetic waves in terms of oscillating electric and magnetic fields, long before there was any experimental evidence for them. He showed that the speed of an electromagnetic wave in a vacuum is the same as the speed of light in free space and he predicted that light is just one form of an electromagnetic wave.

In 1887, the German physicist Heinrich Hertz showed electrical oscillations give rise to transverse waves, verifying the existence of electromagnetic waves travelling at the speed of light. The waves he discovered are known now as radio waves. Bluetooth is just one example of technology we now rely upon that uses radio waves. It uses short wavelength radio waves to allow devices to communicate wirelessly. An example is a cordless telephone, which has one Bluetooth transmitter in the base and another in the handset. A computer communicating with a wireless printer, mouse or keyboard is another.

All electromagnetic radiation exhibits wave properties as it transfers energy through space. All electromagnetic radiation reflects, refracts, diffracts and undergoes interference.

7.4 Permittivity, permeability and the speed of light

Maxwell derived the equation

$$c = \frac{1}{\sqrt{\varepsilon_0 \mu_0}}$$

where
c is the speed of light in free space in m s^{-1};
ε_0 is the permittivity of free space in C^2 N^{-1} m^{-2} or F m^{-1};
μ_0 is the permeability of free space in T m A^{-1} or H m^{-1}.

Using this relationship, and the values $\varepsilon_0 = 8.85 \times 10^{-12}$ C^2 N^{-1} m^{-2} and $\mu_0 = 4\pi \times 10^{-7}$ T m A^{-1}, gives

$$\frac{1}{\sqrt{\varepsilon_0 \mu_0}} = \frac{1}{\sqrt{8.85 \times 10^{-12} \times 4\pi \times 10^{-7}}}$$
$$= 3 \times 10^8$$

Also the dimensions of $\varepsilon_0 \mu_0$ are C^2 N^{-1} m^{-2} × T m A^{-1} and since 1 C = 1 A s and 1 T = 1 N A^{-1} m^{-1}, it can be seen that the dimensions of $\frac{1}{\sqrt{\varepsilon_0 \mu_0}}$ are m s^{-1}. This shows that light is propagated as an electromagnetic wave.

In October 1983 the metre was defined as 'that distance travelled by light, in a vacuum, in a time interval of $\frac{1}{299,792,458}$ seconds'. This means that the speed of light is now a fundamental constant of physics with a value

$$c = 299,792,458 \text{ m s}^{-1}$$

7.5 Summary

Summary

You should now be able to:

- state that the similarities between electricity and magnetism led to their unification i.e. the discovery that they are really manifestations of a single electromagnetic force;
- state that electromagnetic radiation exhibits wave properties i.e. electromagnetic radiation reflects, refracts, diffracts and undergoes interference;
- describe electromagnetic radiation as a transverse wave which has both electric and magnetic field components which oscillate in phase perpendicular to each other and the direction of energy propagation;
- carry out calculations using $c = \frac{1}{\sqrt{\varepsilon_0 \mu_0}}$.

7.6 Extended information

Web links Go online

These web links should serve as an insight to the wealth of information available online and allow you to explore the subject further.

- http://www.cabrillo.edu/~jmccullough/Applets/Flash/Optics/EMWave.swf
- https://www.youtube.com/watch?v=uYz_kQ7UkY8
- https://www.youtube.com/watch?v=R9FVFh3HYaY
- https://www.youtube.com/watch?v=GMnsZuEE_m8
- https://www.youtube.com/watch?v=cy6kba3A8vY
- https://www.youtube.com/watch?v=3_RhISgoXUs

7.7 End of topic 7 test

End of topic 7 test   Go online

Q1: ε_0 is the symbol for the _____ of free space.

1. permeability
2. permittivity

....

Q2: Electromagnetic waves are _____.

1. longitudinal
2. transverse

....

Q3: Electricity and magnetism can be _____ under one theory called electromagnetism.

....

Q4: What is the correct relationship between c, ε_0 and μ_0?

a) $c = \frac{1}{\varepsilon_0 \mu_0}$
b) $c = (\varepsilon_0 \mu_0)^2$
c) $c = \frac{1}{(\varepsilon_0 \mu_0)^2}$
d) $c = \frac{1}{\sqrt{\varepsilon_0 \mu_0}}$
e) $c = \sqrt{\varepsilon_0 \mu_0}$

....

Q5: A student carries out an experiment to determine the permittivity of free space.

It is measured to be 7.7×10^{-12} F m^{-1}.

Use this result and the speed of light in vacuum to determine the permeability of free space.

_____ H m^{-1}

Unit 3 Topic 8

End of section 3 test

End of section 3 test

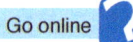

The following data should be used when required:

Quantity	Symbol	Value
Charge on electron	e	-1.60×10^{-19} C
Mass of proton	m_p	1.67×10^{-27} kg
Permittivity of free space	ε_0	8.85×10^{-12} F m^{-1}
Permeability of free space	μ_0	$4\pi \times 10^{-7}$ H m^{-1}

Q1: Two point charges A (+5.95 mC) and B (+7.55 mC) are placed 1.42 m apart.

1. Calculate the magnitude of the Coulomb force that exists between A and B.
   _____ N
2. Calculate the magnitude of the Coulomb force acting on a -1.15 mC charge placed at the midpoint of AB.
   _____ N

....

Q2: A long straight wire carries a steady current I_1.

Calculate the magnetic induction at a perpendicular distance 48 mm from the wire when the current $I_1 = 1.5$ A.

_____ T

....

Q3: An ion carrying charge 2e is accelerated from rest through a potential of 2.5×10^6 V, emerging with a velocity of 5.6×10^6 m s^{-1}.

Calculate the mass of the ion.

_____ kg

....

Q4: Calculate the magnitude of the force on a horizontal conductor 20 cm long, carrying a current of 7.5 A, when it is placed in a magnetic field of magnitude 5.0 T acting at 33° the wire's length.

_____ N

....

TOPIC 8. END OF SECTION 3 TEST

Q5: Consider a capacitor connected in series to an a.c. power supply.

Which one of the following graphs correctly shows how the current in the circuit varies with the frequency of the a.c. supply?

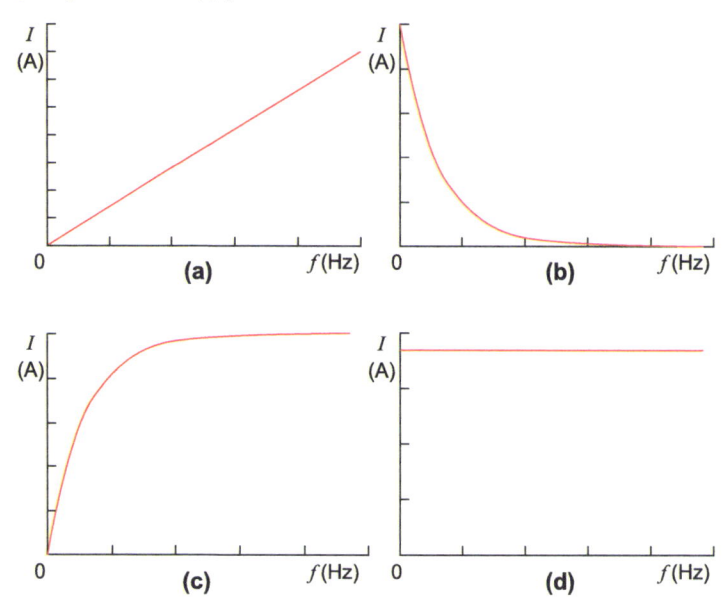

(a) I(A) vs f(Hz): linear increasing from origin

(b) I(A) vs f(Hz): decreasing exponential-like curve

(c) I(A) vs f(Hz): increasing curve leveling off

(d) I(A) vs f(Hz): constant horizontal line

....

Q6: A 200 nF capacitor is connected to 1.0 V a.c. power supply. The frequency of the a.c. supply is 4600 Hz.

Calculate the capacitive reactance of the capacitor.

_____ Ω

....

Q7: Consider the following circuit, in which an ideal inductor L is connected in series to a resistor R and a battery of e.m.f. 6.0 V and zero internal resistance.

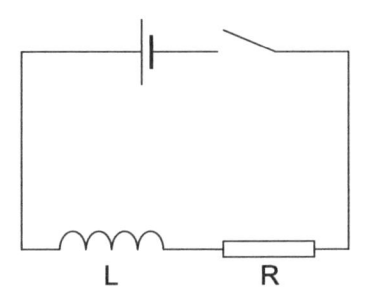

The value of L is 280 mH and the resistance R is 36 Ω.

1. Calculate the initial rate of growth of current in the circuit at the instant the switch is closed.
   _____ A s^{-1}
2. Calculate the energy stored in the magnetic field of the inductor once the current has reached a steady value.
   _____ J

....

Q8: The circuit below shows a 12 V power supply connected to a 1.5 H inductor and a 36 Ω resistor.

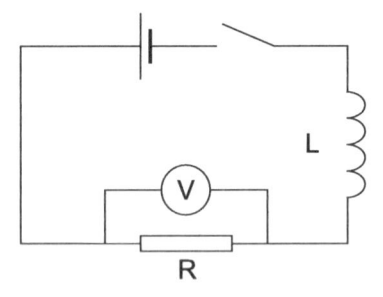

Calculate the potential difference measured by the voltmeter when the current is changing at a rate of 3.3 A s^{-1}.
   _____ V

....

Q9: In the circuit shown below, the voltmeters V_1 and V_2 measure the potential difference across an inductor L and a resistor R respectively.

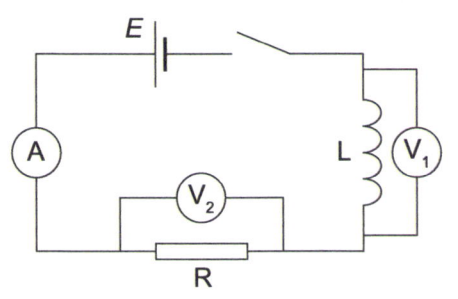

The battery has e.m.f. E. The inductor has an inductance of 660 mH and the resistance of the resistor is 14 Ω.

1. The maximum potential difference in V recorded on the voltmeter V_2 after the switch is closed is 3.2 V.
   State the e.m.f. E of the battery.
   _____ V
2. After the switch has been closed for several seconds, state the value of the potential difference measured by voltmeter V_1.
   _____ V
3. Calculate the maximum current recorded on the ammeter A after the switch is closed.
   _____ A

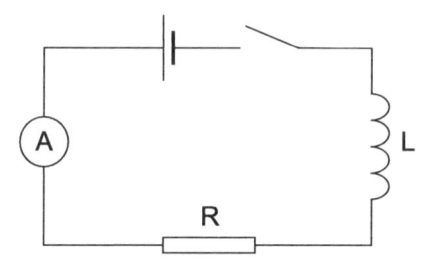

Q10: Consider the circuit below, in which an inductor is connected to a 8.00 V battery of negligible internal resistance.

The resistance R is 40.0 Ω.

1. At the instant the switch is closed, the current in the circuit is changing at a rate of 60.0 A s^{-1}.
   Calculate the inductance L.
   ---------- H
2. Calculate the maximum current in the circuit.
   ---------- A
3. Calculate the energy stored in the inductor when the current has reached its maximum value.
   ---------- J

TOPIC 8. END OF SECTION 3 TEST   501

Q11: Consider an inductor connected in series to an a.c. power supply.

Which one of the following graphs correctly shows how the current in the inductor varies with the frequency of the a.c. supply?

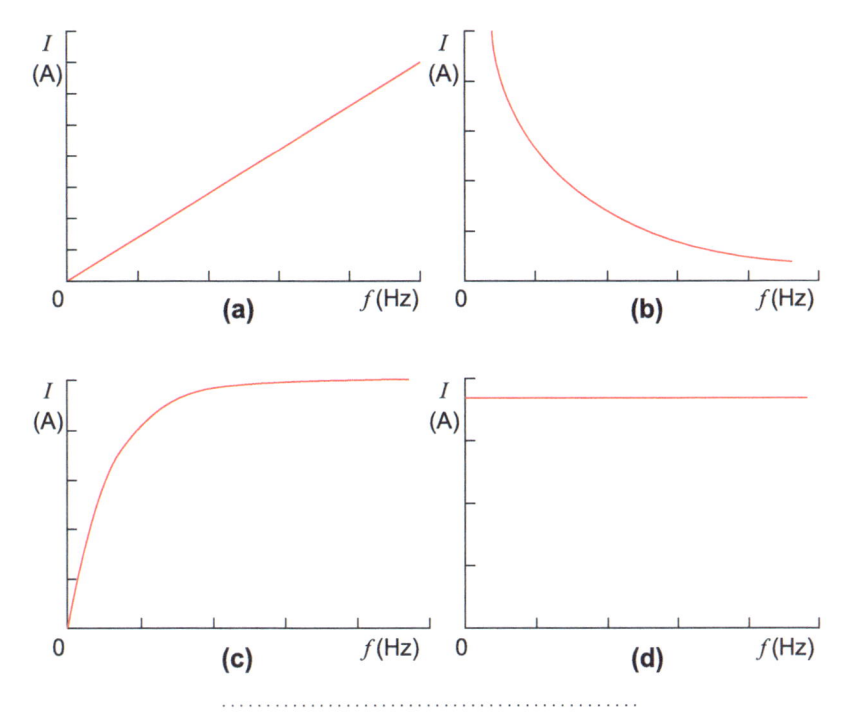

....

Q12: A 150 μF capacitor is connected in series with a 500Ω resistor to a 6.00 V battery. Calculate the time taken for the voltage across the capacitor to increase from 0.00 V to 3.78V.

time = _____ s

Investigating Physics

1   Initial planning, using equipment and recording data **505**
   1.1   Introduction 506
   1.2   The planning cycle and initial plan 508
   1.3   Using equipment and recording experimental data 509
   1.4   Summary 511

2   Measuring and presenting data **513**
   2.1   Result tables 514
   2.2   Summary 517

3   Evaluating findings **519**
   3.1   Evaluating findings 520
   3.2   Summary 522

4   Scientific report **523**
   4.1   Scientific report 524
   4.2   Summary 528

Unit 4 Topic 1

Initial planning, using equipment and recording data

Contents

1.1 Introduction	506
1.2 The planning cycle and initial plan	508
1.3 Using equipment and recording experimental data	509
1.4 Summary	511

Learning objective

By the end of this topic you should be able to:

- be able to make an initial plan and make a timeline for the investigation;
- know how to look up ideas for suitable experiments and who to talk to about accessing and using equipment;
- understand how to take suitable pictures and draw relevant diagrams for your investigation.

1.1 Introduction

As part of the Advanced Higher Physics course you will have to do a project on a suitable Physics topic and complete a written report at the end. This report will be marked by the SQA and is worth a total of 30 marks, (scaled to 40).

The purpose of the project is to allow the you to carry out an in-depth investigation of a physics topic and produce a project report. You are required to plan and carry out the investigation.

Project overview

The project assesses the application of skills of scientific inquiry and related physics knowledge and understanding.

You should choose your topic to investigate. To make sure you are successful you should agree your topic with your teacher. This is to ensure that the project is at a suitable potential level of difficulty and achievable with the resources you can access. This discussion should take place as early in the investigation cycle as possible. As at National 5 and Higher you are asked to investigate/research its underlying physics.

Most of the project will be autonomously, i.e. without your teacher's supervision. The is open-ended, giving you the freedom to pursue your ideas in greater depth. You should keep a record of your work. The record is usually called a daybook. Often the daybook is a jotter, but it can be sheets in a ring binder or an electronic file. This will I form the basis of your report. The daybook should include your thoughts on potential topics, draft procedures and all the details of your research, experiments and recorded data.

Your teacher will need to access your record regularly during the project to make sure your project is your own work.

When planning your project, take time to understand what is being asked of you. The project is an opportunity to demonstrate these skills, knowledge and understanding by:

- extending and applying knowledge of physics to new situations, interpreting and analysing information to solve more complex problems;
- planning and designing physics experiments/investigations, using reference material, to test a hypothesis or to illustrate particular effects;
- recording systematic detailed observations and collecting data;
- selecting information from a variety of sources;
- presenting detailed information appropriately in a variety of forms;
- processing and analysing physics data (using calculations, significant figures and units, where appropriate)
- making reasoned predictions from a range of evidence/information;
- drawing valid conclusions and giving explanations supported by evidence/justification;
- critically evaluating experimental procedures by identifying sources of uncertainty, and suggesting and implementing improvements;
- drawing on knowledge and understanding of physics to make accurate statements, describe complex information, provide detailed explanations, and integrate knowledge;

TOPIC 1. INITIAL PLANNING, USING EQUIPMENT AND RECORDING DATA

- communicating physics findings/information fully and effectively;
- analysing and evaluating scientific publications and media reports.

> *Copyright ©Scottish Qualifications Authority*
> *This may change from year to year, always check the SQA website for the most up to date mark scheme and assessment information.*

To complete the investigative stage of the project you must make independent and rational decisions. These must be based on evidence and interpretation of scientific information. That involves analysing and evaluating your results.

The project report requires you to use your literacy skills. You may need to develop your skills in literature and internet research. The project asks you to write clear, concise final project with a logical structure.

1.2 The planning cycle and initial plan

The first stage in the process is to set up a timeline for the project such as one like this:

Phase	Start date	Tasks	Deadlines
Research and choose a topic	Now!	Check with teacher for suitability and equipment	
Experiment 1			
Experiment 2			
Experiment 3 (+ 4 etc.)			
Write report			
Hand in first draft			
Hand in final draft			

Secondly you need to research topics that you are interested in and perhaps trial some experiments to see if it will make a viable investigation. Research could be from the internet or from textbooks and journals. If you use books you should give the author, title, edition and page numbers and if you use the internet the full URL is needed. Both will need the date you looked at them too. The final report will need a minimum of two correctly referenced references mentioned to achieve the mark.

These are the correct way to record your references:

1. Research found from the internet - a full URL with the date accessed.

    http://www.cyberphysics.co.uk/topics/forces/young_modulus.htm - accessed on 10/10/2018

2. Research found from a textbook.

    Tom Duncan, A Textbook for Advanced Level Students, 2nd Edition, Pages 228 - 229. Read on 06/02/2019

You will have to check with your teacher whether equipment is available Sometimes a visit to a University can be arranged or equipment borrowed, and most schools will have some equipment available. Some students even construct their own equipment for the project but bear in mind this will take up a lot of extra time.

Some ideas for experiments can be found here:

- https://www.sserc.org.uk/subject-areas/physics/physics-advanced-higher/investigation-ideas/

1.3 Using equipment and recording experimental data

Don't be surprised if the practical work takes many more hours than you planned, this is normal. Good planning will help here, making sure all equipment is ready in advance and your daybook is clearly laid out with dates and clearly drawn up tables of collected data. An account of the experimental procedures should also be written and written in the third person form.

Images

A **labelled** diagram of apparatus should be drawn (this can be in rough in the daybook) and take pictures of the apparatus too which can also be used in the final report. With pictures try and avoid background objects or overly crowded pictures which detract from the details in the image.

This image is a bit cluttered and not clear to the reader

"Bomb Calorimeter" by Akshat Goel, licensed under CC BY 3.0

Slightly better though still with a distracting background

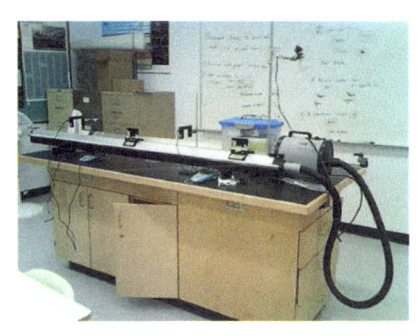

"Air Track with photo-gates and a reverse vacuum" by Bhavesh Chauhan is licensed under CC BY 3.0

Nice clear equipment with no distracting background which can be clearly labelled

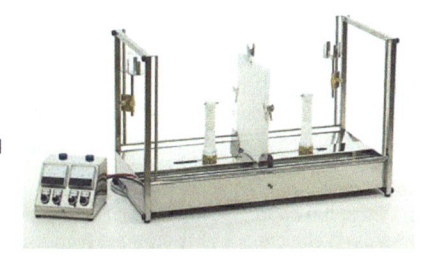

"An Electrospinz Ltd Doris type laboratory electrospinning machine", by Robert Lamberts, licensed under CC BY 3.0

Diagrams

Similarly diagrams should be fully labelled and clear 2D diagrams not 3D works of art, try and think.

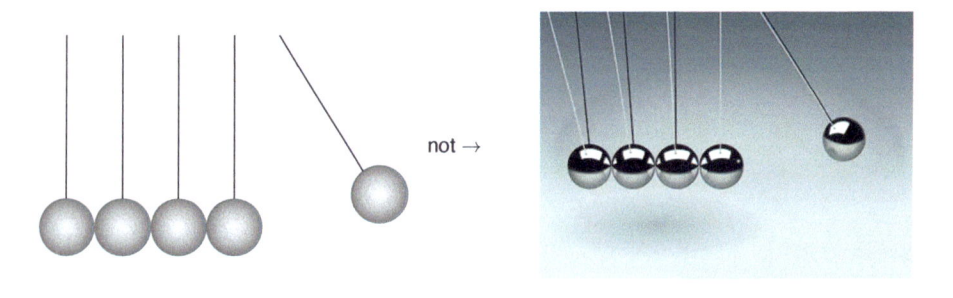

not →

Here are some examples of labelled diagrams that could be used in conjunction with labelled photos.

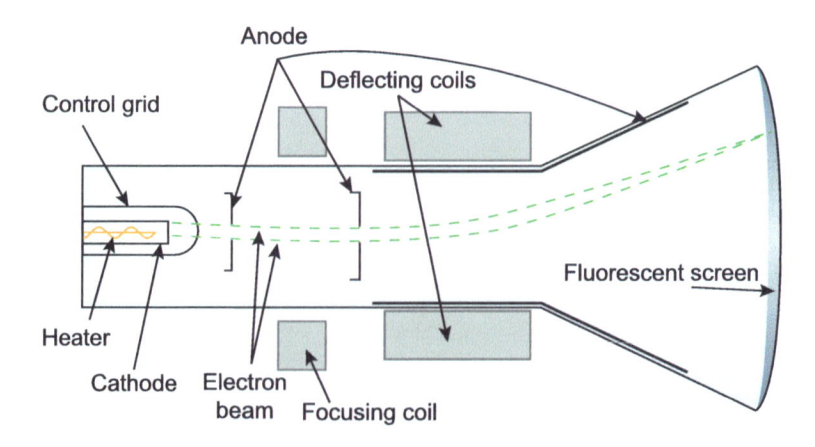

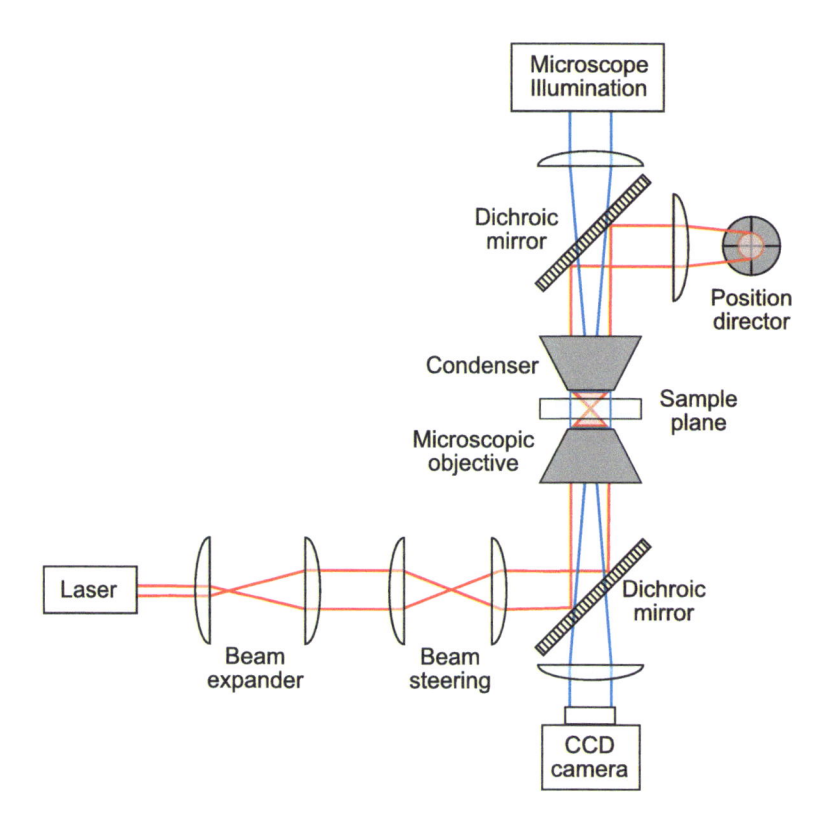

TOPIC 1. INITIAL PLANNING, USING EQUIPMENT AND RECORDING DATA 511

1.4 Summary

Summary

You should now be able to:

- be able to make an initial plan and make a timeline for the investigation;
- know how to look up ideas for suitable experiments and who to talk to about accessing and using equipment;
- how to take suitable pictures and draw relevant diagrams for your investigation.

© HERIOT-WATT UNIVERSITY

Unit 4 Topic 2

Measuring and presenting data

Contents

2.1 Result tables 514
2.2 Summary 517

Learning objective

By the end of this topic you should be able to:

- correctly complete results tables for your investigation;
- produce accurate graphs of the correct size and scale.

2.1 Result tables

Don't forget that tables of data need to have correct units and headings. A rough graph is useful to identify outliers which may affect your results. You may wish to use a spreadsheet program such as Excel or Numbers to produce the graph, but make sure the axes are correctly labelled, the scale is suitable and there are enough gridlines to clearly see the data points. The data points should be suitably small and not obscure the grid lines and ideally should be error bars not dots. (See tips for drawing graphs below for more information.)

A best fit line or curve should be added to aid with analysis of data. Excel can also give you an equation of a best fit line. Uncertainties will need to be taken into account and they will be looked at in more detail later. Also useful is to note down any difficulties encountered and how you dealt with them and any further improvements you can think of or further work you might like to look into. This will help greatly when writing up the evaluation part of the final report. Make sure you get plenty of data. When appropriate, a good scientist repeats readings enough times to calculate the random uncertainty and has a good range of data.

Example

This is an example of a seemingly well presented table of results for a Newton's Second Law experiment involving varying the unbalanced force applied to various masses. Note that this student has made a major mistake in their experiment and a few minor ones as well, can you spot their failings?

Mass (kg)	Force1 (N)	Force2 (N)	Force3 (N)	ForceAVG (N)	Acceleration
1.0	3.4	3.6	3.5	4	4
2.0	2.8	2.2	3.1	3	1.5
3.0	1.9	1.4	1.6	1.6	0.5
4.0	8.6	7.8	8.5	8	2.0
5.0	10.0	11.2	12.8	11.3	2.3
6.0	6.5	6.3	6.5	6.4	1.1
7.0	3.6	4.5	7.2	5	0.7
8.0	7.5	7.8	8.1	8	1.0
9.0	4.5	4.7	3.8	4.3	0.5

Things to note:

- Your results should have the correct number of significant figures/decimal places as appropriate. This will aid graph drawing later.
- You must have all headings labelled with the correct units.
- You may prefer to use a program such as Excel to perform the calculations in the columns for you, e.g. in this table the average force and acceleration were calculated using this method. This can save you a lot of time compared to using a calculator. See Excel Help if your teacher can't help you with this method.

TOPIC 2. MEASURING AND PRESENTING DATA

Mistakes in table:

- The student above has varied the mass **and** the unbalanced force making it difficult to plot a valid graph of results for their experimental data.
- Units for acceleration are missing.
- To calculate the random uncertainty in the Force, you would be advised to have two more repeats.
- Some of the Force and Acceleration values are only quoted to one significant figure.

Tips for drawing graphs

Be careful when using computer programs to draw graphs for you, computers tend not to show minor grid lines, leaving you with floating points which are not at all accurate:

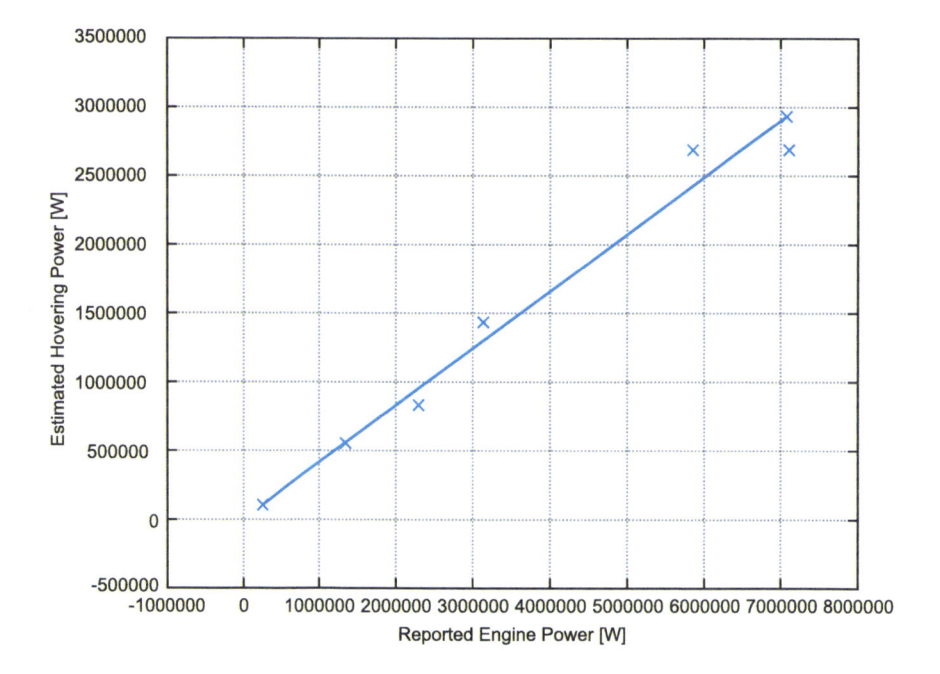

UNIT 4. INVESTIGATING PHYSICS

Another graph shown as follows is way too small on the page, the axes have not been labelled, the points are too large and so on.

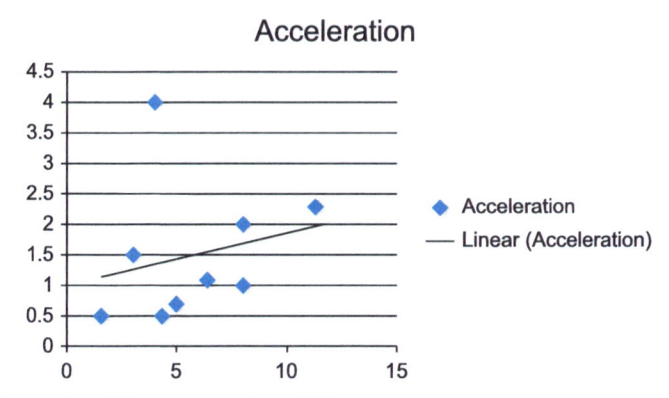

Acceleration

A much better version of the same graph would be:

Newton's Second Law Graph, Acceleration against Force

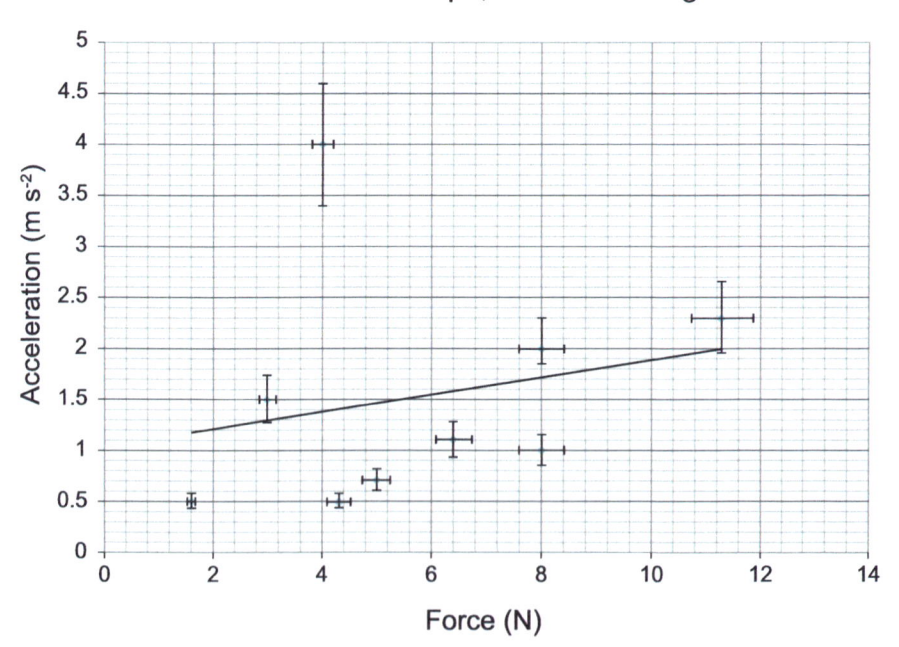

This graph has clearly labelled axes, major and minor gridlines and the large points have been replaced with error bars showing the uncertainties in the mass and acceleration. The point at 4N is an outlier and worthy of comment or even removal and the computer could also be used to calculate the gradient of the line for the conclusion. LINEST is a useful feature on excel for working out the uncertainty in the gradient if desired. Quite often a sharp pencil and a graph paper will enable you to be accurate enough if you prefer the paper and pencil method and the parallelogram method of uncertainties can work out the error in the gradient and intercept. (See Unit 5 for more information on this method).

2.2 Summary

Summary

You should now be able to:

- correctly complete results tables for your investigation;
- produce accurate graphs of the correct size and scale.

Unit 4 Topic 3

Evaluating findings

Contents
3.1 Evaluating findings 520
3.2 Summary 522

Prerequisites
- Evaluating & drawing conclusions (Higher).

Learning objective

By the end of this topic you should be able to:

- evaluate your results and draw conclusions;
- understand some tips for completing the evaluation sections of the report.

3.1 Evaluating findings

Evaluating your scientific findings - general tips.

This is a section that many candidates find it hard to pick up full marks and candidates are advised to use the following points to structure the discussion. Each experiment should have it's own conclusion and evaluation with a final overall evaluation at the end of the report.

Conclusions need to be valid and **related to the aim** of the investigation and they tend to take the form of the calculation of the gradient of a graph with appropriate uncertainties for example.

In the evaluation try and comment on all of the following:

1. **Accuracy of experimental measurements**

    How did you ensure that each result was as accurate as it could be?

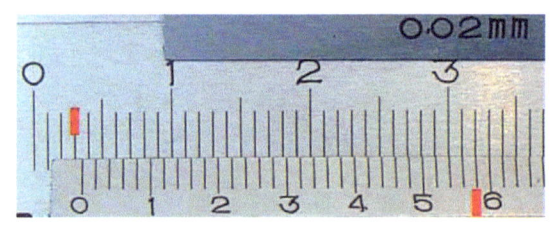

    Zoom-in on "Messschieber.jpg" made by Ultraman,Wikimedia, licensed under CC BY 3.0

    Using a Vernier scale instead of a regular ruler will vastly reduce your reading uncertainty.

    Or being aware of effects such as PARALLAX, where the reading can change depending on how you look at it.

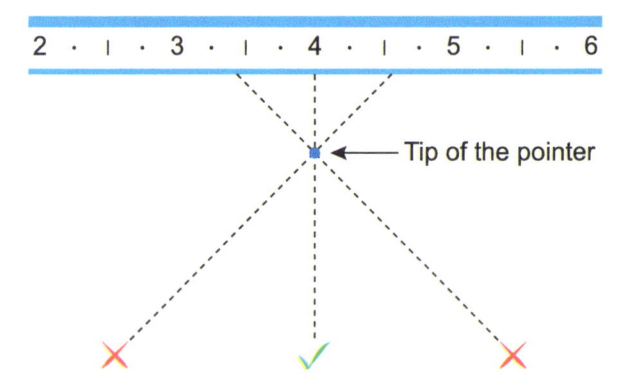

2. **Adequacy of repeated readings**

    Did you have enough readings to calculate an accurate random uncertainty? Was the equipment accurate enough to give you similar results each time?

3. **Adequacy of range over which variables are altered**

    Quite often equipment restraints will prevent you having the range you desire, can you think of any ways to extend the range of your variables or at the least comment on these restrictions?

TOPIC 3. EVALUATING FINDINGS

4. **Adequacy of control of variables**

   How did you reduce uncertainties in each variable? How many fixed variables were there in your experiment and how did you control them?

5. **Limitations of equipment**

   Most school equipment will have some failings, systematic uncertainties are quite common, look out and check for these. For example a Voltmeter that is always 0.2 mV out, a wooden ruler that is 0.5mm shorter than it once was, etc.

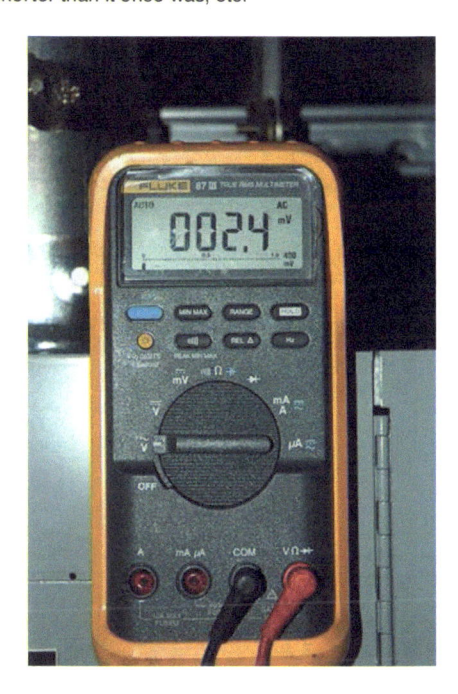

V-O-M (Volt-Ohm-Meter) / Multimeter, Author: Steve C, Source: Flickr

6. **Reliability of methods**

   Be self-critical here, how reliable was your method and how could it be improved?

7. **Sources of errors and uncertainties**

   Think about whether the uncertainties are most likely calibration, scale, reading etc or if human error is a major factor. Looking at the uncertainty calculations will make this much easier to target the largest sources of error.

The overall conclusion and evaluation of the investigation as a whole should mention:

- **Problems overcome** - During all three (or more) experiments.

    Examples:

    - A darkroom was needed to eliminate background light but one wasn't available.
    - The school multimeter/digital balance/data logger only has a limited range/large percentage uncertainty at smaller readings.
    - My school equipment wasn't advanced enough so I had to contact Scotland University for help with my project.
    - I had to limit the current to xA due to wires heating up.
    - etc.

- **Modifications to procedures and suggested future experiments** - If you had an unlimited budget, what would you like to do to improve and enhance your experiments?
- **Significance/interpretation of findings** - Can you find any similar work to compare yours to? Can you see a bigger picture from your findings? Can you link your results back to your background theory?

More detail of what to write in the evaluation can be found in the next topic, which takes you through the process of writing up the scientific report.

3.2 Summary

Summary

You should now be able to:

- evaluate your results and draw conclusions;
- understand some tips for completing the evaluation sections of the report.

Unit 4 Topic 4

Scientific report

Contents

4.1 Scientific report 524
4.2 Summary 528

Learning objective

By the end of this topic you should be able to:

- understand how to approach the project write up and will be familiar with the mark scheme used by the SQA.

4.1 Scientific report

When you come to write up your project report, you need to make sure it is between 2500 and 4500 words in length. Too short and it will not be detailed enough to pick up some of the marks in the mark scheme and too long - over 4950 words (10% over the maximum) and you will suffer a penalty. The total mark is 30 marks and you should make sure you hand in drafts to your teacher well in advance of the submission date for checking and re-editing. The most successful candidates may hand in two or even three rough drafts before the final submission.

The best tip is to pay close attention to the mark scheme and tick off each section as you complete it. The following instructions are taken direct from the SQA website and give detailed instructions on how the marks are awarded with a few extra tips to make sure you get the best out of your report writing process.

> https://www.sqa.org.uk/files_ccc/AHPhysicsCourseSpec.pdf
>
> *This document may be reproduced in whole or in part for assessment purposes provided that no profit is derived from reproduction and that, if reproduced in part, the source is acknowledged. If it needs to be reproduced for any purpose other than assessment, it is the centre's responsibility to obtain copyright clearance. Re-use for alternative purposes without the necessary copyright clearance may constitute copyright infringement.*
>
> *Copyright ©Scottish Qualifications Authority*

Sections you should include in the report

1. Abstract   1 mark
2. Underlying Physics 4 marks
3. Procedures   7 marks
4. Results   8 marks
5. Discussion   8 marks
6. Presentation   2 marks
                Total = **30 marks**

TOPIC 4. SCIENTIFIC REPORT

More detail on each section in the report

1. **Abstract (1 mark)**

   A brief abstract (summary) stating the overall aim(s) and finding(s)/conclusion(s) of the investigation.

   The abstract must be:

   - relevant to the investigation;
   - demonstrating an understanding of the physics theory underpinning the investigation;
   - of an appropriate Advanced Higher Physics level;
   - **immediately following the contents page** and should be under a separate heading. The abstract must be separate from and placed before the 'introduction';
   - the overall findings must be consistent with the conclusion(s) given in the 'discussion' and should relate to the aim(s).

2. **Underlying Physics (4 marks)**

   - Candidates must include an account of the underlying physics that is relevant to the investigation. All terms and symbols used should be clearly defined. Simply stating equations is not sufficient - derivation of formulae should be given and all symbols in the equations must be explained with correct units. Candidates must demonstrate a good understanding of the physics behind these equations.
   - Candidates may (and should) draw on a variety of sources of information when researching their chosen topic. Don't base all your theory on one website of information.
   - Downloading directly from the internet or copying directly from books may suggest that the candidate has not understood the physics involved and will be considered as plagiarism. Where the vast majority is believed to have been copied verbatim then the candidate is not demonstrating understanding.
   - Complicated diagrams copied and pasted from an internet source are perfectly acceptable, especially when the reference is cited in the text and listed at the back of the report.

3. **Procedures (7 marks)**

   Labelled diagrams and/or descriptions of apparatus (2 marks)

   - Candidates must include labelled diagrams and / or descriptions of the apparatus that they used for experimental work. Photographs of assembled apparatus, with appropriate labelling, are acceptable. A satisfactory photograph showing clear detail should be labelled as covered in Topic 3, and if possible use a labelled photograph and a labelled diagram for clarity.

   Clear descriptions of how the apparatus was used to obtain experimental readings (2 marks)

   - Candidates must also give clear descriptions of how they used the apparatus to obtain their experimental results.
   - The report should be written in the *past tense and impersonal voice*. (3rd person)
   - Bulleted / numbered points are not recommended for the method and must be made up of complete sentences. Use proper paragraphs to write your third person description.

© HERIOT-WATT UNIVERSITY

- The procedure should be described well enough for another competent Advanced Higher Physics candidate to be able to repeat the procedure from the description.

Procedures are at an appropriate level for Advanced Higher (3 marks)

Factors to be considered include:

- range of procedures and number of repetitions;
- control of variables;
- accuracy;
- originality of approach and/or experimental techniques;
- degree of sophistication of experimental design and/or equipment.

Some of this is out of your hands due to equipment restraints but reading this list beforehand combined with talking to your teacher should help steer you in the correct direction to attain these marks. Use experiments completed in class as a rough guide to the standard required.

4. **Results (including uncertainties) (8 marks)**

Data sufficient and relevant to the aim(s) of the investigation (1 mark)

- The experimental data that candidates collect must be relevant to the aim(s) of their investigation. Also, the data candidates collect and present in their report must be sufficient in quantity and with a degree of accuracy and precision appropriate to their investigation - ie it must show all readings and not just the mean values. Don't forget too that data in tables is meaningless unless supplied with the correct labels and units.

Appropriate analysis of data, e.g. quality graphs, lines of best fit, calculations (4 marks)

- A candidate's report must include analysis of their experimental data that is appropriate to the investigation. This may involve drawing graphs or calculating and tabulating numerical values. Again further tables, graphs and calculations should still have the correct units and significant figures applied to them.

Uncertainties in individual and final results (3 marks)

- Candidates must include uncertainties in the values of each of the physical quantities that they measure and in the final result(s) of their investigation. Their analysis should show clearly how they have calculated/estimated the uncertainty in their final result(s). The best way to do this is with an example calculation for each method used, e.g. random uncertainty, percentage uncertainty, uncertainty in gradient of graph etc. Uncertainty calculations should all be at Advanced Higher level and all uncertainties (calibration, scale reading and random) that have a bearing on the accuracy on the experimental work should be mentioned. See Unit 5 for more information on uncertainties.

5. **Discussion (8 marks)**

Conclusion(s) is/are valid and relate to the aim(s) of the investigation (1 mark)

- Candidates must include overall conclusion(s) that are relevant to the aims(s) of their investigation and *supported by data* in the report and which are valid for the experimental results obtained.

TOPIC 4. SCIENTIFIC REPORT

Evaluation of experimental procedures (3 marks)

- Candidates must also include a critical evaluation of each experiment. It is often appropriate to include this after the 'procedures' and 'results' of each experiment. This should be a significant part of the candidate's report and should focus on the quality of their experimental work. See topic 4 for some more tips in writing this section. Candidates should include as many factors as possible and suggest improvements to procedures.
  - Accuracy of experimental measurements
  - Adequacy of repeated readings
  - Adequacy of range over which variables are altered
  - Adequacy of control of variables
  - Limitations of equipment
  - Reliability of methods
  - Sources of errors and uncertainties

  One tip here is to make sure you write something relevant for all seven points for each experiment being careful not to repeat yourself.

Coherent discussion of overall conclusion(s) and critical evaluation of the investigation as a whole (3 marks)

- Candidates must include a discussion of their overall conclusion(s) together with a critical evaluation of the investigation as a whole. This should be a more wide-ranging discussion of the investigation. It is an opportunity to explain what the candidate has learned as a result of the investigation and the significance of their findings. Candidates could also demonstrate the depth of their understanding of the physics related to the investigation.
  - Problems overcome
  - Modifications to procedures
  - Significance/interpretation of findings
  - Suggestions for further improvements to procedures
  - Suggestions for further work

Overall quality of the investigation (1 mark)

- This is a final quality mark for the standard of the investigation - not just the 'discussion' part of the report. This is for a good investigation well worked through, taking particular account of the physics involved and synthesis of argument.

6. **Presentation (2 marks)**

Appropriate structure, including informative title, contents page and page numbers (1 mark)

- The report structure should be easy to follow. A *title, contents page* and structure are essential - the contents page must show *page numbers* and the pages throughout the report must be numbered. Occasional missing page numbers (e.g. on hand-drawn graphs) will not be penalised. A title page with a nice picture or diagram to catch the marker's eye and show something of what the report is about is not essential but highly recommended.

References cited in the text and references listed in standard form, acknowledgements, where appropriate (1 mark)

© HERIOT-WATT UNIVERSITY

- At *least three references* must be *cited correctly* in the main body of the report and the same ones also listed correctly at the back of the report. Any additional references cited or listed incorrectly should not be penalised. Referencing must conform to either Vancouver or Harvard referencing systems.
- References must be relevant to the investigation and specific. References must be cited within the text of the candidate's report and in many cases these will occur in the 'Underlying Physics' section. At the end of the report, the candidate must include details on all of the references (e.g. books, journals/periodicals and websites) that they cited. Candidates must include sufficient information to allow a reader to consult the original work to confirm its relevance to the investigation. Candidates should only include details on references; do not include information on materials that were part of background reading but are not cited as references in the report.

Total marks = 30 for the report

> **Top tip**
>
> Quite a lot of the advice above is **Copyright ©Scottish Qualifications Authority** and may change from year to year, always check the website for the most up to date mark scheme.

4.2 Summary

> **Summary**
>
> You should now be able to:
>
> - have a thorough understanding of how to approach the project write up;
> - be familiar with the mark scheme used by the SQA and know how to get as many of the 30 marks available as you can.

Appendix: Units, prefixes, uncertainties and data analysis

A Units, prefixes and scientific notation 531
   A.1 Symbols and units used in Rotational motion and astrophysics 532
   A.2 Symbols and units used in Quanta and waves 534
   A.3 Symbols and units used in Electromagnetism 535
   A.4 Prefixes 536
   A.5 Scientific notation and significant figures 540
   A.6 Summary 543

B Uncertainties 545
   B.1 Introduction 546
   B.2 Random uncertainties 546
   B.3 Scale-reading uncertainties 548
   B.4 Systematic uncertainties 550
   B.5 Calibration uncertainties 554

C Data analysis 555
   C.1 Calculating and stating uncertainties 556
   C.2 Combining uncertainties 563
   C.3 Uncertainties and graphs 564
   C.4 Centroid or parallelogram method 564
   C.5 Accuracy and precision 566

Unit 5 Appendix A

Units, prefixes and scientific notation

Contents

A.1 Symbols and units used in Rotational motion and astrophysics 532
A.2 Symbols and units used in Quanta and waves 534
A.3 Symbols and units used in Electromagnetism 535
A.4 Prefixes 536
    A.4.1 SI Units 536
    A.4.2 Prefixes and Scientific Notation 537
A.5 Scientific notation and significant figures 540
A.6 Summary 543

A.1 Symbols and units used in Rotational motion and astrophysics

Physics Quantity	Symbol	Unit	Unit abbreviation
velocity	v	metre per second	m s^{-1}
displacement	s, x, y	metre	m
acceleration	a	metre per second square	m s^{-2}
time	t	second	s
angular velocity	ω	radian per second	rad s^{-1}
angular displacement	Θ	radian	rad
angular acceleration	α	radian per second square	rad s^{-2}
radius of circle	r	metre	m
torque	T	newton metre	Nm
force	F	newton	N
moment of inertia	I	kilogram metre square	kg m^2
angular momentum	L	kilogram metre square per second	kg m^2 s^{-1}
energy	E	joule	J
universal constant of gravitation	$G\ (= 6.67 \times 10^{-11})$	metre cube per (kilogram second square)	m^3 kg^{-1} s^{-2}
mass of a large object, e.g. planet or star	M	kilogram	kg
mass of a smaller object, e.g. satellite	m	kilogram	kg
gravitational potential	V	joule per kilogram	J kg^{-1}
apparent brightness	b	watt per metre square	W m^{-2}
luminosity	L	watt	W

APPENDIX A. UNITS, PREFIXES AND SCIENTIFIC NOTATION

Physics Quantity	Symbol	Unit	Unit abbreviation
Stefan-Boltzmann constant	$\sigma\ (= 5.67 \times 10^{-8})$	watt per (metre square kelvin quadrupled)	W m^{-2} K^{-4}
temperature	T	kelvin	K
speed of light	$c\ (= 3.0 \times 10^{8})$	metre per second	m s^{-1}

A.2 Symbols and units used in Quanta and waves

Physics Quantity	Symbol	Unit	Unit abbreviation
frequency	f	hertz	Hz
planck's constant	$h\ (= 6.63 \times 10^{-34})$	metre square kilogram per second	m² kg s⁻¹
wavelength	λ	metre	m
momentum	p	kilogram metre per second	kg m s⁻¹
uncertainty in position	Δx	metre	m
uncertainty in momentum, energy, time	$\Delta p, \Delta E, \Delta t$	kilogram metre per second, joule, second	kg m s⁻¹, J, s
charge	q or Q	coulomb	C
magnetic field strength	B	tesla	T
amplitude	A	metre	m
phase difference	φ	radian	rad
fringe widths/spacing	Δx	metre	m
distance between slits and screen	D	metre	m
slit separation/min thickness of lens/width of thin wedge gap	d	metre	m
refractive index	n	no unit	no unit
length of slide (thin wedge)	l	metre	m
Brewster's Angle	i_p	degrees	°

APPENDIX A. UNITS, PREFIXES AND SCIENTIFIC NOTATION

A.3 Symbols and units used in Electromagnetism

Physics Quantity	Symbol	Unit	Unit abbreviation
permittivity of free space (electric constant)	ε_0 ($= 8.85 \times 10^{-12}$)	coulomb square per newton metre square	$C^2\ N^{-1}\ m^{-2}$
electrostatic potential	V	joule per coulomb OR volt	$J\ C^{-1}$ OR V
electric field strength	E	newton per coulomb OR volt per metre	$N\ C^{-1}$ OR $V\ m^{-1}$
current	I	ampere	A
permeability of free space	$\mu_0\ (= 4\pi x 10^{-7})$	metre kilogram per (second square ampere square)	$m\ kg\ S^{-2}\ A^{-2}$
time constant	t	second	s
resistance	R	ohm	Ω
capacitance	C	farad	F
reactance	X	ohm	Ω
self-induced emf	ε	volt	V
inductance	L	henry	H

> **Top tip**
>
> It can be quicker to use $1/4\pi\varepsilon_0\ (= 9 \times 10^9)$ in calculations involving the **permittivity of free space (electric constant)**.
>
> Similarly, it can be quicker to use $\mu_0/2\pi\ (= 2 \times 10^{-7})$ in calculations involving the **permeability of free space**.

© HERIOT-WATT UNIVERSITY

A.4 Prefixes

There are some prefixes that you must know.
These are listed in the following table:

Prefix	Symbol	Equivalent to
pico	p	$\times 10^{-12}$
nano	n	$\times 10^{-9}$
micro	μ	$\times 10^{-6}$
milli	m	$\times 10^{-3}$
kilo	k	$\times 10^{3}$
mega	M	$\times 10^{6}$
giga	G	$\times 10^{9}$

In Advanced Higher Physics you are expected to know and remember the meaning of all of these prefixes. Note that they are from the metric system so are all multiples of 1000 or 1/1000. Hence why centi $\times 10^{-2}$ (eg cm) is not used in the Advanced Higher Physics course.

A.4.1 SI Units

We talk of going 15 miles in a car journey to attend a 5 km race. We use a variety of units in everyday life that can lead to confusion. In science, we have adopted **The International System of Units (SI)**. Sometimes known as the metric system.

SI units all come from seven base units, second (s), metre (m), kilogram (kg), ampere (A), kelvin (K), mole (mol), and candela (cd). The other units derive from these constants.

Symbol	Quantity	Symbol and unit	
a	acceleration	m s^{-2}	metres per second per second
A	activity	Bq	Becquerel
d	distance	m	metre
E	energy	J	Joule
E_h	heat energy	J	Joule
E_k	kinetic energy	J	Joule
E_p	potential energy	J	Joule
E_W	work done	J	Joule
F	force	N	Newton
g	gravitational field strength	N kg^{-1}	Newtons per kilogram
h	height	m	metre
m	mass	kg	kilogram

APPENDIX A. UNITS, PREFIXES AND SCIENTIFIC NOTATION

Symbol	Quantity	Symbol and unit	
p	pressure	Pa	Pascals
P	power	W	Watt
s	displacement	m	metre
t	time	s	second
u	initial velocity	m s^{-1}	metres per second
v	velocity (or final velocity)	m s^{-1}	metres per second
$\bar{v}$	average velocity	m s^{-1}	metres per second
W	weight	N	Newton
Q	charge	C	Coulomb
R	resistance	Ω	Ohm
R_T	total resistance	Ω	Ohm

Symbol	Quantity	Symbol and unit	
T	temperature	K	Kelvin
ΔT	change in temperature	°C	degrees Celsius
V	volume	m^3	metres cubed
V	voltage	V	Volt
V_s	supply voltage	V	Volt
c	specific heat capacity	J kg^{-1} °C^{-1}	Joules per kilogram per degree Celsius
I	current	A	Amperes
l	specific latent heat	J kg^{-1}	Joules per kilogram
λ	wavelength	m	metres
ω_R	radiation weighting factor		(no units)
D	absorbed dose	Gy	Gray
f	frequency	Hz	Hertz
H	equivalent dose	Sv	Sievert
$\dot{H}$	equivalent dose rate	Sv s^{-1}	(many possible units)
T	period	s	seconds

A.4.2 Prefixes and Scientific Notation

Scientific Notation is a way of abbreviating, shortening, numbers e.g. 700 becomes 7×10^2 or 29000000 becomes 2.9×10^8.

Scientific Notation makes it simpler to use large and small values.

How do we use it?

700 is the same as 7×10, remember that 100 is 10^2.

So $700 = 7 \times 10^2$, they are the same value, just written differently.

When a large number has more than one digit such as 29000000 we use a decimal point after the first digit. The $\times 10$ power is how many places we have to move the decimal point.

So 29000000 is written 2.9×10^8 because 2.9×10000000 the decimal point has moved eight places.

© HERIOT-WATT UNIVERSITY

UNIT 5. UNITS, PREFIXES, UNCERTAINTIES AND DATA ANALYSIS

For numbers less than zero, the movement of the decimal point is labelled as negative. So 0.000005 becomes 5×10^{-6} as it equals $5 \div 1000000$.

Quiz

Go online

Q1: Write the following numbers in scientific notation.

a) 64
b) 658423
c) 2345
d) 0.0026
e) 0.000056
f) 0.2304

....

Q2: Write the following numbers in decimals.

a) 1.92×10^3
b) 3.051×10^1
c) 4.29×10^7
d) 1.03×10^{-2}
e) 8.862×10^{-7}
f) 9.512×10^{-5}

Prefixes

To deal with the range of numbers that can appear in physics and allow to make more sense standard prefixes are used to specify multiples and fractions of the units. These are based on Engineering Notation, like Scientific Notation but going up in multiples of 3.

Prefix	Symbol	Multiple	Multiple in full
Giga	G	$\times 10^9$	$\times$ 1 000 000 000
Mega	M	$\times 10^6$	$\times$ 1 000 000
Kilo	k	$\times 10^3$	$\times$ 1 000
Milli	m	$\times 10^{-3}$	$\div$ 1 000
Micro	μ	$\times 10^{-6}$	$\div$ 1 000 000
Nano	n	$\times 10^{-9}$	$\div$ 1 000 000 000

Examples

1. An ultrasound pulse last a time of 852 μs, how many seconds is this?

852 μs = 852 microseconds = 852×10^{-6} s = $852 \div 1\,000\,000$ = 0.000852 seconds.

....

© HERIOT-WATT UNIVERSITY

APPENDIX A. UNITS, PREFIXES AND SCIENTIFIC NOTATION

2. A wave has a period of 70 ms. How many seconds is this?

70 ms = 70 milliseconds = 70 × 10^{-3} s = 70 ÷ 1 000 = 0.070 seconds.

3. A boat travels 54.1 km in an hour. What is the distance the boat travelled in metres?

54.1 km = 54.1 kilometres = 54.1 × 10^3 m = 54.1 × 1 000 = 54 100 metres

4. In Back to the Future, the Flux Capacitor required 1.21 GW at 88mph to send Marty back. How much Power is required in Watts?

1.21 GW = 1.21 Giga Watts = 1.21 × 10^9 W = 1 210 000 000 Watts

Significant Figures

Often in Maths, they ask you to answer a question to a certain number of significant figures or decimal places.

In Physics, the number of significant figures in the answer is determined by the values in the questions.

Your answer should have the number of significant figures as the smallest number of significant figures given in the question.

Example A greyhound runs round a 515 m field in 12 seconds. Find their average speed?

d = 515 m

t = 12 s

$$d = v \times t$$
$$515 = v \times 12$$
$$v = \frac{515}{12}$$
$$v = 42.91666667 \; m \; s^{-1}$$

The 7 in this answer is × 10^{-8} m every second. Less than the width of a human hair on a run of half a kilometre. It makes no sense to go to that level of precision, especially as our measurements are in metres.

The smallest number of significant figures is 2, for 12 seconds.

The answer becomes $v = 43 \; m \; s^{-1}$.

A.5 Scientific notation and significant figures

Scientific notation

When carrying out calculations, you should be able to use scientific notation. This type of notation has been used throughout the topics where necessary, so you should already be familiar with it.

Scientific notation is used when writing very large or very small numbers. When a number is written in scientific notation there is usually one, non-zero number, before the decimal point.

Examples

1. The speed of light is often written as 3.0×10^8 m s^{-1}.

This can be converted into a number in ordinary form by moving the decimal point 8 places to the right, giving 300 000 000 m s^{-1}.

....

2. The capacitance of a capacitor may be 0.000 022 F.

This very small number would often be written as 2.2×10^{-5} F. The x 10^{-5} means move the decimal point 5 places to the left.

> **Top tip**
>
> Make sure you know how to enter numbers written in scientific notation into your calculator.

You must give final answers to calculations to an appropriate number of significant figures. In general, this means that your final answer should be given to no more than the least number of significant figures used for the data that was used to obtain the answer.

Significant figures

As a general rule, the final numerical answer that you quote should be to the same number of significant figures as the data given in the question. The above rule is the key point but you might like to note the following points:

1. The answer to a calculation cannot increase the number of significant figures that you can quote.
2. If the data is not all given to the same number of significant figures, identify the least number of significant figures quoted in the data. This least number is the number of significant figures that your answer should be quoted to.
3. When carrying out sequential calculations carry many significant figures as you work through the calculations. At the end of the calculation, round the answer to an appropriate number of significant figures.

Examples

1. The current in a circuit is 6.7 A and the voltage across the circuit is 21 V. Calculate the resistance of the circuit.

APPENDIX A. UNITS, PREFIXES AND SCIENTIFIC NOTATION 541

Note: Both of these pieces of data are given to two sig. figs. so your answer must also be given to two sig figs.

I = 6.7 A
V = 21 V
R = ?

$$V = IR$$
$$21 = 6.7 \times R$$
$$R = 3.1343$$
$$R = 3.1 \; \Omega$$

round to 2 sig figs

....

2. A 5.7 kg mass accelerates at 4.36 m s^{-2}.
Calculate the unbalanced force acting on the mass.

Note: The mass is quoted to two sig. figs and the acceleration is quoted to three sig. figs. so the answer should be quoted to two sig figs.

m = 5.7 kg
a = 4.36 m s^{-2}
F = ?

$$F = ma$$
$$F = 5.7 \times 4.36$$
$$F = 24.852$$
$$F = 25 \; N$$

round to 2 sig figs

....

A car accelerates from 0.5037 m s^{-1} to 1.274 m s^{-1} in a time of 4.25 s.
The mass of the car is 0.2607 kg.
Calculate the unbalanced force acting on the car.

Note: The time has the least number of sig figs, three, so the answer should be quoted to three sig figs.

u = 0.5037 m s^{-1}
v = 1.274 m s^{-1}
t = 4.25 s
m = 0.2607 kg

© HERIOT-WATT UNIVERSITY

Step 1: calculate a

$$a = \frac{v - u}{t}$$
$$a = \frac{1.274 - 0.5037}{4.25}$$
$$a = 0.181247 \; m \; s^{-2}$$

Step 2: calculate F

$$F = m \, a$$
$$F = 0.2607 \times 0.18147$$
$$F = 0.0472511$$
$$F = 0.0473 \; N$$

round to 3 sig figs

Finally, it is good practice to check your answers to calculations. Is the answer about what was expected (at least an acceptable order of magnitude)? Do you know a different relationship that can be used to confirm an answer? Does a check on the units confirm that the correct relationship was used (this is called 'dimensional analysis')?

Quiz questions

Q3: A car travels a distance of 12 m in a time of 9.0 s.
The average speed of the car is:

a) 1.3333
b) 1.33
c) 1.3
d) 1.4
e) 1

Q4: A mass of 2.26 kg is lifted a height of 1.75 m. The acceleration due to gravity is 9.8 m s^{-2}.
The potential energy gained by the mass is:

a) 38.759 J
b) 38.76 J
c) 38.8 J
d) 39 J
e) 40 J

Q5: A trolley of 5.034 kg is moving at a velocity of 4.03 m s^{-1}.
The kinetic energy of the trolley is:

a) 40.878 J
b) 40.88 J
c) 40.9 J
d) 41 J
e) 40 J

A.6 Summary

Summary

You should now be able to:

- have a good knowledge of all the units and prefixes needed for Advanced Higher Physics;
- correctly use scientific notation and significant figures at this level.

Unit 5 Appendix B

Uncertainties

Contents

B.1 Introduction 546
B.2 Random uncertainties 546
B.3 Scale-reading uncertainties 548
B.4 Systematic uncertainties 550
B.5 Calibration uncertainties 554

B.1 Introduction

Uncertainties are covered in some detail at Higher level, but a quick review follows and the main differences at Advanced Higher are that you need to be able to combine uncertainties and work out the uncertainty in the gradient and intercept of a graph. This is covered in detail in the 'Data analysis' section. These skills are highly useful in writing up your project but also could be assessed in the examination.

Whenever a physical quantity is measured, there is always an uncertainty in the measurement - no measurement is ever exact. Uncertainties can never be eliminated but must be reduced as far as possible if experimental results are to be valid.

If an experiment 'does not work' - i.e. the expected result is not obtained - this usually means that the uncertainties in the experimental measurements are very high - so high that the anticipated result may be only obtained by chance. Uncertainties can be reduced by careful experimental design and by experimenters exercising care in the way in which they carry out the experiment and take the measurements. Uncertainties must be taken into account when stating the results of experimental investigation.

Quoting a numerical result of an experiment as (value ± uncertainty) allows us to check the validity of our experimental method. In addition it enables comparison of the numerical result of one experiment with that of another.

If the result of an experiment to measure a physical quantity of known value (e.g. the speed of light *in vacuo*) leads to a range of values that does not include the accepted value then either the experiment is not valid or, more commonly, the uncertainties have been underestimated. An experiment that leads to a smaller range of uncertainties is more valid than an experiment that has a wider range.

When undertaking experiments you should be prepared to discard or to repeat any measurement that is obviously 'wrong' - i.e. not consistent with the other measurements that you have taken.

There are several causes of uncertainty in experimental measurements and these may be random, scale-reading or systematic.

B.2 Random uncertainties

The effects of random uncertainties are not predictable. For example, when an experimental measurement is repeated several times, the result may not be the same each time. It is likely that some of the readings will be slightly higher than the true value and some will be slightly lower than the true value. Examples could include measurements of time using a stop-watch, measuring an angle using a protractor, measuring length using a measuring tape or ruler.

Random uncertainties are due to factors that cannot be completely eliminated by an experimenter. For example, when taking a measurement of length using a metre stick there may be small variations in the exact positioning of the metre stick from one reading to the next; similarly when reading an analogue meter there may be slight variations in the positions of the experimenter's eyes as readings are taken.

The effects of random uncertainties can be reduced by repeating measurements and finding the mean. The mean value of a number of measurements is the best estimate of the true value of the quantity being measured.

APPENDIX B. UNCERTAINTIES

Where a quantity Q is measured n times, the measured value is usually quoted as the mean Q_{mean} of the measurements taken ± the approximate random uncertainty in the mean. Q_{mean} is the best estimate of the true value and is given by:

$$Q_{mean} = \frac{\Sigma Q_i}{n}$$

The approximate random uncertainty in the mean is given by:

$$approximate\ random\ uncertainty = \frac{Q\ maximum - Q\ minimum}{n}$$

Notes:

1. A random uncertainty can only be calculated from measured data that you would expect to be the same value.
2. A random uncertainty must not be found in calculated values.
3. The above relationship is an approximation; it is not statistically rigorous, but it is sufficiently accurate at this level when at least 5 readings have been taken.

Example A student uses a computer program to measure their reaction time. The following values are obtained for the reaction time of the student.

Attempt number	1	2	3	4	5
Reaction time /s	0.273	0.253	0.268	0.273	0.238

a) Calculate the mean reaction time of the student.

b) Calculate the approximate random uncertainty in the mean.

a)

$$mean = \frac{total\ of\ values}{number\ of\ values}$$
$$mean = \frac{(0.273 + 0.253 + 0.268 + 0.273 + 0.238)}{5}$$
$$mean = \frac{1.305}{5}$$
$$mean = 0.261\ \text{s}$$

b)

$$random\ uncertainty = \frac{(max\ value - min\ value)}{number\ of\ values}$$
$$random\ uncertainty = \frac{(0.273 - 0.238)}{5}$$
$$random\ uncertainty = 0.007\ \text{s}$$

© HERIOT-WATT UNIVERSITY

Interpretation of these calculations

These are often written as:

best estimate = mean value ± uncertainty

best estimate of reaction time = 0.261 s ± 0.007 s

This means that if the reaction time was measured again it is likely, not guaranteed, that the value would be with the range of 0.261 s plus or minus 0.007 s.

⇒ Likely that measured value of time would lie between 0.254 s and 0.268 s.

Increasing the reliability

In order to increase the reliability of a measurement, increase the number of times that the quantity is measured. It is likely that the random uncertainty will decrease.
In the above example this would mean that instead of finding the mean reaction time based on 5 attempts, repeat the measurement so that the calculation is based on 10 attempts.

If you repeat a measurement 5 times and you measure exactly the same value on each occasion then the random uncertainty will be zero. Making further repeated measurements is unnecessary as this will not reduce the random uncertainty so it will not increase the reliability.

B.3 Scale-reading uncertainties

A scale reading uncertainty is a measure of how well an instrument scale can be read. This type of uncertainty is generally random and is due to the finite divisions on the scales of measuring instruments. For example, the probable uncertainty in a measurement of length, using a metre stick graduated in 1 mm divisions, is 0.5 mm. If more precision is needed then a different measuring instrument (e.g. a metal ruler or a micrometer) or a different technique must be used.

For instruments with analogue scales, the scale-reading uncertainty is usually taken as ± half of the smallest scale division. In some cases, it may be possible to make reliable estimates of smaller fractions of scale divisions.

For instruments with digital scales the reading uncertainty is 1 in the last (least significant) digit.

APPENDIX B. UNCERTAINTIES

Examples

1. Analogue scale

This approach is used for rulers, metre sticks, liquid in glass thermometer and meters which have a pointer.

The length of metal is measured with the ruler shown below.

Length 6 cm
Scale reading uncertainty = half of one scale division = 0.5 cm

Often expressed as 6.0 cm $\pm$ 0.5 cm

This means that the best estimate of the length is 6.0 cm and it would be expected that the "true" length would be between 5.5 cm and 6.5 cm.

2. Digital display

This approach is used whenever a seven segment digital display is present.

The following image shows a digital ammeter.

Current = 12.9 A
Scale reading uncertainty = one in smallest scale division = 0.1 A

Often expressed as 12.9 A $\pm$ 0.1 A

This means that the best estimate of the current is 12.9 A and it would be expected that the "true" current would be between 12.8 A and 13.0 A

© HERIOT-WATT UNIVERSITY

B.4 Systematic uncertainties

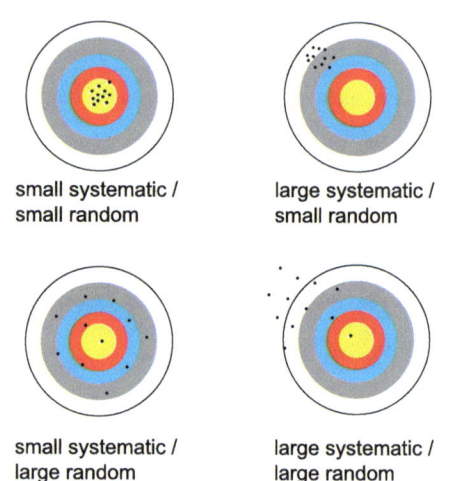

small systematic / small random

large systematic / small random

small systematic / large random

large systematic / large random

Systematic uncertainties have consistent effects on the quantities being measured.

Systematic uncertainties often arise due to experimental design or issues with the equipment.

The following example shows a ruler being used to measure the length of a metal bar.

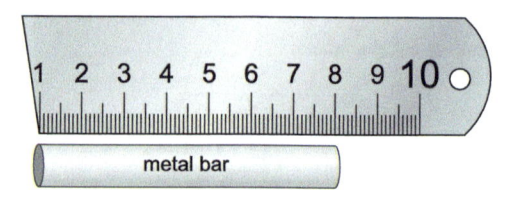

At first sight the length of the metal bar is 8 cm.

However, on closer inspection the actual length is only 7 cm as the ruler starts at 1 cm rather than 0 cm.

This ruler could easily cause all measured values to be too long by 1 cm. This would be a systematic uncertainty.

This systematic uncertainty could have been noticed by the experimenter and corrected but often the presence of a systematic uncertainty is not detected until data is analysed.

Example A student is investigating how the distance between a loudspeaker and microphone affects the time it takes a pulse of sound to travel from the loudspeaker to the microphone. The equipment used is shown below.

APPENDIX B. UNCERTAINTIES

Electronic timer

switch — start timer — stop timer

loudspeaker   microphone

distance measured

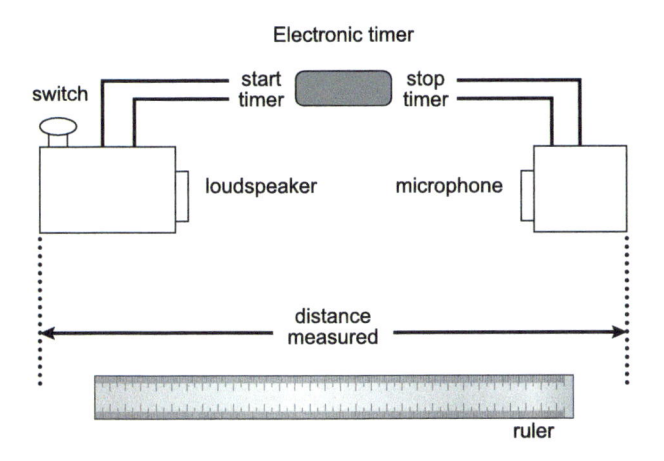

ruler

When the switch is pressed the loudspeaker produces a sound and the timer starts. When the sound reaches the microphone the timer is stopped.

The distance shown is measured with a ruler. The distance is altered by moving the microphone to a greater distance from the loudspeaker and further measurements are taken.

The results obtained are displayed on the following graph.

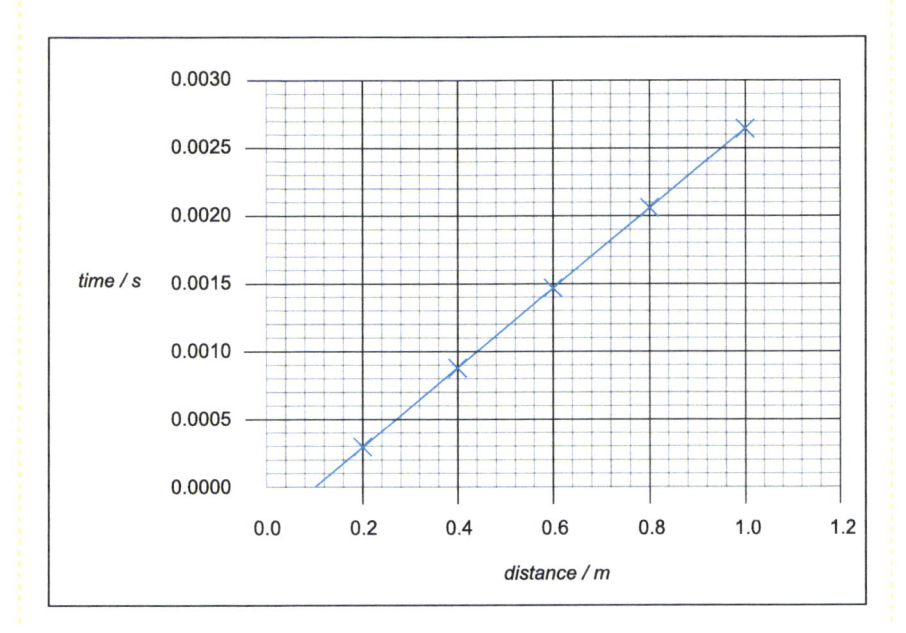

time / s vs *distance / m*

The expected graph is a straight line through the origin.
Here a straight line is obtained but it does not go through the origin.
This shows that there is a systematic uncertainty in the investigation.

The line is too far to the right so **all** of the distance measurements are too big by the same value. There is a systematic uncertainty of 0.1 m. This value is found by finding the intercept on the distance axis.

What has caused this systematic uncertainty?

Look at the labelled diagram and notice that the distance is between the extreme edges of the loudspeaker and the microphone.
The sound will be made inside the loudspeaker box and the microphone will be inside the microphone box. This means that the sound does not have to travel this distance and all the distances measured are too big by 0.1 m.

Further thoughts on this investigation

1. The gradient of this graph can lead to an estimate of the speed of sound.

$$gradient = \frac{\Delta y}{\Delta x} = \frac{y_2 - y_1}{x_2 - x_1} = \frac{rise}{run}$$

$$gradient = \frac{\Delta time}{\Delta distance}$$

$$gradient = \frac{(0.0015 - 0)}{(0.6 - 0.1)}$$

$$gradient = 3 \times 10^{-3}$$

since

$$speed = \frac{\Delta distance}{\Delta time}$$

and here

$$gradient = \frac{\Delta time}{\Delta distance}$$

hence

$$speed = \frac{1}{gradient}$$

$$speed = \frac{1}{3 \times 10^{-3}}$$

$$Speed\ of\ sound = 333\ m\ s^{-1}$$

2. It may be suggested that the systematic uncertainty could be removed by measuring the distance between the inside edges of the loudspeaker and microphone as shown in the diagram below.

APPENDIX B. UNCERTAINTIES

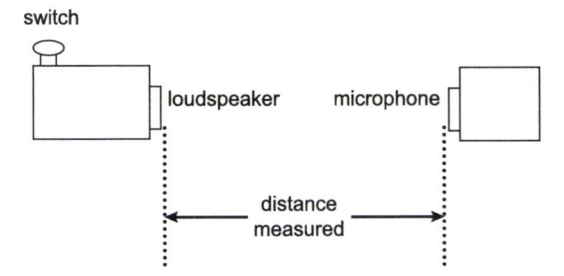

This would result in the following graph.

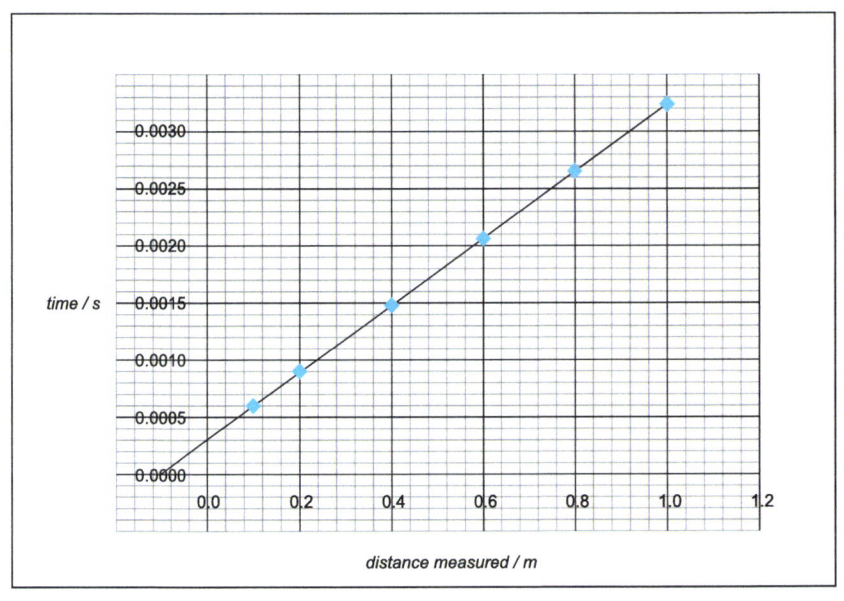

Using this approach, a straight line is obtained but again does not pass through the origin indicating the presence of a systematic uncertainty. The line is too far to the left.

The distance measured is too short and the underestimate is always 0.1 m. This value is found from the intercept on the distance axis. This means that all the distance measurements are too small by 0.1 m.

It is impossible to remove the systematic uncertainty unless the actual positions of where the sound is produced and where the sound is detected are known. This cannot be done if the components are mounted inside "boxes".

The gradient of this graph would again give an estimate of the speed of sound.

Identifying systematic effects is often an important part of the evaluation of an experiment.

© HERIOT-WATT UNIVERSITY

B.5 Calibration uncertainties

A Calibration uncertainty is a manufacturer's claim for the accuracy of a instrument compared with an approved standard. It is usually found in the instructions that are supplied with the instrument. Calibration uncertainties are often systematic in nature. Calibration uncertainties may be predictable or unpredictable. For example the drift of the time base of an oscilloscope due to temperature changes may not be predictable but it is likely to have a consistent effect on experimental results. Other examples of calibration uncertainties are a clock running consistently fast or consistently slow, an ammeter reading 5% higher than the true reading and a balanced incorrectly zeroed at the start of an experiment reading consistently too high or too low.

Unit 5 Appendix C

Data analysis

Contents
C.1 Calculating and stating uncertainties 556
C.2 Combining uncertainties 563
C.3 Uncertainties and graphs 564
C.4 Centroid or parallelogram method 564
C.5 Accuracy and precision 566

556 UNIT 5. UNITS, PREFIXES, UNCERTAINTIES AND DATA ANALYSIS

C.1 Calculating and stating uncertainties

Single measurements may be quoted as ± measurement absolute uncertainty, for example 53.20 ± 0.05 cm. When measured quantities are combined (e.g. when the quantities are multiplied, divided or raised to a power) to obtain the final result of an experiment it is often more useful to quote measurement ± percentage uncertainty, where

$$percentage\ uncertainty = \frac{actual\ uncertainty}{measurement} \times 100$$

In an experiment where more than one physical quantity has been measured, the largest percentage uncertainty in any individual quantity is often a good estimate of the percentage uncertainty in the final numerical result of the experiment.

When comparing the uncertainty in two or more measured values it is necessary to compare percentage uncertainties not absolute uncertainties.

In an investigation the distance travelled and the time taken are measured and the results are expressed in the form.

Best estimate ± absolute uncertainty

distance, d	=	125 mm ± 0.5 mm (metre stick, analogue device)
time, t	=	5. 2 s ± 0.1 s (stop watch, digital device)

$$\%uncert\ in\ d = \frac{absolute\ uncert\ in\ d}{measurement\ of\ d} \times 100$$

$$\%uncert\ in\ d = \frac{0.5}{125} \times 100$$

$$\%uncert\ in\ d = 0.4\%$$

$$\%uncert\ in\ t = \frac{absolute\ uncert\ in\ t}{measurement\ of\ t} \times 100$$

$$\%uncert\ in\ t = \frac{0.1}{5.2} \times 100$$

$$\%uncert\ in\ t = 2\%$$

In order to compare the precision of these two measurements the percentage uncertainty in each measurement must be calculated.

Comparing these two percentage uncertainties it can be seen that the percentage uncertainty in time is much greater than the percentage uncertainty in the distance.

Finding the uncertainty in a calculated value

The uncertainty in a calculated value can be estimated by comparing the percentage uncertainties in the measured values. At Higher level, normally one percentage uncertainty was three or more times larger than all the others and as a result this largest percentage uncertainty was a good estimate of the uncertainty in the calculated value. This may again be the case at Advanced Higher level.

Evaluating an experimental method

In order to improve the precision of an experiment it is necessary to find the measurement with the largest percentage uncertainty and consider how this percentage uncertainty could be reduced. Using the figures given above for distance and time the percentage uncertainty in time is greatest therefore an improvement method of measuring the time is required. Using two light gates connected to an electronic timer would enable the time to be measured with a smaller scale reading uncertainty. This would improve the precision in the measurement of time and hence in average speed.

Example Using the measured values of distance and time given, calculate the average speed of the moving object. In order to carry this out the percentage uncertainties in distance and time must be know.

distance, d = 125 mm ± 0.4%
time, t = 5. 2 s ± 2%

$$averagespeed = \frac{distance\ gone}{time\ taken}$$
$$averagespeed = \frac{125}{5.2}$$
$$averagespeed = 24 \text{mm s}^{-1}$$

The percentage uncertainty in the average speed will be 2%. The percentage uncertainty in t is more than three time the percentage uncertainty in d.

$$averagespeed = 24 \text{mms}^{-1} \pm 2\%$$

When measured quantities are combined it is usual to ignore any percentage uncertainty that is not significant.

> A percentage uncertainty in an individual measured quantity can be regarded as insignificant if it is less than one third of any other percentage uncertainty.

> Absolute uncertainty should always be rounded to one significant figure.

Quiz

Go online

Q1: State the scale reading uncertainty in the following voltmeter reading.

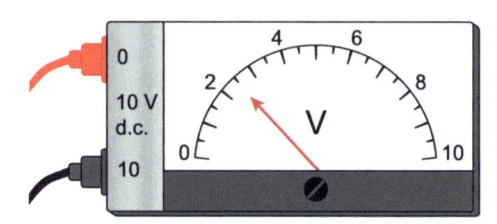

a) ± 0.25 V
b) ± 0.5 V
c) ± 1.0 V
d) ± 2.0 V
e) ± 5.5 V

....

Q2: A student carries out an investigation to measure the time taken for ten complete swings of a pendulum.
The following values are obtained for the time for ten complete swings.

| 3.1 s | 3.8 s | 3.3 s | 4.1 s | 3.4 s |

What is the random uncertainty in the time for ten complete swings?

a) ± 0.01 s
b) ± 0.02 s
c) ± 0.1 s
d) ± 0.2 s
e) ± 1.0 s

....

Q3: A student carries out three investigations into the variation of voltage and current. The results obtained are shown in the Graphs A, B and C.

APPENDIX C. DATA ANALYSIS 559

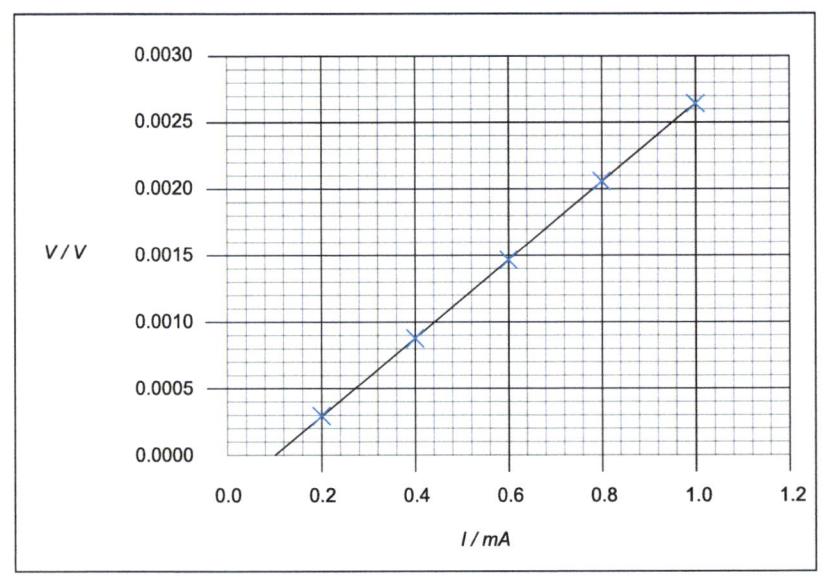

Graph A

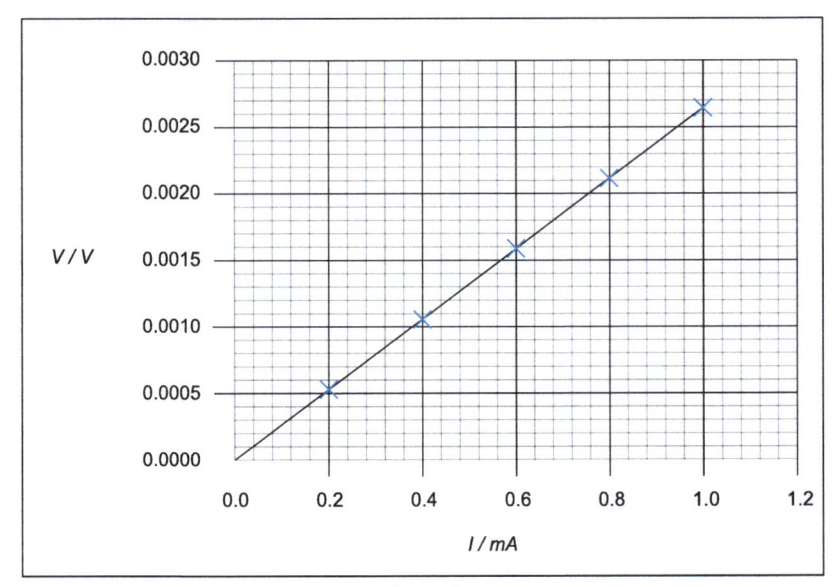

Graph B

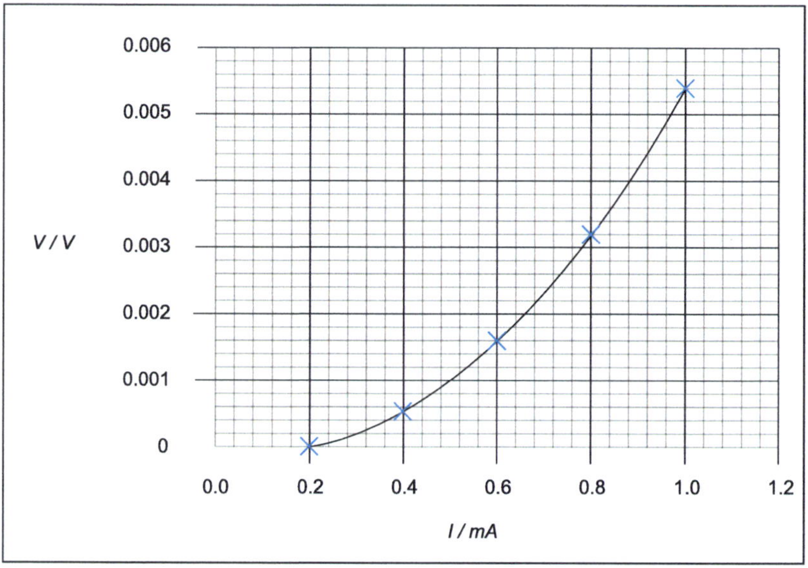

Graph C

Which of the following statements is/are true?

- I Graph A shows a systematic uncertainty
- II Graph B shows a proportional relationship
- III Graph C shows a systematic uncertainty

a) I only
b) II only
c) I and II only
d) I and III only
e) I, II and III

....

Q4: In an experiment the following measurements and uncertainties are recorded.

Temperature rise	=	$10°$ C $\pm$ 1°C
Heater current	=	5.0 A $\pm$ 0.2 A
Heater voltage	=	12.0 V $\pm$ 0.5 V
Time	=	100 s $\pm$ 2 s
Mass of liquid	=	1.000 kg $\pm$ 0.005 kg

APPENDIX C. DATA ANALYSIS

The measurement which has the largest percentage uncertainty is the:

a) Temperature rise
b) Heater current
c) Heater voltage
d) Time
e) Mass of liquid

....

Q5: In an investigation the acceleration of a trolley down a slope is found to be 2.5 m s^{-2} ± 4%.

The absolute uncertainty in this value of acceleration is:

a) ± 0.04 m s^{-2}
b) ± 0.1 m s^{-2}
c) ± 0.4 m s^{-2}
d) ± 1.0 m s^{-2}
e) ± 4.0 m s^{-2}

....

Q6: In an investigation the voltage across a resistor is measured as 20 V ± 2 V and the current through it is 5.0 A ± 0.1 A.

The percentage uncertainty in the power is:

a) 0.1%
b) 2%
c) 3%
d) 10%
e) 12%

....

Q7: Specific heat capacity can be found from the experimental results given below.

Which one of the following measurements creates most uncertainty in the calculated value of the specific heat capacity?

a) Power = 2000 ± 10 W
b) Time = 300 ± 1 s
c) Mass = 5.0 ± 0.2 kg
d) Final temperature = 50 ± 0.5 °C
e) Change in temperature = 30 ± 1 °C

....

© HERIOT-WATT UNIVERSITY

Q8: The light coming from a spectral lamp is investigated and the following data is obtained.

$$\lambda = 450 \text{ nm} \pm 10\%$$
$$f = 6.7 \times 10^{14} \text{Hz} \pm 2\%$$

This data is used to estimate the speed of light. The absolute uncertainty in this estimate of the speed of light is:

a) ± 2.0 m s^{-1}
b) ± 10 m s^{-1}
c) $\pm 6.0 \times 10^6$ m s^{-1}
d) $\pm 3.0 \times 10^7$ m s^{-1}
e) $\pm 3.6 \times 10^7$ m s^{-1}

....

Q9: Two forces P and Q act on an object X as shown.

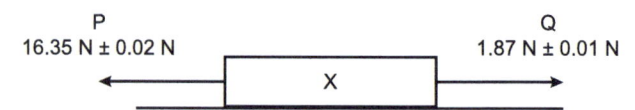

P
16.35 N ± 0.02 N

Q
1.87 N ± 0.01 N

X

The value of the unbalanced force acting on the object X and the percentage uncertainty in this value, expressed in the form value ± absolute uncertainty is:

a) 14.48 N ± 0.03 N
b) 14.48 N ± 0.08 N
c) 14.48 N ± 0.5 N
d) 18.22 N ± 0.03 N
e) 18.22 N ± 0.08 N

....

Q10: A student measures their reaction time using the digital stop watch on a computer.

The following measurements of their reaction time are displayed on the computer's digital stop watch.

| 0.29 s | 0.25 s | 0.22 s | 0.26 s | 0.24 s |

When evaluating this set of measurements the student makes the following statements.

- I Increasing the number of attempts from 5 to 10 would make the mean value more reliable.
- II The scale reading uncertainty in this set of measurements is ± 0.01 s.
- III You can tell by reviewing the measurements that there is no systematic uncertainty present.

Which of the above statements is/are correct?

a) I only
b) II only
c) III only
d) I and II only
e) I, II and III

C.2 Combining uncertainties

Uncertainties in combinations of quantities are calculated as follows.

Addition and subtraction

When two quantities P with absolute uncertainty ΔP, and Q with absolute uncertainty ΔQ, are added or subtracted to give a further quantity S, the absolute uncertainty ΔS in S, is given by

$$\Delta S = \sqrt{\Delta P^2 + \Delta Q^2}$$

Multiplication and division

When two quantities P with absolute uncertainty ΔP, and Q with absolute uncertainty ΔQ, are multiplied or divided to give a further quantity S, the percentage uncertainty in S is given by

$$\% \text{ uncertainty in } S = \sqrt{(\% \text{ uncertainty in } P)^2 + (\% \text{ uncertainty in } Q)^2}$$

$$\frac{\Delta S}{S} \times 100 = \sqrt{\left(\frac{\Delta P}{P} \times 100\right)^2 + \left(\frac{\Delta Q}{Q} \times 100\right)^2}$$

Powers

When a quantity P is raised to a power n to give a further quantity Q, then

$$\% \text{ uncertainty in } Q = n \times \% \text{ uncertainty in } P$$

When measured quantities are combined it is usual to ignore any percentage uncertainty that is not significant; a percentage uncertainty in an individual measured quantity can be regarded as insignificant if it is less than one third of any other percentage uncertainty.

© HERIOT-WATT UNIVERSITY

Example

Calculate the kinetic energy and corresponding absolute uncertainty of a car moving at (25.0 ± 0.1)ms^{-1} with a mass of (2200 ± 10)kg.

First you'd find the kinetic energy:

$$E_k = \frac{1}{2}mv^2$$
$$E_k = 0.5 \times 2200 \times 25.0^2$$
$$E_k = 687,500 \text{ J} = 690,000 \text{ J to the correct two significant figures.}$$

Next you should work out the percentage uncertainties in each of the quantities:

$$\% \text{ uncertainty in mass} = 10/2200 \times 100 = 0.45\%$$
$$\% \text{ uncertainty in velocity} = 0.1/25 \times 100 = 0.40\%$$

which due to being v^2 is doubled to get 0.80% (see powers equation above)

Next step is to combine the uncertainties using the multiplication rule:

$$\% \text{ uncertainty in kinetic energy} = \sqrt{(0.45^2 + 0.8^2)} = 0.92\%$$

Finally work out the absolute uncertainty which is 0.92% of 690kJ = 6000J

(Remember absolute uncertainties are always quoted to one significant figure.)

Your answer is best expressed as:

Kinetic Energy = (690 ± 6)kJ

C.3 Uncertainties and graphs

When graphing quantities that include uncertainties, note the following points.

- Each individual point on the graph should include error bars, on one or both of the quantities being plotted, as appropriate.
- The error bars on each point could indicate either the absolute uncertainty or the percentage uncertainty in the quantities being plotted, as appropriate.
- The error bars are used to draw the best straight line or the best fit curve, as appropriate.

C.4 Centroid or parallelogram method

The following method (known as the 'centroid' method) can be used to estimate the uncertainty in the gradient and the uncertainty in the y-intercept of a straight-line graph.

APPENDIX C. DATA ANALYSIS

1. Plot the points and error bars.
2. Calculate the centroid of the points. The x co-ordinate of the centroid is the mean of the values plotted on the x-axis and the y co-ordinate of the centroid is the mean of the values plotted on the y-axis.
3. Draw the best fitting straight line through the centroid.
4. Construct a parallelogram by drawing lines, which are parallel to this line, and which pass through the points furthest above and furthest below this line.
5. Calculate the gradients, m_1 and m_2, of the diagonals of the parallelogram. (These two diagonals represent the greatest and least values that the gradient could have.)
6. Calculate the uncertainty in the gradient, Δm using
$$\Delta m = \frac{m_1 - m_2}{2\sqrt{(n-2)}}$$
where n is the number of points (not including the centroid) plotted on the graph.
7. Read off the intercepts, c_1 and c_2, on the y-axis, of the diagonals of the parallelogram.
8. Calculate the uncertainty in the intercept, Δc using
$$\Delta c = \frac{c_1 - c_2}{2\sqrt{(n-2)}}$$
Again, n is the number of points (not including the centroid) plotted on the graph.

There are various methods possible using computers to calculate the error in the gradient and the intercept. One option is to use one of the functions available in graph drawing software e.g. LINEST and Trendline functions in Excel. Use the Help feature on Excel to find out more or ask your teacher for guidance.

LINEST in exel - (Takes you through the steps to use LINEST)

http://www.colby.edu/chemistry/PChem/notes/linest.pdf

(*last accessed in October 2019*)

© HERIOT-WATT UNIVERSITY

C.5 Accuracy and precision

"Dart180" by Sebastian Kamper is licensed under CC BY 3.0

The accuracy of a measurement compares how close the measurement is to the 'true' or accepted value. The darts player in the picture above is clearly accurate! The precision of a measurement gives an indication of the uncertainty in the measurement, the lower the uncertainty the more precise the reading. The darts player is also very precise and has achieved the maximum 180 with three darts.

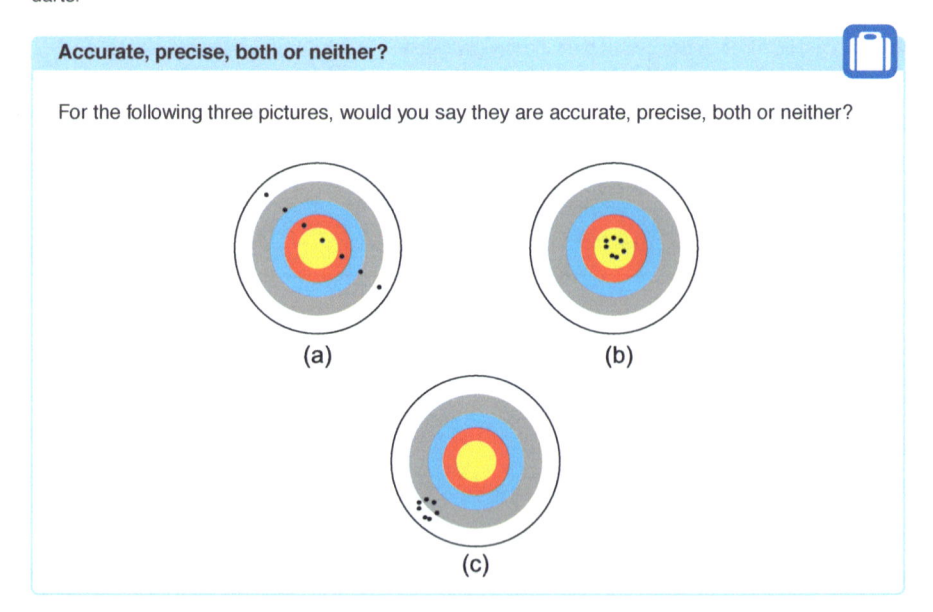

Accurate, precise, both or neither?

For the following three pictures, would you say they are accurate, precise, both or neither?

(a)

(b)

(c)

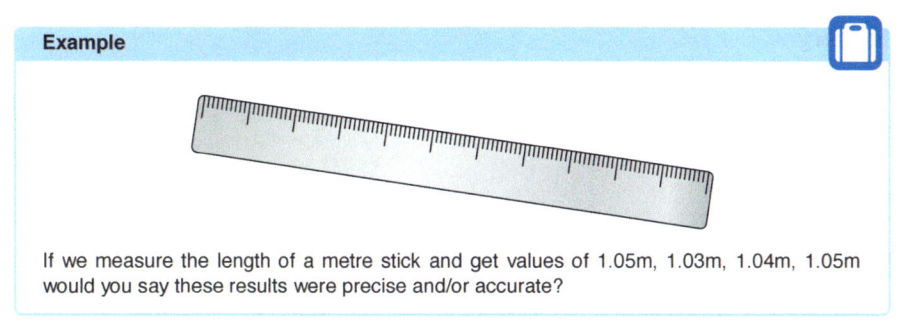

> **Example**
>
> If we measure the length of a metre stick and get values of 1.05m, 1.03m, 1.04m, 1.05m would you say these results were precise and/or accurate?

The following link will help you to understand the Accuracy and Precision, Systematic Error and Random Uncertainty

https://www.youtube.com/watch?v=icWY7nICrfo

(*last accessed in October 2019*)

Glossary

Amplitude
   the maximum displacement of an oscillating object from the zero displacement (equilibrium) position.

Angular acceleration
   the rate of change of angular velocity, measured in rad s^{-2}.

Angular displacement
   the angle, measured in radians, through which a point or line has been rotated about an axis, in a specified direction.

Angular momentum
   the product of the angular velocity of a rotating object and its moment of inertia about the axis of rotation, measured in kg m^2 s^{-1}.

Angular velocity
   the rate of change of angular displacement, measured in rad s^{-1}.

Apparent brightness
   the amount of energy per second reaching a detector per unit area.

Back e.m.f.
   an induced e.m.f. in a circuit that opposes the current in the circuit.

Beat frequency
   is the frequency of the changes between loud and quiet sounds heard by the listener - for example when a tuning fork and a piano note of similar frequencies are played at the same time.

Birefringence
   the property of some materials to split an incident beam into two beams polarised at right angles to each other.

Black body
   an object which absorbs all incident radiation and then re-emits all this energy again. Can be visualised as standing waves inside the black body cavity.

Black body emitter
   a perfect absorber and emitter of radiation at all wavelengths of the electromagnetic spectrum.

Black dwarf
   the remains of a white dwarf after it has cooled.

Black hole
   a region in space from which no matter or radiation can escape, resulting from the extreme curvature of spacetime caused by its compact mass.

Bloomed lens
   a lens that has been given a thin coating to make it anti-reflecting at certain wavelengths.

GLOSSARY

Bohr model
: a model of the hydrogen atom, in which the electron orbits numbered $n = 1, 2, 3....$ have angular momentum $nh/2\pi$.

Brewster's law
: a beam of light travelling in a medium of refractive index n_1 will be 100% polarised by reflection from a medium of refractive index n_2 if the angle of incidence i_p obeys the relationship $\tan i_p = \frac{n_2}{n_1}$

Capacitive reactance
: the opposition which a capacitor offers to current.

Centripetal acceleration
: the acceleration of an object moving in a circular path, which is always directed towards the centre of the circle.

Centripetal force
: a force acting on an object causing it to move in a circular path.

Coherent waves
: two or more waves that have the same frequency and wavelength, and similar amplitudes, and that have a constant phase relationship.

Compton scattering
: the reduction in energy of a photon due to its collision with an electron. The scattering manifests itself in an increase in the photon's wavelength.

Conical pendulum
: a pendulum consisting of a bob on a string moving in a horizontal circle.

Conservative field
: a field in which the work done in moving an object between two points in the field is independent of the path taken.

Coulomb's law
: the electrostatic force between two point charges is proportional to the product of the two charges, and inversely proportional to the square of the distance between them.

Current-carrying conductor
: exactly as its name suggests - a conductor of some sort in which there is a current.

Damping
: a decrease in the amplitude of oscillations due to the loss of energy from the oscillating system, for example the loss of energy due to work against friction.

De Broglie wavelength
: a particle travelling with momentum p has a wavelength λ associated with it, the two quantities being linked by the relationship $\lambda = h/p$. λ is called the de Broglie wavelength.

© HERIOT-WATT UNIVERSITY

Dichroism
    the property of some materials to absorb light waves oscillating in one plane, but transmit light waves oscillating in the perpendicular plane.

Eddy currents
    an induced current in any conductor placed in a changing magnetic field, or in any conductor moving through a fixed magnetic field.

Electric field
    a region in which an electric charge experiences a Coulomb force.

Electric potential
    the electric potential at a point in an electric field is the work done per unit positive charge in bringing a charged object from infinity to that point.

Electromagnetic braking
    the use of the force generated by eddy currents to slow down a conductor moving in a magnetic field.

Equivalence principle
    Einstein's principle which states that there is no way of distinguishing between the effects on an observer of a uniform gravitational field and of constant acceleration.

Escape velocity
    the minimum speed required for an object to escape from the gravitational field of another object, for example the minimum speed a rocket taking off from Earth would need to escape the Earth's gravitational field.

Event horizon
    the boundary of a black hole; no matter or radiation can escape from within the event horizon. Time appears to be frozen at the event horizon of a black hole.

Faraday's law of electromagnetic induction
    the magnitude of an e.m.f. produced by electromagnetic induction is proportional to the rate of change of magnetic flux through the coil or circuit.

Ferromagnetic
    materials in which the magnetic fields of the atoms line up parallel to each other in regions known as magnetic domains.

Frequency
    the rate of repetition of a single event, in this case the rate of oscillation. Frequency is measured in hertz (Hz), equivalent to s^{-1}.

Fundamental unit of charge
    the smallest unit of charge that a particle can carry, equal to 1.60×10^{-19} C.

General relativity
    the study of non-inertial frames of reference.

Geodesic path
: the shortest path between two events in spacetime.

Geometrical path difference
: the difference in physical length between two sources.

Gravitational equilibrium
: when the outward force on a star due to thermal pressure from fusion balances the inwards gravitational pull.

Gravitational field
: the region of space around an object in which any other object with a mass will have a gravitational force exerted on it by the first object.

Gravitational field strength
: the gravitational field strength at a point in a gravitational field is equal to the force acting per unit mass placed at that point in the field.

Gravitational lensing
: the bending of light as it passes through curved spacetime.

Gravitational potential
: at a particular point in a gravitational field, the gravitational potential is the work done by external forces in bringing a unit mass from infinity to that point.

Gravitational time dilation
: the effect whereby time runs more slowly close to an object with a large gravitational field strength.

Gravitational waves
: general relativity predicts that the Big Bang created ripples in the curvature of spacetime. Such ripples are also thought to be generated during astronomical events, such as the collision of two black holes.

Heliosphere
: a massive bubble-like volume surrounded by the empty space.

Helix
: the shape of a spring, with constant radius and stretched out in one dimension.

Hertzsprung-Russell (H-R) diagram
: a plot of absolute magnitude or luminosity against spectral class or surface temperature. By convention the surface temperature is in descending order.

High-pass filter
: an electrical filter that allows high frequency signals to pass, but blocks low frequency signals.

Induced e.m.f.
: the e.m.f. induced in a conductor by electromagnetic induction.

Induction heating
> the heating of a conductor because of the eddy currents within it.

Inductive reactance
> the opposition which an inductor offers to current.

Inductor
> acoil that generates an e.m.f. by self-inductance. The inductance of an inductor is measured in henrys (H).

Inertial frame of reference
> a frame of reference that is stationary or has a constant velocity.

Interference by division of amplitude
> a wave can be split into two or more individual waves, for example by partial reflection at the surface of a glass slide, producing two coherent waves. Interference by division of amplitude takes place when the two waves are recombined.

Interference by division of wavefront
> all points along a wavefront are coherent. Interference by division of amplitude takes place when waves from two such points are superposed.

Irradiance
> the rate at which energy is being transmitted per unit area, measured in $W\ m^{-2}$ or $J\ s^{-1}\ m^{-2}$.

Lenz's law
> the induced current produced by electromagnetic induction is always in such a direction as to oppose the change that is causing it.

Liquid crystal displays
> a display unit in which electronically-controlled liquid crystals are enclosed between crossed polarisers. If no signal is supplied to the unit, it transmits light. When a signal is applied, the liquid crystal molecules take up a different alignment and the cell does not transmit light, hence appearing black.

Luminosity
> a measure of the total power a star emits i.e. the total energy emitted per second.

Magnetic domains
> regions in a ferromagnetic material where the atoms are aligned with their magnetic fields parallel to each other.

Magnetic flux
> a measure of the quantity of magnetism in a given area. Measured in weber (Wb), equivalent to $T\ m^2$.

Magnetic induction
> a means of quantifying a magnetic field.

Magnetic poles
one way of describing the magnetic effect, especially with permanent magnets. There are two types of magnetic poles - north and south. Opposite poles attract, like poles repel.

Magnetosphere
the region near to an object such as a planet or a star where charged particles are affected by the object's magnetic field.

Main sequence
stars that are in their long lived stable phase where they are fusing hydrogen into helium in their cores. They form the long diagonal band on a H-R diagram.

Moment of inertia
the moment of inertia of an object about an axis is the sum of mass × distance from the axis, for all elements of the object.

Nebula
a large cloud of gas and dust from which a star is formed.

Neutron star
a high density small star that is composed almost entirely of neutrons and is leftover from a supernova explosion.

Non-inertial frame of reference
a frame of reference that is accelerating.

Optical activity
the effect of some materials of rotating the plane of polarisation of a beam of light.

Optical path difference
the optical path between two points is equal to the distance between the points multiplied by the refractive index. An optical path difference will exist between two rays travelling between two points along different paths if they travel through different distances or through media with different refractive indices.

Periodic time
the time taken for an object rotating or moving in a circle to complete exactly one revolution.

Permeability of free space
a constant used in electromagnetism. It has the symbol μ_0 and a value of $4\pi \times 10^{-7}$ H m^{-1} (or T m A^{-1}).

Phase
a way of describing how far through a cycle a wave is.

Phase difference
if two waves that are overlapping at a point in space have their maximum and minimum values occurring at the same times, then they are in phase. If the maximum of one does not occur at the same time as the maximum of the other, there is a phase difference between them

Photoelasticity
: the effect of a material becoming birefringent when placed under a mechanical stress.

Photoelectric effect
: the emission of electrons from a substance exposed to electromagnetic radiation.

Photoelectrons
: an electron emitted by a substance due to the photoelectric effect.

Photons
: a quantum of electromagnetic radiation, with energy $E = hf$, where f is the frequency of the radiation and h is Planck's constant.

Pitch
: the distance travelled along the axis of a helix per revolution.

Planetary nebula
: the outer layers of a low or medium mass star that are shed to leave behind a white dwarf.

Polarisation
: the alignment of all the oscillations of a transverse wave in one direction.

Polarised light
: tight in which all the electric field oscillations are in the same direction.

Potential difference
: the potential difference between two points is the difference in electric potential between the points. Since electric potential tells us how much work is done in moving a positive charge from infinity to a point, the potential difference is the work done in moving unit positive charge between two points. Like electric potential, potential difference V is measured in volts V.

Precess
: the motion of a spinning body in which its axis of rotation changes.

Principle of superposition
: states that the total disturbance at a point due to the presence of two or more waves is equal to the algebraic sum of the disturbances that each of the individual waves would have produced.

Principle of superposition of forces
: the total force acting on an object is equal to the vector sum of all the forces acting on the object.

Prominences
: arcs of gas that erupt from the surface of the sun launching charged particles into space, sometimes called a filament.

Proton-proton chain
: the process whereby a star fuses hydrogen into helium.

Quantum mechanics
a system of mechanics, developed from quantum theory, used to explain the properties of particles, atoms and molecules.

Radian
a unit of measurement of angle, where one radian is equivalent to $180/\pi$ degrees. One radian is defined in terms of the angle at the centre of a circle made by the sector of the circle in which the length along the circumference is equal to the radius of the circle.

Red giants
high luminosity and low surface temperature stars, lying to the right of the main sequence.

Red supergiant
a large mass star will progress beyond that of a red giant to that of a red supergiant, expanding due to hydrogen depletion. This is the stage before it undergoes a supernova explosion.

Rotational kinetic energy
the kinetic energy of an object due to its rotation about an axis, measured in J. A rotating object still has kinetic energy even if it is not moving from one place to another. The total kinetic energy of a rotating object is the sum of the rotational and translational kinetic energies.

Saccharimetry
a technique that uses the optical activity of a sugar solution to measure its concentration.

Schwarzschild radius
the distance from the centre of a black hole at which not even light can escape.

Self-inductance
the generation of an e.m.f. by electromagnetic induction in a coil owing to the current in the coil.

Simple harmonic motion (SHM)
motion in which an object oscillates around a fixed (equilibrium) position. The acceleration of the object is proportional to its displacement, and is always directed towards the equilibrium point.

Singularity
the point of infinite density at the centre of a black hole to which all mass would collapse.

Spacetime
a representation of four dimensional space, in which the three spatial dimensions and time are blended.

Special relativity
the study of inertial frames of reference.

Spectral class
stars can be put into the classes O,B,A,F,G,K and M, where O is the hottest and M is the coolest.

Speed

the speed of a wave is the distance travelled by a wave per unit time, measured in m s^{-1}.

Stationary wave

a wave in which the points of zero and maximum displacement do not move through the medium (also called a standing wave).

Stellar nucleosynthesis

process whereby a star carries out nuclear fusion to produce new elements heavier than hydrogen.

Strong nuclear force

the force that acts between nucleons (protons and neutrons) in a nucleus, binding the nucleus together.

Supernova

the explosion of a red supergiant, leaving behind a black hole or a neutron star.

Tangential acceleration

the rate of change of tangential speed, measured in m s^{-2}.

Tangential speed

the speed in m s^{-1} of an object undergoing circular motion, which is always at a tangent to the circle.

Time constant

the time taken for the charge stored by a capacitor to increase by 63% of the difference between initial charge and full charge, or the time taken to discharge a capacitor to 37% of the initial charge.

Torque

also called the moment or the couple, the torque due to a force is the turning effect of the force, equal to the size of the force times the perpendicular distance from the point where the force is applied to the point of rotation. Torque is measured in units of N m.

Torsion balance

very small forces can be measured using a torsion balance. The forces are applied to the end of a light rod suspended at its centre by a vertical thread. As the forces turn the rod, the restoring torque (turning force) in the thread increases until the turning forces balance.

Transmission axis

the transmission axis of a sheet of Polaroid is the direction in which transmitted light is polarised.

Travelling wave

a periodic disturbance in which energy is transferred from one place to another by the vibrations.

Universal Law of Gravitation

also known as Newton's law of gravitation, this law states that there is a force of attraction between any two massive objects in the universe. For two point objects with masses m_1 and m_2, placed a distance r apart, the size of the force F is given by the equation:

$$F = \frac{Gm_1m_2}{r^2}$$

Unpolarised light

light in which the electric field oscillations occur in random directions.

Van Allen radiation belts

regions around the Earth, caused by the Earth's magnetic field, that trap charged particles. Discovered in 1958 after the Explorer 1 mission.

Wave function

a mathematical function used to determine the probability of finding a quantum mechanical particle within a certain region of space.

Wavelength

the distance between successive points of equal phase in a wave, measured in metres.

Wave-particle duality

the concept that, under certain conditions, waves can exhibit particle-like behaviour and particles can exhibit wave-like behaviour.

Weak nuclear force

a nuclear force that acts on particles that are not affected by the strong force.

Weight

the weight of an object is equal to the force exerted on it by the Earth.

White dwarfs

faint hot stars that are left behind when the outer layers of a red giant drift away. They are the remnants of a medium or low mass star at the end of its life. They lie below and to the left of the main sequence.

Work function

in terms of the photoelectric effect, the work function of a substance is the minimum photon energy which can cause an electron to be emitted by that substance.

Worldline

a line on a spacetime diagram mapping a particle's spatial location at every instant in time.

Hints for activities for Unit 1

Topic 1: Kinematic relationships

Horizontal motion

Hint 1:

List the data you are given in the question

$u = ...$ m s^{-1}
$v = ...$ m s^{-1}
$s = ...$ m
$a = ?$

Once you have the data listed, decide on the appropriate kinematic relationship.

Quiz: Motion in one dimension

Hint 1: First sketch the graph - or see the examples in the section Motion in one dimension.

Hint 2: This is a straight application of the second equation of motion.

Hint 3: This is a straight application of the second equation of motion.

Hint 4: First, work out the displacement of the stone from its starting position. Initial velocity is zero, $a = g$ - use the third equation of motion to find v.

Hint 5: You must choose a positive direction 'up' or 'down' - the initial vertical velocity is up and the final displacement is down - be careful with the sign of the acceleration.

Topic 2: Angular motion

Quiz: Radian measurement

Hint 1: $360° = 2\pi$ radians.

Hint 2: $360° = 2\pi$ radians.

Hint 3: $360° = 2\pi$ radians.

Hint 4: 1 complete rotation = 2π radians

Hint 5: First find out the fraction of the circumference that the object has moved.

Quiz: Angular velocity and angular kinematic relationships

Hint 1: $\omega = 2\pi f$

Hint 2: This is the angular equivalent of a speed, displacement, time calculation.

Hint 3: This is the angular equivalent of a speed, displacement, time calculation.

Hint 4: First work out the change in angular velocity.

Hint 5: The angular displacement after one complete revolution = 2π.

© HERIOT-WATT UNIVERSITY

HINTS: UNIT 1 TOPIC 3

Quiz: Angular velocity and tangential speed

Hint 1: This is a straight application of $v = r\omega$.

Hint 2: This is a straight application of $v = r\omega$.

Hint 3: First work out the circumference of the circle.

Hint 4: First work out the angular deceleration then use $a = r\alpha$.

Hint 5: First work out the angular displacement.

Quiz: Centripetal acceleration

Hint 1: This is a straight application of $a_\perp = r\omega^2$.

Hint 2: See the section titled Centripetal acceleration.

Hint 3: This is a straight application of $a_\perp = \frac{v^2}{r}$

Hint 4: First, work out either ω or $a_\perp = \frac{v^2}{r}$.

Hint 5: Substitute $r = 2r$ in $a_\perp = \frac{v^2}{r}$

Quiz: Horizontal and vertical motion

Hint 1: This is a straight application of $F = \frac{mv^2}{r}$

Hint 2: This is a straight application of $F = \frac{mv^2}{r}$

Hint 3: Consider each statement in turn while referring to $F = \frac{mv^2}{r}$

Hint 4: Calculate the speed for a central force equal to the maximum tension.

Hint 5: At what point is the tension in the string greatest? See the section on vertical motion.

Quiz: Conical pendulum and cornering

Hint 1: See the section titled Conical pendulum.

Hint 2: See the section titled Conical pendulum for the derivation of the relationship $\cos\theta = \frac{g}{l\omega^2}$. Remember to use SI units.

Hint 3: Start from $\cos\theta = \frac{g}{l\omega^2}$ and remember

$$\omega = 2\pi f$$

Hint 4: See the section titled Cars cornering

Hint 5: See the activity banked corners.

© HERIOT-WATT UNIVERSITY

Topic 3: Rotational dynamics

Quiz: Torques

Hint 1: Consider the units on both sides of the relationship $T = Fr$.

Hint 2: This is a straight application of $T = Fr$.

Hint 3: This is a straight application of $T = Fr$.

Hint 4: First, calculate the component of the force perpendicular to the radius.

Hint 5: First calculate the torque and use this to find the component of the force perpendicular to the radius.

Topic 4: Angular momentum

Quiz: Conservation of angular momentum

Hint 1: This is a straight application of $L = I\omega$.

Hint 2: Use the relationship given to find the moment of inertia of the disc. Then apply $L = I\omega$.

Hint 3: Angular momentum is conserved. What effect does adding the clay have on the total moment of inertia?

Quiz: Angular momentum and rotational kinetic energy

Hint 1: Use the relationship given to find the moment of inertia of the sheet. Then apply rotational $E_k = \frac{1}{2}I\omega^2$.

Hint 2: First, work out the value of ω.

Topic 5: Gravitation

Quiz: Gravitational force

Hint 1: This is a straight application of $F = \frac{Gm_1m_2}{r^2}$

Hint 2: This is an example of Newton's Third Law which is sometimes stated as "To every action there is an equal an opposite reaction."

Hint 3: Apply Weight $= \frac{Gm_1m_2}{r^2}$

Hint 4: What happens to the distance between the object and the centre of the planet?

Hint 5: Consider the weight of a mass of 1 kg on the surface of Venus

Quiz: Gravitational fields

Hint 1: This is a straight application of $g = \frac{GM}{r^2}$

Hint 2: Think of Newton's Second Law!

Hint 3: The gravitational strength is halved - this means the distance from the centre of the planet is increased by a factor $\sqrt{2}$ - see the relationship $g = \frac{GM}{r^2}$

Hint 4: The two gravitational forces must be equal and opposite.

Hint 5: This is a straight application of $g = \frac{GM}{r^2}$

Quiz: Gravitational potential

Hint 1: Consider the relationship $V = -\frac{GM}{r}$.

Hint 2: This is a straight application of $V = -\frac{GM}{r}$.

Hint 3: This is an application of $V = -\frac{GM}{r}$.
Remember r is the distance from the centre of the Earth.

Hint 4: This is a straight application of $E_P = -\frac{Gm_1m_2}{r}$.

Hint 5: Consider the options in the context of the relationship $E_P = -\frac{Gm_1m_2}{r}$.

Quiz: Escape velocity

Hint 1: See the section titled Escape velocity.

Hint 2: Consider the relationship $v = \sqrt{\frac{2GM_E}{r_E}}$.

Hint 3: This is an application of the relationship $v = \sqrt{\frac{2GM_{\text{planet}}}{r_{\text{planet}}}}$.

Hints to questions and activities for Unit 2

Topic 1: Introduction to quantum theory

Quiz: Atomic models

Hint 1: See the section titled The Bohr model of the hydrogen atom.

Hint 2: See the section titled The Bohr model of the hydrogen atom - in particular the relationship for angular momentum.

Hint 3: See the section titled The Bohr model of the hydrogen atom.

Hint 4: See the section titled Atomic spectra.

Topic 2: Wave particle duality

Quiz: Wave-particle duality of waves

Hint 1: This is a straight application of $E = hf$.

Hint 2: The minimum photon energy equals the work function.

Hint 3: See the section titled Compton scattering.

Hint 4: $E = hf$ and $p = \frac{h}{\lambda}$.

Hint 5: This is a straight application of $p = \frac{h}{\lambda}$.

Quiz: Wave-particle duality of particles

Hint 1: See the section titled Wave-particle duality of particles.

Hint 2: This is a straight application of $\lambda = \frac{h}{p}$.

Hint 3: This is a straight application of $\lambda = \frac{h}{p}$.

Hint 4: See the section titled Diffraction.

Hint 5: See the section titled The electron microscope.

Topic 3: Magnetic fields and particles from space

Quiz: The force on a moving charge

Hint 1: How does the charge on a proton compare with the charge on an electron?

Hint 2: This is a straight application of $F = q\,v\,B$.

Hint 3: What is the size of the magnetic force acting on a charged particle moving parallel to magnetic field lines?

Hint 4: What is the direction of a magnetic force relative to the direction of motion of a charged particle?

Hint 5: This is an application of $qvB = \frac{mv^2}{r}$.

HINTS: UNIT 2 TOPIC 5

Quiz: Motion of charged particles in a magnetic field

Hint 1: See the section titled Helical motion.

Hint 2: See the section titled Helical motion.

Hint 3: This is a straight application of $r = \frac{mv\sin\theta}{qB}$.

Hint 4: See section entitled The Solar Wind.

Hint 5: See the section Charged particles in the Earth's magnetic field.

Topic 4: Simple harmonic motion

Quiz: Defining SHM and equations of motion

Hint 1: See the section titled Defining SHM.

Hint 2: $\omega = 2\pi f$.

Hint 3: See the section titled Defining SHM.

Hint 4: See the relationships in the section titled Equations of motion in SHM. Calculate ω. Maximum acceleration occurs when displacement equals amplitude.

Hint 5: See the relationships in the section titled Equations of motion in SHM.

Quiz: Energy in SHM

Hint 1: See the relationships in the section titled Energy in SHM.

Hint 2: The total energy of the system is constant.

Hint 3: Consider the relationship $E_k = \frac{1}{2}m\omega^2(a^2 - y^2)$.

Hint 4: $PE = KE$ when half of the total energy is kinetic and half is potential - find the displacement when the kinetic energy is half its maximum value.

Hint 5: Consider the relationship $E_k = \frac{1}{2}m\omega^2(a^2 - y^2)$ with $y = \frac{1}{2}a$.

Quiz: SHM Systems

Hint 1: See the section titled Mass on a spring - vertical oscillations. First, find ω and hence find the period.

Hint 2: See the section titled Simple pendulum. First, find ω and hence find the frequency.

Hint 3: See the section titled Mass on a spring - vertical oscillations. First, find ω and hence find the spring constant.

Hint 4: See the section titled Simple pendulum.

Hint 5: See the section titled Simple pendulum - how does ω vary with l ?

© HERIOT-WATT UNIVERSITY

Topic 5: Waves

Quiz: Properties of waves

Hint 1: See the section titled Definitions.

Hint 2: Use $v = f\lambda$ twice.

Hint 3: See the figure The electromagnetic spectrum in the section titled Definitions.

Hint 4: Use $v = f\lambda$.

Hint 5: See the section titled Definitions.

Quiz: Travelling waves

Hint 1: Substitute the values in the equation.

Hint 2: Compare the options with the general expression for a travelling wave $y = A \sin 2\pi(ft - \frac{x}{\lambda})$.

Hint 3: Substitute the values in the equation.

Hint 4: Compare the equation with the general expression for a travelling wave $y = A \sin 2\pi(ft - \frac{x}{\lambda})$.

Hint 5: First work out the values of f and λ by comparing the equation with the general expression for a travelling wave $y = A \sin 2\pi(ft - \frac{x}{\lambda})$.

Quiz: Superposition

Hint 1: The amplitude of the disturbance at this point is equal to the difference between amplitudes of the two waves.

Hint 2: Consider the phase of the waves arriving at the listener from the two loudspeakers.

Hint 3: Divide the wavelength into the distance.

Hint 4: See the section titled Fourier Series.

Hint 5: Use the Phase Angle Equation and be careful with the wavelength.

Quiz: Stationary waves

Hint 1: How does the frequency of the third harmonic compare with the fundamental frequency?

Hint 2: See the section titled Stationary waves.

Hint 3: How many wavelengths are there on the string when the fundamental note is played?

Hint 4: See the section titled Stationary waves.

Hint 5: To find out how to calculate the wavelength see the activity titled Longitudinal stationary waves. Then use $v = f\lambda$.

Hint 6: What is the difference between the frequencies? What is the Beat Frequency? Read the section on Beats for more information.

Topic 6: Interference
Quiz: Coherence and optical paths

Hint 1: See the section titled Coherence.

Hint 2: Divide the wavelength into the distance.

Hint 3: The optical path length in the glass is equal to *(actual path length in the glass $\times$ refractive index)*.

Hint 4: Be careful to work out the optical path **difference**.

Hint 5: The optical path **difference** is equal to 4λ; remember that the ray in air travelled the thickness of the plastic.

Topic 7: Division of amplitude
Quiz: Thin film interference

Hint 1: See the section titled Optical path difference.

Hint 2: See the section titled Thin film interference.

Hint 3: Find the values of λ that give constructive interference.

Hint 4: See the end of the section titled Thin film interference.

Hint 5: See the end of the section titled Thin film interference for the correct relationship.

Wedge fringes

Hint 1: Make sure you are measuring all the distances on the same scale - use m throughout.

Quiz: Wedge fringes

Hint 1: See the relationship derived in the section titled Wedge fringes.

Hint 2: Consider the expression for constructive interference derived in the section titled Wedge fringes.

Hint 3: See the relationship derived at the end of the section titled Wedge fringes.

Hint 4: See the relationship derived in the section titled Wedge fringes. Find the thickness of the air wedge for $m = 1$.

Hint 5: Consider the effect on the optical path and the condition for constructive interference.

© HERIOT-WATT UNIVERSITY

Topic 8: Division of wavefront

Quiz: Young's slits

Hint 1: See the title of this topic.

Hint 2: Apply the relationship derived in the section titled Young's slit experiment.

Hint 3: Consider the relationship for Δx derived in the section titled Young's slit experiment.

Hint 4: This is a straight application of the relationship derived in the section titled Young's slit experiment.

Hint 5: Work out 5 times Δx.

Topic 9: Polarisation

Quiz: Polarisation

Hint 1: See the introduction to this topic.

Hint 2: The plane of polarisation of the emergent beam is the same as the transmission axis of the last polarising filter through which it travelled.

Hint 3: Only transverse waves can be polarised.

Quiz: Brewster's law and applications of polarisation

Hint 1: This is a straight application of Brewster's Law.

Hint 2: See the section titled Brewster's Law.

Hint 3: This is a straight application of Brewster's Law.

Hint 4: See the section titled Applications of polarisation.

Hint 5: See the section titled Applications of polarisation.

Hints to questions and activities for Unit 3

Topic 1: Electric force and field

Quiz: Coulomb force

Hint 1: Remember Newton's Third law.

Hint 2: The number of electrons in 1 C is equal to the inverse of the fundamental charge.

Hint 3: This is a straight application of Coulomb's Law.

Hint 4: This is a straight application of Coulomb's Law.

Hint 5: Work out the size and direction of the force exerted by X on Y. Then work out the size and direction of the force exerted by Z on Y. Then add the two vectors.

Quiz: Electric field

Hint 1: How does the strength of the electric force exerted by a point charge vary with distance?

Hint 2: Electric field strength is the force per unit positive charge.

Hint 3: Electric field strength is the force per unit positive charge.

Hint 4: Work out the size and direction of the electric field due to the 30 nC. Then work out the size and direction of the electric field due to the 50 nC. Then add the two vectors.

Hint 5: Electric field is zero at the point where the magnitude of the field due to the 1.0 μC charge is equal to the magnitude of the field due to the 4.0 μC charge.

Topic 2: Electric potential

Quiz: Potential and electric field

Hint 1: This is a straight application of $V = Ed$.

Hint 2: Make E the subject of the relationship $V = Ed$; then consider units on both sides of the equation.

Hint 3: This is a straight application of $E_W = QV$.

Hint 4: This is a straight application of $E_W = QV$.

Quiz: Electrical potential due to point charges

Hint 1: This is a straight application of $V = \frac{Q}{4\pi\varepsilon_0 r}$.

Hint 2: The charge of an alpha particle is double the charge of an electron. Use $E_p = E_W = QV$.

Hint 3: Find the potential due to each charge using $V = \frac{Q}{4\pi\varepsilon_0 r}$. Don't forget to include the minus sign for negative charges here.

Hint 4: To find out how $\frac{E}{V}$ depends on r, substitute $E = \frac{Q}{4\pi\varepsilon_0 r^2}$ and $V = \frac{Q}{4\pi\varepsilon_0 r}$ in $\frac{E}{V}$.

Topic 3: Motion in an electric field

Quiz: Acceleration and energy change

Hint 1: Is the velocity of the electron increasing, decreasing or staying the same?

Hint 2: This is a straight application of $E_W = QV$.

Hint 3: The electrical energy QV is converted to kinetic energy $\frac{1}{2}mv^2$.

Hint 4: Use $V = Ed$ and then $E_W = QV$.

Quiz: Charged particles moving in electric fields

Hint 1: Electric field strength is the force per unit (positive) charge.

Hint 2: First find the electrical force, then use Newton's Second Law.

Hint 3: Electric field strength is the force per unit **positive** charge.

Hint 4: Electric field strength is the force per **unit** positive charge.

Hint 5: What is the initial value of the vertical velocity of the electron? Find the vertical electrical force and use this to calculate the vertical acceleration of the electron. Then use the first equation of motion.

Topic 4: Magnetic fields

Quiz: Magnetic fields and forces

Hint 1: See the section titled Magnetic forces and fields.

Hint 2: See the section titled Magnetic forces and fields.

Hint 3: See the section titled Magnetic forces and fields.

Quiz: Current-carrying conductors

Hint 1: See the section titled Force on a current-carrying conductor in a magnetic field.

Hint 2: See the section titled Force on a current-carrying conductor in a magnetic field.

Hint 3: See the section titled Magnetic Field Around a current-carrying conductor.

Hint 4: See the section titled Magnetic induction.

The hiker

Hint 1: (a) What is the expression for the magnetic field due to a current-carrying conductor?

Hint 2: (b) What is the maximum value of B at the new position?

Hint 3: (b)

© HERIOT-WATT UNIVERSITY

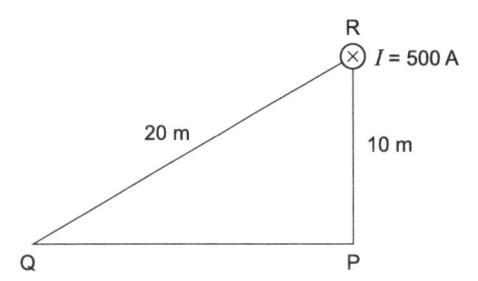

What is the horizontal distance PQ in relation to QR and PR?

Topic 6: Inductors

Quiz: Self-inductance

Hint 1: This is a straight application of $\varepsilon = -L\frac{dI}{dt}$.

Hint 2: What is the rate of change of current?

Hint 3: This is an application of Lenz's law.

Hint 4: This is a straight application of $E = \frac{1}{2}LI^2$

Hint 5: This is a straight application of $E = \frac{1}{2}LI^2$

Quiz: Inductors in d.c. circuits

Hint 1: The maximum current is the steady value reached when the induced e.m.f. is zero.

Hint 2: The maximum potential difference across the inductor is the value when current in the circuit is zero.

Hint 3: The current is steady!!

Hint 4: For 'growth' read 'variation of current with time'.

Quiz: a.c. circuits

Hint 1: $X_L = 2\pi f L$.

Hint 2: See the section titled Inductors in a.c. circuits.

Hint 3: $X_L = 2\pi f L$.

Hints to questions and activities for Unit 5

Appendix A: Units, prefixes and scientific notation

Quiz questions

Hint 1: Data is quoted to 2 sig figs so answer must be quoted to 2 sig figs.

Hint 2: The acceleration due to gravity is quoted to only 2 sig figs so the answer must be given to 2 sig figs.

Hint 3: The mass of the trolley is given to 4 sig figs and the velocity is given to 3 sig figs.

Answers to questions and activities for Unit 1

Topic 1: Kinematic relationships

Horizontal motion (page 12)

Expected answer

List the data you are given in the question

$u = 12.0$ m s^{-1}

$v = 0$ m s^{-1}

$s = 30.0$ m

$a = ?$

The appropriate kinematic relationship is $v^2 = u^2 + 2as$.

Putting the values into this equation

$$v^2 = u^2 + 2as$$
$$\therefore 0^2 = 12.0^2 + (2 \times a \times 30.0)$$
$$\therefore 0 = 144 + 60a$$
$$\therefore -60a = 144$$
$$\therefore a = -\frac{144}{60}$$
$$\therefore a = -2.40 \text{ m s}^{-2}$$

So to stop the car in exactly 10.0 m, the car must have an acceleration of -2.40 m s^{-2}, equivalent to a deceleration of 2.40 m s^{-2}.

Quiz: Motion in one dimension (page 14)

Q1: a) increases with time.

Q2: d) 45.0 m

Q3: c) 5.05 s

Q4: c) 18.8 m s^{-1}

Q5: d) 13 m s^{-1}

End of topic 1 test (page 30)

Q6: a = 1.5 m s^{-2}

Q7: Stopping distance = 41 m

Q8: t = 3.3 s

Q9: t = 1.8 s

Q10: The instantaneous acceleration of a body can be found by calculating the **gradient** of the tangent to the velocity-time graph.

Q11:
The displacement can be found from a velocity-time graph by determining the area under the graph. This process is equivalent to **integrating/integration** between limits.

Q12:
$$\frac{dv}{dt} = a$$
$$\int_0^v dv = \int_0^5 4.0t - 1.2 \; dt$$
$$[v]_0^v = \left[\frac{4.0t^2}{2} - 1.20t\right]_0^5$$
$$v - 0 = \left(\left(\frac{4.0 \times 5^2}{2}\right) - (1.2 \times 5)\right) - 0$$
$$v = 44 \text{ m s}^{-1}$$

Q13:
$$\frac{dv}{dt} = a$$
$$\int_5^v dv = \int_0^3 4.2t - 0.6 \; dt$$
$$[v]_5^v = \left[\frac{4.2t^2}{2} - 0.6t\right]_0^3$$
$$v - 5 = \left(\frac{4.2 \times (3)^2}{2} - (0.6 \times 3)\right) - 0$$
$$v = 22.1 \text{ m s}^{-1}$$

Q14:
$$\frac{ds}{dt} = v$$
$$\int_{5.0}^s ds = \int_0^{2.0} 9.0t^2 - 0.5t \; dt$$
$$[s]_{5.0}^s = \left[\frac{9.0t^3}{3} - \frac{0.5}{2}t^2\right]_0^{2.0}$$
$$s - 5.0 = \left(\frac{9.0 \times (2.0)^3}{3} - \frac{0.5 \times (2.0)^2}{2}\right) - 0$$
$$s = 28 \text{ m}$$

Topic 2: Angular motion

Quiz: Radian measurement (page 37)

Q1: c) 2.53 rad

Q2: c) 68.75°

Q3: d) $2\pi/3$ rad

Q4: d) 6π rad

Q5: a) 0.208 rad

Orbits of the planets (page 40)

Expected answer

Planet	Orbit radius (m)	Period (days)	Period (s)	Angular velocity (rad s^{-1})
Mercury	5.79×10^{10}	88.0	7.60×10^6	8.27×10^{-7}
Venus	1.08×10^{11}	225	1.94×10^7	3.24×10^{-7}
Earth	1.49×10^{11}	365	3.15×10^7	1.99×10^{-7}

Quiz: Angular velocity and angular kinematic relationships (page 45)

Q6: e) 50.3 rad s^{-1}

Q7: a) 0.45π rad s^{-1}

Q8: d) 60.0 rad

Q9: b) 0.70 rad s^{-2}

Q10: c) 7.9 s

Quiz: Angular velocity and tangential speed (page 49)

Q11: c) 4.80 m s^{-1}

Q12: c) 1.60 m

Q13: e) 4.58 m s^{-1}

Q14: a) 0.108 m s^{-2}

Q15: b) 3.2 m

Quiz: Centripetal acceleration (page 54)

Q16: c) 8.4 m s^{-2}

Q17: b) The centripetal acceleration is always directed towards the centre of the circle.

Q18: b) 5.6 m s^{-2}

Q19: e) 140 m s^{-2}

Q20: a) The centripetal acceleration halves in value.

Motion in a vertical circle (page 58)

Expected answer

1. The rope is most likely to go slack when the mass is at the top of the circle. Compare Equation 2.16 and Equation 2.17 to understand why.
2. At the top of the circle, Equation 2.16 tells us

$$T_{top} = mr\omega^2 - mg$$

Expressing this in terms of the tangential speed v,

$$T_{top} = \frac{mv^2}{r} - mg$$

When the rope goes slack, the tension in it must have dropped to zero, so

$$0 = \frac{mv^2}{r} - mg$$
$$\therefore \frac{mv^2}{r} = mg$$
$$\therefore v^2 = gr$$
$$\therefore v = \sqrt{gr}$$
$$\therefore v = \sqrt{9.8 \times 0.80}$$
$$\therefore v = 2.8 \text{ m s}^{-1}$$

ANSWERS: UNIT 1 TOPIC 2

Quiz: Horizontal and vertical motion (page 59)

Q21: c) 40 N

Q22: c) 6.0 rad s^{-1}

Q23: d) If the radius of the circle is increased, the centripetal force decreases.

Q24: b) 14.4 m s^{-1}

Q25: a) When the mass is at the bottom of the circle.

Quiz: Conical pendulum and cornering (page 67)

Q26: d) $T \times \sin \theta$

Q27: c) 57°

Q28: e) $2\pi \sqrt{l \cos \theta / g}$

Q29: a) centripetal force is provided by the frictional force.

Q30: d) a component of the normal reaction force contributes to the centripetal force.

End of topic 2 test (page 69)

Q31: 1.15

Q32: Time taken = 1.8 s

Q33: Angular velocity = 3.03 × 10^{-6} rad s-1

Q34: t = 1.53 s

Q35: Angular deceleration = 4.20 rad s^{-2}

Q36: Speed = 0.69 m s^{-1}

Q37: Average angular acceleration = 0.075 rad s^{-2}

Q38: Centripetal acceleration = 5.3 m s^{-2}

Q39: Radius = 8.6 cm

Q40: Centripetal force = 23.6 N

Q41: Minimum radius = 6.3 m

Q42:
 1. Tension = 3.7 N
 2. Angular velocity = 3.1 rad s^{-1}

Q43: Maximum speed = 14.0 m s^{-1}

Q44: Maximum speed = 18.5 m s^{-1}

© HERIOT-WATT UNIVERSITY

Topic 3: Rotational dynamics

Quiz: Torques (page 77)

Q1: e) N m

Q2: d) 49 N m

Q3: a) 3.75 m

Q4: c) 8.46 N m

Q5: d) 111 N

Torque and static equilibrium (page 81)

Q6: $(450 \times 4) = (600 \times \text{distance})$
Distance = 3 m on the right.

Q7: $(450 \times 3) = (600 \times \text{distance})$
Distance = 2.25 m on the right.

Q8: $(450 \times 3) + (480 \times 2) = (600 \times \text{distance})$
Distance = 3.85 m on the right.

Q9: $(450 \times 3) + (480 \times 2) = (360 \times 4) + (600 \times \text{distance})$
Distance = 1.45 m on the right.

Combinations of rotating bodies (page 88)

Q10: 0.16 kg m^2

Q11: 20 cm from the centre

Q12: 0.40 kg m^2

Q13: 0.43 kg m^2

Q14: 28 cm

End of topic 3 test (page 90)

Q15: Moment = 25 Nm

Q16: Angular acceleration = 3.8 rad s^{-2}

Q17: I = 0.25 kg m^2

Q18: I = 4.18 kg m^2

Q19: I = 0.12 kg m^2

Q20: The moment of inertia of an object is a measure of its resistance to **angular** acceleration about a given axis.

The moment of inertia of an object about an axis depends on the **mass** of the object, and the distribution of the **mass** about the axis.

Q21: I = 6.0 $\times$ 10^{-5} kg m^2

Q22: m = 15.0 g

Q23: r = 0.951 cm

Topic 4: Angular momentum

Quiz: Conservation of angular momentum (page 97)

Q1: d) 7.20 kg m^2 s^{-1}

Q2: b) 0.25 kg m^2 s^{-1}

Q3: a) The turntable would slow down.

Quiz: Angular momentum and rotational kinetic energy (page 99)

Q4: c) 0.015 J

Q5: e) 187 J

End of topic 4 test (page 102)

Q6: Rotational E_K = 0.235 J

Q7: ω = 3.4 rad s^{-1}

Q8: ω = 3.5 rad s^{-1}

Q9:
1. ω = 4.5 rad s^{-1}
2. Total E_K = 76.5 J

Topic 5: Gravitation

Quiz: Gravitational force (page 113)

Q1: d) 1.0×10^{-10} N

Q2: a) $F_S = F_E$

Q3: d) 8.4 N

Q4: b) Its mass remains constant but its weight decreases.

Q5: e) 8.87 m s^{-2}

Quiz: Gravitational fields (page 118)

Q6: d) 26.5 N kg^{-1}

Q7: a) m s^{-2}

Q8: a) 7.1×10^5 m

Q9: c) 3.0 m from P

Q10: d) 1.03×10^{26} kg

Quiz: Gravitational potential (page 123)

Q11: c) the distance of A from the centre of the Earth.

Q12: c) -2.9×10^6 J kg^{-1}

Q13: c) -6.0×10^7 J kg^{-1}

Q14: d) -4.9×10^{10} J

Q15: a) The satellite has moved closer to the Earth.

Quiz: Escape velocity (page 127)

Q16: e) escape from the Earth's gravitational field.

Q17: b) the mass and radius of the Earth.

Q18: c) 5020 m s^{-1}

End of topic 5 test (page 129)

Q19: F = 2.16×10^{20}

Q20: F = 8.09×10^{-10}

© HERIOT-WATT UNIVERSITY

Q21: Weight on Neptune = 74.2 N

Q22: g = 8.9 m s^{-2}

Q23: 3.8 N

Q24: 2.75 N kg^{-1}

Q25: 0.378 N kg^{-1}
Note: It is possible to solve this problem without having to calculate the mass of the planet.
Gravitational field strength equation is $g = \frac{GM}{r^2}$.
Rearranging this we get $gr^2 = GM$.
The GM term remains constant and $gr^2 =$ constant.
This means the problem can be solved with the following expression:

$$g_1 r^2{}_1 = g_2 r^2{}_2$$
$$6.45 \times 2.26 \times (10^6)^2 = g_2 \times 9.12 \times (10^6)^2$$
$$g_2 = 0.403 N kg^{-1}$$

Q26: 7.29 × 10^{-3} N kg^{-1}

Q27: V = -3.2 × 10^7 J kg^{-1}

Q28: E$_p$ = -7.2 × 10^{13}

Q29: V = -2.6 × 10^7

Q30: E$_p$ = -3.2 × 10^{11}

Q31: 2.6 × 10^5 m

Q32: 1.87 × 10^4 m s^{-1}

Q33: 3.27 × 10^{30} kg

Topic 6: General relativity and spacetime
End of topic 6 test (page 156)

Q1: b) non-inertial

Q2: c) accelerating

Q3: equivalence principle

Q4: a) slowly

Q5: a) slowly

Q6: b) False

Q7: a) True

Q8: b) False

Q9: b) False

Q10: a) True

Q11: b) No

Q12: a) True

Q13: spacetime

Q14: light

Q15: light

Q16: worldline

Q17: accelerating

Q18: a) X: constant speed, Y: accelerating, Z: stationary

Q19: gradient

Q20: b) R and S

Q21:

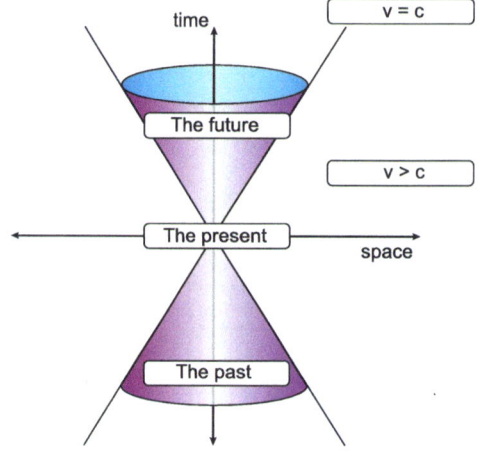

Q22: e) I, II and III

Q23:

$$r_{Schwarzschild} = \frac{2GM}{c^2}$$

$$r_{Schwarzschild} = \frac{2 \times 6.67 \times 10^{-11} \times 5.97 \times 10^{31}}{(3 \times 10^8)^2}$$

$$r_{Schwarzschild} = 8.85 \times 10^4 \text{ m}$$

Q24:

$$r_{Schwarzchild} = \frac{2GM}{c^2}$$

$$4.31 \times 10^8 = \frac{2 \times 6.67 \times 10^{-11} \times M}{(3 \times 10^8)^2}$$

$$M = 2.91 \times 10^{35} \text{ kg}$$

Topic 7: Stellar physics
Quiz: Properties of stars (page 173)

Q1: $\lambda_{max}T_{max} = constant$ (3000000)

Q2:

Class	Temperature (K)	Colour
O	> 33000	**Blue**
B	10000 - 33000	Blue to blue-white
A	**7500 - 10000**	White
F	6000 - 7500	Yellowish-white
G	**5200 - 6000**	Yellow
K	3700 - 5200	**Orange**
M	< **3700**	Red

Q3:
$P = \sigma T^4$
$P = 5.67 \times 10^{-8} \times 4970^4$
$P = 3.46 \times 10^7 Wm^{-2}$

Q4:
$L = 4\pi r^2 \sigma T^4$
$L = 4\pi \times (8.33 \times 10^{10})^2 \times 3.46 \times 10^7$
$L = 3.02 \times 10^{30} W$

Q5:
$b = \dfrac{L}{4\pi d^2}$
$b = \dfrac{3.02 \times 10^{30}}{4\pi(42.9 \times 365 \times 24 \times 60 \times 60 \times 3 \times 10^8)^2}$
$b = 2.07 \times 10^{36} Wm^{-2}$

© HERIOT-WATT UNIVERSITY

Hertzsprung-Russell diagram matching task (page 181)

Q6:

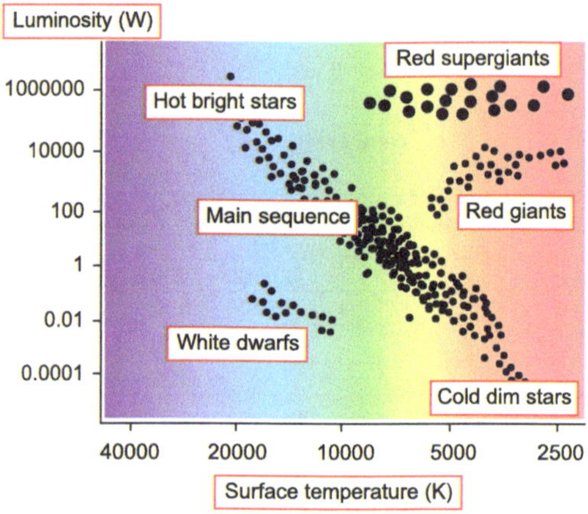
Luminosity (W)
1000000
10000
100
1
0.01
0.0001
Red supergiants
Hot bright stars
Main sequence
Red giants
White dwarfs
Cold dim stars
40000
20000
10000
5000
2500
Surface temperature (K)

Stellar matching task (page 184)

Q7: 1D; 2E; 3B; 4F; 5C; 6A

The life cycle of a low or medium mass star (page 184)

Q8:

1. Nebula contracts due to gravitational attraction.
2. The centre of the nebula's temperature increases.
3. Nuclear fusion starts.
4. Hydrogen runs out in the core and a red giant forms.
5. The outer layers of the red giant drift off into space.
6. A white dwarf is formed.

ANSWERS: UNIT 1 TOPIC 7

The life cycle of a large mass star (page 185)

Q9:

1. Nebula contracts due to gravitational attraction.
2. The centre of the nebula's temperature increases.
3. Nuclear fusion starts.
4. Hydrogen runs out in the core and a red supergiant forms.
5. The core collapses and a supernova explosion occurs.
6. A dense neutron star or a black hole is formed.

End of topic 7 test (page 187)

Q10: e) its surface temperature and its radius.

Q11: a) True

Q12: a) True

Q13: $L = 2.7 \times 10^{29}$

Q14: $L_{Procyon\ A} = 5.8 \times L_{Sun}$

Q15: $r_{Sirius\ A} = 8.1 \times 10^{16}$ m

Q16: d) blue and hot.

Q17: This is called the **main sequence**.

Q18: a) up

Q19: a) red giant / red supergiant.

Q20: e) main sequence star.

Q21: d) fusion balances

Q22: White dwarf

Q23: a) True

Q24: b) False

Q25: b) False

Q26: b) its surface temperature decreases and luminosity increases.

© HERIOT-WATT UNIVERSITY

Q27:

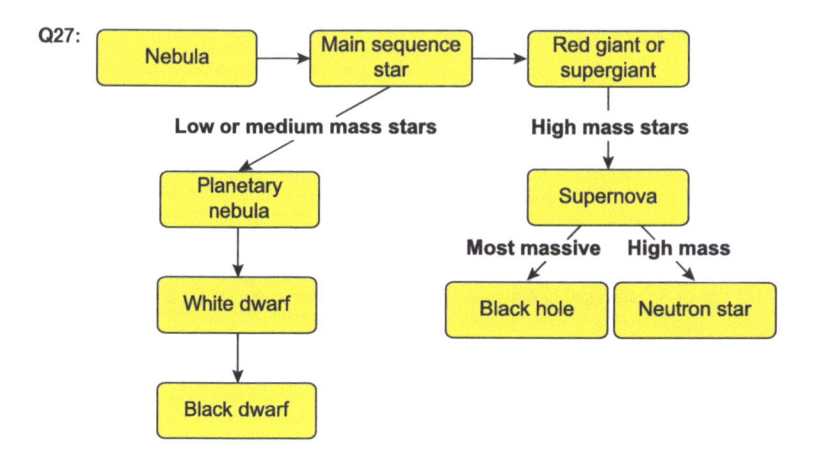

Nebula → Main sequence star → Red giant or supergiant

Low or medium mass stars: Main sequence star → Planetary nebula → White dwarf → Black dwarf

High mass stars: Red giant or supergiant → Supernova
- **Most massive:** Black hole
- **High mass:** Neutron star

Topic 8: End of section 1 test

End of section 1 test (page 192)

Q1: $v = 53$ m s^{-1}

Q2:

1. 3.96 rad s^{-1}
2. 3.79 N
3. 7.90 N

Q3: 2.4×10^3 m s^{-1}

Q4: $r_{Schwarzschild} = 3.72 \times 10^4$ m

Q5: $r = 9.0 \times 10^6$ m

Q6: T = 6250 K

Q7: a) upper right.

Q8: c) C

Q9: b) shorter

Q10: b) larger mass and have higher temperatures.

Answers to questions and activities for Unit 2

Topic 1: Introduction to quantum theory

Hydrogen line spectrum (page 207)

Q1: Photon is absorbed as the electron needs to gain energy to move up an energy level.

Q2: The larger difference in energy levels causes the photon to have more energy and hence a higher frequency due to $E = hf$. UV is higher frequency than visible light.

Quiz: Atomic models (page 207)

Q3: b) the electron's angular momentum is quantised.

Q4: d) 4.22×10^{-34} kg m² s⁻¹

Q5: e) $r = n\lambda/2\pi$

Q6: a) photons can only be emitted with specific energies.

End of topic 1 test (page 215)

Q7: 4.22×10^{-34} kg m² s⁻¹

Q8: 3.32×10^{-10} m

Q9: 9.79×10^{-11} m

Q10: 3.19×10^{-19} J

Q11: 6.24×10^{-7} m

Q12: A, C and E

Q13: 2.51×10^{-32} m

Topic 2: Wave particle duality

Quiz: Wave-particle duality of waves (page 223)

Q1: a) 3.88×10^{-19} J

Q2: b) 9.85×10^{14} Hz

Q3: d) the wavelength of the scattered photons increases.

Q4: a) The blue photons have greater photon energy and greater photon momentum.

Q5: c) 2.65×10^{-32} kg m s^{-1}

Quiz: Wave-particle duality of particles (page 228)

Q6: a) particles exhibit wave-like properties.

Q7: d) 1.14×10^{-10} m

Q8: c) 1.7×10^{-27} kg

Q9: b) particles can exhibit wave-like properties.

Q10: b) the electrons have a shorter wavelength than visible light.

End of topic 2 test (page 232)

Q11: $f = 5.66 \times 10^{14}$ Hz

Q12: $\lambda = 3.72 \times 10^{-6}$ m

Q13: $E_k = 1.89 \times 10^{-19}$ J

Q14: p = 1.56×10^{-27} kg m s^{-1}

Q15: p = 2.21×10^{-27} kg m s^{-1}

Q16: a) The proton

Q17: b) The electron

Q18: p = 5.78×10^{-24} kg m s^{-1}

Q19: $\lambda_e = 1.15 \times 10^{-10}$ m

Q20: Quantum tunnelling is when an incident **electron** is thought of as a **wave** and part of the wave crosses a **barrier** . Due to the **uncertainty** of the wave's position, this allows the electron to pass through the barrier. This is how nuclear **fusion** is possible in the sun and how modern **transistors** have become so efficient and tiny.

Topic 3: Magnetic fields and particles from space

Quiz: The force on a moving charge (page 242)

Q1: b) The particles experience the same magnitude of force but in opposite directions.

Q2: c) 2.4×10^{-13} N

Q3: e) all of them

Q4: a) constant speed.

Q5: a) 3.16×10^6 m s^{-1}

Quiz: Motion of charged particles in a magnetic field (page 251)

Q6: a) (i) only

Q7: c) helical, with the axis in the magnetic field direction

Q8: e) 5.69×10^{-6} m

Q9: e) Van Allen belts

End of topic 3 test (page 253)

Q10: 2.97×10^{-16} N

Q11:
1. 7.41×10^{-15}
2. 20.1 mm

Q12: 3.97×10^{-8} seconds

Q13: 2

Q14:
1. 4.2×10^{-4} m
2. 2.2×10^{-3} m

Q15: B,D and E

Q16: d) plasma

Topic 4: Simple harmonic motion

Quiz: Defining SHM and equations of motion (page 263)

Q1: b) displacement.

Q2: a) 1.27 Hz

Q3: d) Acceleration

Q4: e) 98.7 ms^{-2}

Q5: c) 0.84 ms^{-1}

Energy in simple harmonic motion (page 265)

Expected answer

1. Note that the sum of kinetic and potential energies is constant, so that energy is conserved in the SHM system

$$PE + KE$$
$$= \tfrac{1}{2}m\omega^2 y^2 + \tfrac{1}{2}m\omega^2 \left(a^2 - y^2\right)$$
$$= \tfrac{1}{2}m\omega^2 a^2$$

The total energy is independent of the displacement y.

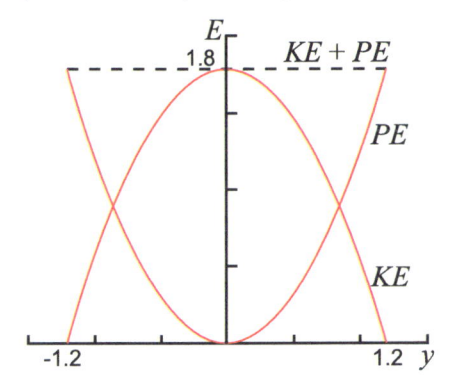

2. Again the sum of kinetic and potential energies is constant.

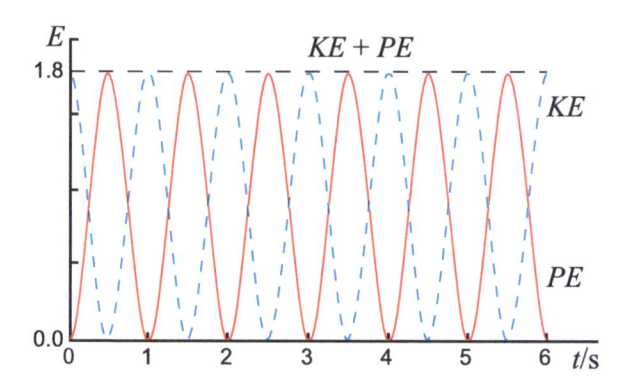

Quiz: Energy in SHM (page 266)

Q6: b) 0.037 J

Q7: d) 45 J

Q8: c) When its displacement from the rest position is zero.

Q9: c) $\pm a/\sqrt{2}$

Q10: d) 75 J

Quiz: SHM Systems (page 271)

Q11: d) 3.14 s

Q12: b) 0.910 Hz

Q13: e) 37 N m^{-1}

Q14: b) 0.75 s

Q15: a) Decrease its length by a factor of 4.

ANSWERS: UNIT 2 TOPIC 4

End of topic 4 test (page 275)

Q16: 1.0 m s^{-2}

Q17: 0.84 s

Q18:

1. 3.1 m s^{-2}
2. 0.86 m s^{-1}

Q19: 2.20 J

Q20: 0.25 m

Q21: 0.73 Hz

Q22: 1.5 × 10^{-3} J

Q23: 0.35 m

Topic 5: Waves

Quiz: Properties of waves (page 281)

Q1: e) amplitude

Q2: b) 4.3×10^{14} - 7.5×10^{14} Hz

Q3: a) X-rays, infrared, microwaves.

Q4: c) 4.74×10^5 GHz

Q5: d) 180 W m^{-2}

Quiz: Travelling waves (page 285)

Q6: c) 0.25 m

Q7: b) $y = 2\sin 2\pi (20t - x)$

Q8: e) 4 m

Q9: e) 12 Hz

Q10: b) 1.25 m s^{-1}

Quiz: Superposition (page 288)

Q11: d) 3.0 cm

Q12: a) a loud signal, owing to constructive interference?

Q13: c) 22.5 wavelengths

Q14: b) any periodic wave is a superposition of harmonic sine and cosine waves.

Q15: Phase angle = 3.7 rad

Superposition of two waves (page 289)

Q16: The resulting amplitude is equal to the sum of the amplitudes of A and B.

Q17: Once again the waves are back in phase, and so they interfere constructively again.

Q18: The amplitude is equal to the difference in the amplitudes of A and B, since the two waves are interfering destructively.

Longitudinal stationary waves (page 297)

Expected answer

a) (i)

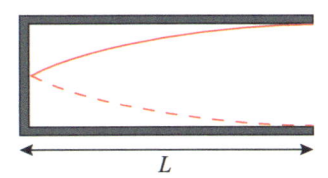

The fundamental has wavelength 4L. Remember that λ is twice the distance between adjacent nodes, and so four times the distance between adjacent node and antinode.

a) (ii)

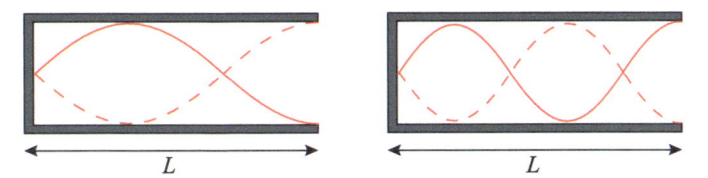

The wavelengths of the harmonics of a tube or pipe with one end open are $4L/3$, $4L/5$...

Note that only odd harmonics are possible for a tube with one end open.

b) (i)

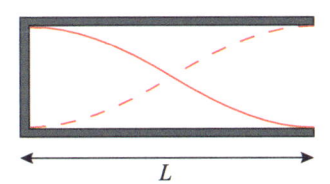

The fundamental has wavelength 2L. Remember that λ is twice the distance between adjacent nodes or antinodes.

b) (ii)

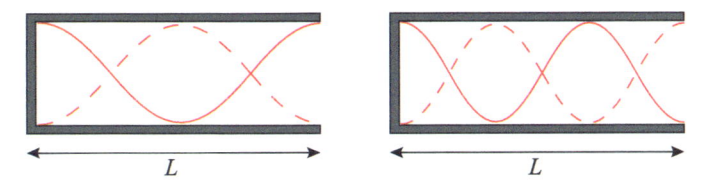

The wavelengths of the harmonics of a tube or pipe with both ends open are L, $2L/3$...

Quiz: Stationary waves (page 299)

Q19: c) 143 Hz

Q20: b) Every point between adjacent nodes of a stationary wave oscillates in phase.

Q21: e) 3.00 m

Q22: c) 4

Q23: d) 408 Hz

Q24: b) 25 beats

End of topic 5 test (page 302)

Q25: 5.54×10^{14} Hz

Q26: 6.52×10^{6} W m^{-2}

Q27: 272 cm

Q28: 8.8 m s^{-1}

Q29: 0.44 m

Q30: 3.1 cm

Q31: 0.61 m

Q32: 148 Hz

Q33: 0.24 m

Q34: 0.750 m

Q35: 26 m s^{-1}

Q36: 450 Hz

Topic 6: Interference

Quiz: Coherence and optical paths (page 309)

Q1: d) their phase difference is constant.

Q2: e) 9 wavelengths

Q3: c) 0.180 m

Q4: a) 8.75×10^{-3} m

Q5: b) 0.0682 m

End of topic 6 test (page 314)

Q6: 32.7 mm

Q7: 12.3 mm

Topic 7: Division of amplitude

Quiz: Thin film interference (page 321)

Q1: c) their optical path difference is an integer number of wavelengths.

Q2: b) 2.09×10^{-7} m

Q3: c) 1120 nm and 373 nm

Q4: d) The coating is only anti-reflecting for the green part of the visible spectrum.

Q5: a) 9.06×10^{-8} m

Wedge fringes (page 325)

Expected answer

Use Equation 7.7, and make sure you have converted all the lengths into metres.

1. In the first case, the wavelength is 6.33×10^{-7} m.

$$\Delta x = \frac{\lambda l}{2y}$$
$$\therefore \Delta x = \frac{6.33 \times 10^{-7} \times 0.08}{2 \times 2 \times 10^{-5}}$$
$$\therefore \Delta x = \frac{5.064 \times 10^{-8}}{4 \times 10^{-5}}$$
$$\therefore \Delta x = 1.3 \times 10^{-3} \text{m} = 1.3 \,\text{mm}$$

2. Now, the argon laser has wavelength 5.12×10^{-7} m.

$$\Delta x = \frac{\lambda l}{2y}$$
$$\therefore \Delta x = \frac{5.12 \times 10^{-7} \times 0.08}{2 \times 2 \times 10^{-5}}$$
$$\therefore \Delta x = \frac{4.096 \times 10^{-8}}{4 \times 10^{-5}}$$
$$\therefore \Delta x = 1.0 \times 10^{-3} \text{m} = 1.0 \,\text{mm}$$

Quiz: Wedge fringes (page 326)

Q6: d) 480 nm

Q7: a) Blue

Q8: d) 3.15×10^{-3} m

Q9: b) 162 nm

Q10: b) The fringes would move closer together.

ANSWERS: UNIT 2 TOPIC 7

End of topic 7 test (page 328)

Q11: 1.14×10^{-7} m

Q12:
1. 718 nm
2. 359 nm

Q13: 543 nm

Q14: 2.25×10^{-6} m

Q15: 4.21×10^{-3} m

Q16: 525 nm

Topic 8: Division of wavefront
Quiz: Young's slits (page 337)

Q1: b) interference by division of wavefront.

Q2: d) 333 nm

Q3: b) The fringes move closer together.

Q4: a) 2.4 mm

Q5: e) 0.016 m

End of topic 8 test (page 339)

Q6: 1.21×10^{-3} m

Q7: 491 nm

Q8: 441 nm

Q9: 2.23 mm

Q10: 535 nm

Topic 9: Polarisation

Quiz: Polarisation (page 345)

Q1: e) transverse waves.

Q2: c) At 20° to the *y*-axis.

Q3: b) No, because the oscillations are parallel to the direction of travel.

Quiz: Brewster's law and applications of polarisation (page 352)

Q4: e) 56.7°

Q5: a) polarised parallel to the reflecting surface.

Q6: d) 1.80

Q7: b) The material can rotate the plane of polarisation of a beam of light.

Q8: a) vertical.

End of topic 9 test (page 354)

Q9: 60°

Q10: a) In the z-direction.

Q11: 1.61

Q12: 49.2°

Q13:
 1. 90°
 2. 1.58

Q14:
 a) 90°
 b) It increases to a maximum when their head is at 90°.

Topic 10: End of section 2 test

End of section 2 test (page 356)

Q1:
1. 5.02×10^{-10} m
2. 3.17×10^{-34} kg m^2 s^{-1}

Q2: 2.74×10^{-12} m

Q3: 2, 4 and 7 are the correct answers

Q4:
1. 0.40 m s^{-2}
2. 0.090 m s^{-1}

Q5:
1. 0.4 m
2. 5.1 m s^{-1}

Q6: 6 beats

Q7:
1. 9.94×10^{-7} m
2. 6.05 m

Q8: 1.83×10^{-6} metres

Q9: 673 nm

Q10:
1. 1.22
2. a) Horizontally.

Answers to questions and activities for Unit 3

Topic 1: Electric force and field

Three charged particles in a line (page 366)

Expected answer

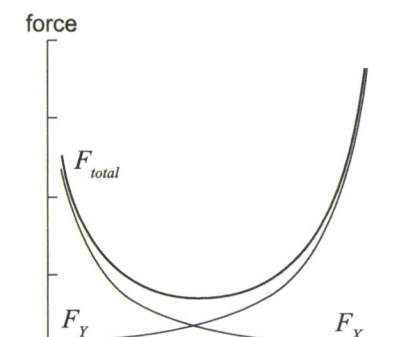

force

The graph shows the forces on the third charge due to charges X and Y, and the total force. As the third charge is moved from X to Y, the magnitude of the force due to X decreases whilst the magnitude of the force due to Y increases. The two forces both act in the **same** direction.

Quiz: Coulomb force (page 366)

Q1: c) F N

Q2: d) 6.25×10^{18}

Q3: d) 9.0 N towards B.

Q4: a) 7.7 cm

Q5: b) 14.4 N towards Z

Quiz: Electric field (page 372)

Q6: a) $E/4$

Q7: e) 1.25×10^{-8} N

Q8: d) 8.00×10^{-18} N in the -x-direction

Q9: d) 180 N C^{-1}

Q10: b) 17 cm

End of topic 1 test (page 377)

Q11: 2.0×10^{17}

Q12: 3.06 N

Q13: 9.2 N

Q14: 25 N

Q15: 5.7 N C^{-1}

Q16: 1.5×10^3 N C^{-1}

Q17: 3.2×10^5 N C^{-1}

Q18: 2.3×10^9 m s^{-2}

Topic 2: Electric potential

Quiz: Potential and electric field (page 382)

Q1: c) 8.00 V

Q2: d) V m^{-1}

Q3: a) 0.24 J

Q4: e) 8000 V

Quiz: Electrical potential due to point charges (page 386)

Q5: c) 1.4×10^5 V

Q6: e) 4.48×10^{-14} J

Q7: a) -7.2×10^4 V

Q8: b) 50 m^{-1}

End of topic 2 test (page 389)

Q9: 270 V

Q10: 0.063 J

Q11: 0.050 m

Q12: 26 J

Q13: 3.8×10^5 V

Q14: 59 V

Q15: 1.35×10^4 V

Q16:
1. 2.3 m
2. 1.6×10^{-6} C

Topic 3: Motion in an electric field

Quiz: Acceleration and energy change (page 395)

Q1: a) The electron gains kinetic energy.

Q2: b) 1.2×10^{-4} J

Q3: c) 3.10×10^5 m s^{-1}

Q4: b) 9.6×10^{-17} J

Quiz: Charged particles moving in electric fields (page 399)

Q5: c) 4.00×10^{-16} N

Q6: d) 7.03×10^{14} m s^{-2}

Q7: a) accelerated in the direction of the electric field.

Q8: c) 2.41×10^9 m s^{-2} downwards

Q9: d) 3.51×10^6 m s^{-1}

Rutherford scattering (page 405)

Expected answer

Use the formula

$$E_W = QV$$
$$\therefore E_W = eV_{gold}$$
$$\therefore E_W = e \times \frac{Q_{gold}}{4\pi\varepsilon_0 r}$$
$$\therefore r = e \times \frac{79e}{4\pi\varepsilon_0 E_W}$$

Now put in the values given in the question

$$r = \frac{79e^2}{4\pi\varepsilon_0 \times 8.35 \times 10^{-14}}$$
$$\therefore r = \frac{2.02 \times 10^{-36}}{9.29 \times 10^{-24}}$$
$$\therefore r = 2.18 \times 10^{-13} \text{ m}$$

End of topic 3 test (page 407)

Q10: 4.48×10^{-17} J

Q11: 5.1×10^6 m s^{-1}

Q12: 2.0×10^4 m s^{-2}

Q13:
1. 3.63×10^5 m s^{-1}
2. 2.34×10^{-3} m

Q14: 6.67×10^{-14} J

Topic 4: Magnetic fields

Quiz: Magnetic fields and forces (page 414)

Q1: e) All magnets have two poles called north and south.

Q2: e) (i) and (iii) only

Q3: c) (iii) only

Oersted's experiment (page 415)

Expected answer

1. A current through the wire produces a circular magnetic field centred on the wire.
2. The greater the current, the stronger is the magnetic field. This is shown by the separation of the field lines.
3. If the direction of the current is reversed, the direction of the magnetic field is also reversed.

Quiz: Current-carrying conductors (page 424)

Q4: e) field the same, current doubled, length halved

Q5: d) 60°

Q6: e) circular, decreasing in magnitude with distance from the wire

Q7: b) N A^{-1} m^{-1}

The hiker (page 427)

Expected answer

a)
$$B = \frac{\mu_0 I}{2\pi r}$$
$$\therefore B = \frac{4\pi \times 10^{-7} \times 500}{2\pi \times 10}$$
$$\therefore B = 1.0 \times 10^{-5} \text{ T}$$

ANSWERS: UNIT 3 TOPIC 4

b)

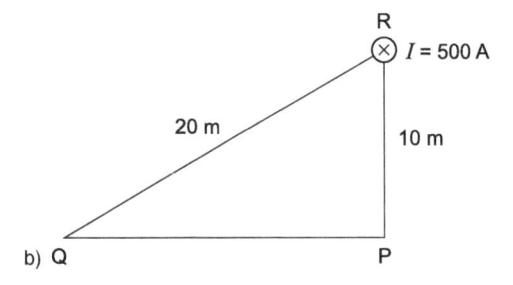

```
         R
         ⊗ I = 500 A
    20 m   |
            | 10 m
            |
   Q        P
```

$B_{\text{Earth}} = 0.5 \times 10^{-4}$ T
$\therefore 10\%$ of $B_{\text{Earth}} = 0.5 \times 10^{-5}$ T

$$B = \frac{\mu_0 I}{2\pi r}$$
$$\therefore 0.5 \times 10^{-5} = \frac{4\pi \times 10^{-7} \times 500}{2\pi \times \text{QR}}$$
$$\therefore \text{QR} = \frac{4\pi \times 10^{-7} \times 500}{2\pi \times 0.5 \times 10^{-5}}$$
$$\therefore \text{QR} = 20 \text{ m}$$

$$\text{QP} = \sqrt{20^2 - 10^2}$$
$$\therefore \text{QP} = \sqrt{400 - 100}$$
$$\therefore \text{QP} = \sqrt{300}$$
$$\therefore \text{QP} = 17.3 \text{ m}$$

Electrostatic and gravitational forces (page 428)

Expected answer

The Coulomb force is given by

$$F_C = \frac{Q_1 Q_2}{4\pi\varepsilon_0 r^2}$$

$$\therefore F_C = \frac{(1.6 \times 10^{-19})^2}{4\pi \times 8.85 \times 10^{-12} \times (10^{-15})^2}$$

$$\therefore F_C \sim \frac{2.5 \times 10^{-38}}{1 \times 10^{-40}}$$

$$\therefore F_C \sim 250 \text{ N}$$

The gravitational force is given by

$$F_G = G \frac{m_1 m_2}{r^2}$$

$$\therefore F_G = 6.67 \times 10^{-11} \times \frac{(1.67 \times 10^{-27})^2}{(10^{-15})^2}$$

$$\therefore F_G \sim 6.67 \times 10^{-11} \times \frac{2.8 \times 10^{-54}}{10^{-30}}$$

$$\therefore F_G \sim 1.9 \times 10^{-34} \text{ N}$$

We can combine these two results to find the ratio F_C/F_G

$$F_C/F_G \sim \frac{250}{1.9 \times 10^{-34}}$$

$$\therefore F_C/F_G \sim 10^{36}$$

End of topic 4 test (page 433)

Q8: 0.0254 N

Q9: 276 N

Q10: 4.37 m s^{-2}

Q11:
1. b) added to QR.
2. 0.21 T

Q12: 2.3 A

Q13: 4.03 × 10^{-5} T

Q14:
1. 1.7 × 10^{-3} T
2. 4.5 × 10^3 A

Topic 5: Capacitors

End of topic 5 test (page 457)

Q1: 45 s

Q2: 150 Ω

Q3: 51 mA

Q4: d)

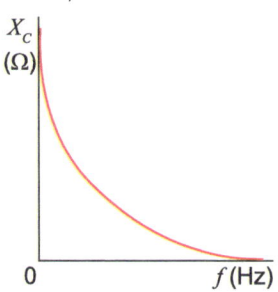

X_C (Ω) vs f (Hz): decreasing curve from high value at low frequency approaching zero at high frequency.

Q5:

1. 152 Ω
2. 0.092 A

Topic 6: Inductors

Quiz: Self-inductance (page 472)

Q1: b) moved across a magnetic field.

Q2: d) the induced current is always in such a direction as to oppose the change that is causing it.

Q3: b) 0.29 H

Q4: a) 0 V

Q5: d) The self-induced e.m.f. in an inductor always opposes the change in current that is causing it.

Q6: a) 0.18 J

Q7: d) 0.85 H

Quiz: Inductors in d.c. circuits (page 478)

Q8: b) 25 mA

Q9: d) 1.5 V

Q10: a) 0 V

Q11: d) stopwatch

Quiz: a.c. circuits (page 481)

Q12: c) $X \propto f$

Q13: e) An inductor can smooth a signal by filtering out high frequency noise and spikes.

Q14: a) $I \propto 1/f$

End of topic 6 test (page 485)

Q15: 8.3 V

Q16: 0.48 J

Q17: c) 1 V s A^{-1}

Q18: 0.048 V

Q19: 8.2 V

Q20: 8.4 V

Q21:
1. 2.6 V
2. 0 V
3. 0.19 A

Q22:
1. 0.65 V
2. 0.043 A
3. 3.1×10^{-4} J

Q23:
1. 0.12 H
2. 0.18 A
3. 1.9×10^{-3} J

Q24:
1. 2.8 V
2. 2.8 V
3. 4.9×10^{-3} J

Topic 7: Electromagnetic radiation
End of topic 7 test (page 494)

Q1: ε_0 is the symbol for the **permittivity** of free space.

Q2: Electromagnetic waves are **transverse**.

Q3: Electricity and magnetism can be **unified** under one theory called electromagnetism.

Q4: d) $c = \frac{1}{\sqrt{\varepsilon_0 \mu_0}}$

Q5: 1.44×10^{-6} H m^{-1}

Topic 8: End of section 3 test

End of section 3 test (page 496)

Q1:

1. 2.00×10^5 N
2. 3.28×10^4 N

Q2: 6.3×10^{-6} T

Q3: 5.1×10^{-26} kg

Q4: 4.1 N

Q5:

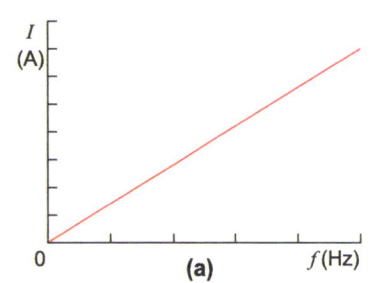

(graph of I (A) vs f (Hz), linear increasing line) **(a)**

Q6: 173 Ω

Q7:

1. 21 A s^{-1}
2. 3.9×10^{-3} J

Q8: 7.1 V

Q9:

1. 3.2 V
2. 0 V
3. 0.23 A

Q10:

1. 0.133 H
2. 0.200 A
3. 2.67×10^{-3} J

Q11:

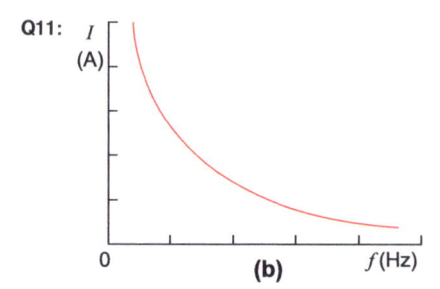

[Graph: I (A) vs f (Hz), decreasing curve, labeled (b)]

Q12:
time = 0.075 s

Answers to questions and activities for Unit 5

Appendix A: Units, prefixes and scientific notation

Quiz (page 538)

Q1:
- a) 6.4×10^1
- b) 6.58423×10^5
- c) 2.345×10^3
- d) 2.6×10^{-3}
- e) 5.6×10^{-5}
- f) 2.304×10^{-1}

Q2:
- a) 192
- b) 30.51
- c) 42900000
- d) 0.0103
- e) 0.00000008862
- f) 0.000009512

Quiz questions (page 543)

Q3: c) 1.3

Q4: d) 39 J

Q5: c) 40.9 J

Appendix C: Data analysis

Quiz (page 558)

Q1: a) ± 0.25 V

Q2: d) ± 0.2 s

Q3: c) I and II only

Q4: a) Temperature rise

Q5: b) ± 0.1 m s^{-2}

Q6: d) 10%

Q7: c) Mass = 5.0 ± 0.2 kg

Q8: d) $\pm 3.0 \times 10^7$ m s^{-1}

Q9: b) 14.48 N ± 0.08N

Q10: d) I and II only

Accurate, precise, both or neither? (page 566)

Expected answer

a) Is neither precise nor accurate.

b) Is precise and accurate.

c) Is precise but inaccurate.

Example (page 567)

Expected answer

These numbers are clearly precise enough for us to believe that if we measure it again we would get (1.04 ± 0.01)m. The small uncertainty shows how precise our results are. These meaurements may be precise but are not accurate as the mean value is 0.04m larger than a metre stick should be.

The SCHOLAR programme offers award-winning online courses for Scottish schools covering subjects from SQA's Curriculum for Excellence (CfE) and provides a route into careers in science, engineering, business, design and languages. These printed Study Guides are designed as personal and affordable student learning resources and are intended to be used in conjunction with the SCHOLAR online e-learning resources at https://scholar.hw.ac.uk.

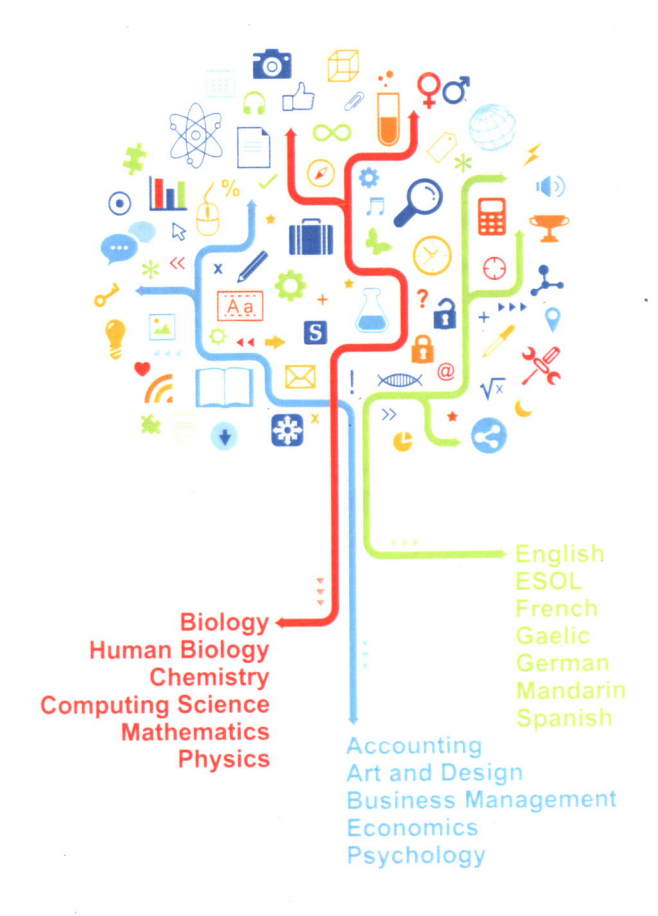

Biology
Human Biology
Chemistry
Computing Science
Mathematics
Physics

English
ESOL
French
Gaelic
German
Mandarin
Spanish

Accounting
Art and Design
Business Management
Economics
Psychology

SCHOLAR Forum

A programme of Heriot-Watt University
Prògram Oilthigh Heriot-Watt

SCHOLAR Unit
Heriot-Watt University
Edinburgh Campus
EH14 4AS

9 781911 057789

Telephone: +44 (0)131 451 4002
Email: info@scholar.hw.ac.uk
Web: https://scholar.hw.ac.uk

The full range of SCHOLAR Study Guides are available at: https://books.scholar.hw.ac.uk